Functional Foods

As widely reported, functional foods are highly nutritious and associated with a number of powerful health benefits. They may protect against disease, prevent nutrient deficiencies and promote proper growth and development. *Functional Foods: Technological Challenges and Advancement in Health Promotion* presents information related to the bioactive compounds present in functional foods and derived from fruits and vegetables, cereals and pulses, dairy and meat, herbs and spices, and other foods. The text describes novel techniques and methodologies used in the extraction, isolation, encapsulation, identification and characterisation of bioactive compounds.

Key features of the book

- Covers the most recent research related to the bioactive compounds present in functional foods
- Presents the latest information on extraction, isolation, encapsulation, identification and characterisation
- Discusses formulation challenges with an emphasis on stability and safety evaluations of functional foods

Finally, the book includes substantial scientific research and innovation for new product development with health benefits. This text will serve as a valuable resource for researchers, academicians and students interested in many aspects of functional foods.

Functional Foods

Technological Challenges and Advancement in Health Promotion

Edited by
Sajad Ahmad Wani,
Mohamed S. Elshikh,
Mona S. Al-Wahaibi, and
Haroon Rashid Naik

CRC Press
Taylor & Francis Group
Boca Raton London New York

CRC Press is an imprint of the
Taylor & Francis Group, an **informa** business

Designed cover image: Shutterstock

First edition published 2024
by CRC Press
6000 Broken Sound Parkway NW, Suite 300, Boca Raton, FL 33487-2742

and by CRC Press
4 Park Square, Milton Park, Abingdon, Oxon, OX14 4RN
CRC Press is an imprint of Taylor & Francis Group, LLC

ISBN: 9781032316758 (hbk)
ISBN: 9781032324524 (pbk)
ISBN: 9781003315100 (ebk)

DOI: 10.1201/9781003315100

Typeset in Times LT Std
by Deanta Global Publishing Services, Chennai, India

Contents

Preface

Nature offers us foods from different sources. Awareness of food can help prevent diseases. Knowledge of functional foods and their relationship with health is currently required. The presence of bioactive compounds in foods makes them interesting from health and scientific points of view. Functional foods and nutraceuticals are taken as dietary supplements in most countries. Functional foods are trusted products, and their ingestion on a daily basis helps in the prevention of various types of diseases.

Researchers and scientists are interested in nutraceuticals and functional foods. Different industries, such as the food and pharmaceutical industries, are interested in the use of bioactive compounds for the production of drugs and nutraceuticals. Bioactive compounds are functional ingredients because they are beneficial to health. Bioactive compounds such as peptides, carotenoids and polyphenols have been studied most. Some foods are rich sources of bioactive compounds. Therefore, even though it is challenging, developing food that is rich in bioactive compounds is exceptionally important to develop a quality and healthy life. Collecting such information in a book is a great achievement.

Adequate intake of functional foods has been linked to reductions in the risk of chronic diseases and to maintenance of body weight. There is a recent trend of developing novel food products using functional ingredients. This book has the potential to add to current scientific knowledge with respect to recent advances in functional foods and bioactive compounds found in functional foods derived from fruits and vegetables, cereals and pulses, dairy and meat products, herbs and spices, and other foods. This text covers the most recent research related to nutraceutical and functional foods. The main aim of the book is to assemble information related to nutraceuticals in order to make functional foods attractive to consumers and to make the job of finding information related to functional foods and bioactive compounds easy for researchers and scientists.

About the Editors

Sajad Ahmad Wani is currently working as a Dr. D.S. Kothari Post-Doctoral Fellow in the Department of Food Science and Technology, SKUAST-Kashmir, J&K, India. He completed his master's in food technology at IUST, Awantipora, J&K, India and his PhD at Sant Longowal Institute of Engineering and Technology, Punjab, India. He passed the National Eligibility Test (NET) (2015) conducted by the Indian Council of Agricultural Research (ICAR) in 2015. He has published more than 40 research/review articles, ten book chapters and five books. He also writes articles for the popular magazine, *Food & Beverage News* (India's first magazine for the food and beverage industry). He has attended 50 national and 20 international conferences, seminars and workshops. Dr. Wani has participated in various faculty development programmes. He serves as Honorary Associate Editor of the esteemed *International Journal of Food Science and Technology* (Wiley). He is a potential peer reviewer for reputed international journals related to food science and technology published by Elsevier, Taylor & Francis, Wiley and Springer. He is also a member of various associations such as IFT, IFERP, International Association for Agricultural Sustainability, AFSTI, the Asia Society of Researchers and Asian Council of Science Editors. Dr. Wani is a recipient of the prestigious Dr. D.S. Kothari Post-Doctoral Fellowship and the Maulana Azad National Fellowship (MANF 2013–14) from the University Grant Commission, New Delhi, India.

Mohamed S. Elshikh is a PhD researcher at King Saud University, College of Science, Botany and Microbiology Department, Riyadh, Saudia Arabia. He completed his master's in 2016 and is currently pursuing a PhD in plant taxonomy and bioinformatics from King Saud University, Riyadh, Saudi Arabia. He has attended 15 national and ten international conferences, seminars and workshops. He has published more than 120 papers in ISI journals. He is a potential peer reviewer for various reputed international journals related to botany and biotechnology published by Elsevier, Taylor & Francis, Wiley and Springer.

Mona S. Al-Wahaibi, is an associate professor at King Saud University, College of Science, Botany and Microbiology Department, Riyadh, Saudia Arabia. She completed her master's in 2008 in systematic studies on the species of the genus heliotropium (boragineaceae) in Saudi Arabia. She was awarded her PhD in 2013 from King Saud University, Riyadh, Saudi Arabia. Her research title was, "The Genetic Diversity of Populations of Heliotropium digynum in the Central Region of the Kingdom of Saudi Arabia by Using Some Molecular and Non-Molecular Aspects," and her work has been published in reputed journals with high impact factors in eminent publishing houses. She has presented her research and review papers at various national and international conferences and workshops. She has authored and co-authored numerous scientific articles published by Elsevier and Springer. She is also a potential peer reviewer for reputed international journals related to botany and biotechnology that are published by popular publishing houses Elsevier, Taylor & Francis, Wiley and Springer.

Haroon Rashid Naik currently works as Director of Planning and Monitoring at SKUAST-Kashmir, J&K, India. He previously worked as Dean of Research at IUST, Kashmir, professor and Head of Division of Post Harvest Technology at SKUAST-Kashmir from 2013–2016, and Student Welfare Officer from 2011–2016. He was a professor, Head of Department of Food Technology and Dean of Students at the Islamic University of Science & Technology, Kashmir, India from 2007–2009. Dr. Naik is a graduate in agriculture from GNDU Amritsar (1987), received a master's in food technology from CFTRI Mysore (1990) and a PhD in food technology from PAU Ludhiana (2000). He pioneered establishment of the Department of Food Technology at IUST Awantipura and created food technology programmes in SKUAST-Kashmir. Dr. Naik has supervised more than 20 MSc students and ten PhD students to date. Currently, eight PhD scholars are pursuing their programme under his supervision. He has published more than 100 research/review articles, ten book chapters and ten books. He served as the principal investigator of the project "All India Coordinated Research on PHET" from 2002–2010 and 2013–2016, and under his supervision many useful technologies for commercialisation were developed, particularly for temperate crops. Prof. Naik has successfully established food processing labs and training centres through various funded projects from the MOFPI, Government of India. He has more than 25 years of experience in teaching, research and extension education to his name. He has attended hundreds of national and international conferences, seminars and workshops. Prof. Naik serves as an editorial board member of a number of national and international journals. He is also a member of various associations such as the IFT, IFERP, the International Association for Agricultural Sustainability, AFSTI, the Asia Society of Researchers and the Asian Council of Science Editors.

Contributors

Gholamreza Abdi
Persian Gulf University
Bushehr, Iran

Samuel Ayofemi Olalekan Adeyeye
Hindustan Institute of Technology & Science
Chennai, India

Divya Aggarwal
CSIR-Central Food Technological Research Institute
Mysuru, Karnataka

Tahmeed Aijaz
Islamic University of Science and Technology
Awantipora, India

Akhila
GADVASU
Punjab, India

Ayman Younes Allam
Menofiya University
Shibin El-Kom, Egypt

Bindvi Arora
ICAR – Indian Agricultural Research Institute
New Delhi, India

Oyas Asimi
SKUAST
Kashmir, India

Ayenampudi Surendra Babu
Malla Reddy University
Hyderabad, India

Nur Alim Bahmid
National Research and Innovation Agency (BRIN)
Yogyakarta, Indonesia

Mukul Machhindra Barwant
Commerce and Science College Kopargaon
Maharashtra, India

Anu Chandel
Defense Food Research Laboratory (DFRL) Defense Research and Development Organization (DRDO)
Mysore, India

Monika Choudhary
Punjab Agricultural University
Ludhiana, India

Sunakshi Gautam
Shoolini University
Solan, India

Gousia
University of Kashmir
Kashmir, India

Hamid
Lovely Professional University
Phagwara, India

Heru Pitria Hastuti
National Research and Innovation Agency (BRIN)
Yogyakarta, Indonesia

Syed Zameer Hussain
Sher-e-Kashmir University of Agricultural Sciences and Technology
Kashmir, India

S. Padma Ishwarya
Polymer Engineering and Colloid Science Laboratory
Tamil Nadu, India

Makdud Islam
Sant Longowal Institute of Engineering & Technology
Punjab, India

Nusrat Jan
Sher-e-Kashmir University of Agricultural Sciences and Technology
Kashmir, India

Surbhi Kapoor
Punjab Agricultural University
Ludhiana, India

Ikawati Karim
Universitas Sulawesi Barat
Majene, Indonesia

Deepika Kathuria
ICAR – National Dairy Research Institute
Karnal, Haryana, India

Amarjeet Kaur
Punjab Agricultural University
Ludhiana, India

Zakir Showkat Khan
Uttaranchal University
Uttarakhand, India

Srishti Khurana
KU Leuven
Leuven, Belgium

Supriya Kumari
Guru Nanak Dev University
Amritsar, India

Toiba Majeed
Islamic University of Science and Technology
Kashmir, India

Nusrat Maqbool
Islamic University of Science and Technology
Awantipora, India

Muhpidah
Hasanuddin University
Makassar, Indonesia

Kamlesh Prasad
Sant Longowal Institute of Engineering & Technology
Punjab, India

Andi Rahmayanti
Hasanuddin University
Makassar, Indonesia

Preethi Ramachandran
GB Pant University of Agriculture and Technology
Pantnagar, Uttarakhand, India

Faisal Rashid
SKUAST
Kashmir, India

Vijay Reddy
Nitte University Center for Science Education and Research
Manglore, India

Sarvesh Rustagi
Uttaranchal University
Uttarakhand, India

Kawkabul Saba
SKUAST
Kashmir, India

Abu Saeid
NPI University of Bangladesh
Manikganj, Bangladesh

Praveen Saini
Sant Longowal Institute of Engineering & Technology
Punjab, India

Asima Sajad
Islamic University of Science and Technology
Awantipora, India

Loveleen Sharma
Amity University Uttar Pradesh (AUUP)
Noida, India

Vinita Sharma
Gautam Buddha University
Uttar Pradesh, India

Pallavi Singh
ICAR – Indian Agricultural Research Institute
New Delhi, India

Shakeel Ahmad Sofi
Central University of Kashmir (J&K)
Ganderbal, India

Shyam Sundar
Guru Nanak Dev University
Amritsar, India

Abhimanyu Thakur
Dr YS Parmar University of Horticulture and Forestry
Solan, India

NS Thakur
Dr YS Parmar University of Horticulture and Forestry
Solan, India

Sajad Mohd Wani
Sher-e-Kashmir University of Agricultural Sciences and Technology
Kashmir, India

Upasana Yadav
Manav Rachna International Institute of Research and Studies
Faridabad, Haryana, India

CHAPTER 1

Introduction to Functional Foods

S. Padma Ishwarya*

1.1 INTRODUCTION

Food has always remained a classic subject of interest, as it quenches the hunger of both palate and perceptions! Food is defined as any substance consumed by living beings to acquire nutrients such as carbohydrates, fats, proteins, vitamins and minerals, which facilitate bodily functions such as breathing, energy production and immune response (Granato et al., 2020). With the advancement in nutritional science and as the need for affordable healthcare emerged, scientists began to explore foods through the lens of public health. This led to the emergence of the functional food concept to realise the potential of foods beyond providing basic nutrition. Consequently, the term "functional food" was coined in 1993 in an article published in *Nature News* magazine, entitled, "Japan explores the boundary between food and medicine" (Swinbanks & O'Brien, 1993). Generally, functional foods are defined as "food products to be taken as part of a diet that provides health benefits beyond traditional nutritional effects" (Roberfroid, 2002). The health benefits include prevention of nutrition deficiencies, promoting optimal growth and development, and reducing the risk of lifestyle diseases and disorders such as obesity, type 2 diabetes, cancer, hypertension and cardiovascular diseases, which are avoidable by dietary modifications (World Health Organization, 2003). Thus, the concept of functional foods is the outcome of scientific advancement in understanding the relationship between diet and disease (Singletary & Morganosky, 2004). It is perceived as the union between food and health/medical sciences, based on the famous quote of Hippocrates, "Let food be thy medicine and medicine be thy food."

The increasing healthcare costs that prevailed in Japan in the 1980s have been stated as the major driving force that led to the advent of functional foods. However,

* Corresponding author E-mail: padmaishwarya.s@gmail.com

DOI: 10.1201/9781003315100-1

Table 1.1 Trends of scientific publications on functional foods

Keyword (In title)	Year range	Google Scholar	Science.gov	PubMed Central
"Functional food"	Total	45,40,000*	2,134#	578,186
	1993–2022	17,60,000*	2,134#	560,135

**including patents; #year range:* 2002–2022; data as of July 2022

the medicinal properties of foods have been realised since pre-historic times in the Vedic texts of India and in Chinese traditional medicine (Henry, 2010). Yet, the formalisation and compilation of technical evidence on the health benefits of functional foods have evolved across years hand in hand with the scientific advancements in this field. The large number of scientific publications (Table 1.1) and the growing market trend are the respective testimonials for the escalating interest of researchers and consumers in functional foods. A literature search carried out using the Google Scholar search engine shows that the different formats of scholarly and peer-reviewed publications on functional foods focus on diverse perspectives: the science behind functional foods and their relationship with physiology, functionality and health, product development, marketing, consumer preference and the regulations regarding safety of functional foods.

With respect to market potential and trends, the preference for functional foods is driven by the health awareness and nutritional knowledge of consumers (Ares, Gimenez & Gambaro, 2008). A recent report states that the international functional food market size is expected to reach $137.1 billion by 2026 from the value of $98.9 billion observed in 2021. The market size is projected to record a CAGR of 6.8% during the forecast period (Markets and Markets Research Private Ltd., 2020). This upward trend is envisaged to continue with advancements in food processing technologies, the discovery of novel food ingredients from natural sources rich in bioactive substances and the demand for fortified food and beverage products. Asia Pacific and Western Europe are the two major regions dominating the functional food market (Reportlinker, 2022). Owing to the stringent regulations that govern the certifications pertaining to functional food, the major players in this field are large enterprises that have global operations and are capable of managing resources and distributing products worldwide. Accordingly, the major players in the functional food global market are Amway (USA), Archer Daniels Midland Company (USA), Arla Foods (Denmark), BASF SE (Germany), Cargill Inc. (USA), DSM NV (Netherlands), General Mills Inc. (USA), Herbalife (USA), Ingredion Incorporated (USA), Kerry Group PLC (Ireland), Nestlé (Switzerland) and Tate & Lyle PLC (UK) (Farid et al., 2019).

The functional foods that we encounter in our daily routine include herbs, spices, fruits, vegetables, cereals, legumes, nuts, oilseeds, green tea and fish that inherently contain bioactive substances such as antioxidants, omega-3 fatty acids and fibre (Wani et al., 2022a). For example, oats are rich in β-glucan, a type of fibre that alleviates inflammation, boosts immune function and improves heart health (Bashir & Choi,

2017). Food products fortified with vitamins (e.g. milk fortified with vitamin D), minerals, probiotics (e.g. dairy products) or fibre also fall under the category of functional foods (Ellis, 2022; Pem & Jeewon, 2015). The various health benefits of plant-based and animal-based functional foods are schematically represented in Figure 1.1.

With the above background, this introductory chapter aims at presenting an overview of the concept of functional foods. The initial part of this chapter presents the most recent formal definition and classification of functional foods. The subsequent sections discuss the guidelines that govern the development of functional foods and the validation of their health claims. The latter sections of this chapter provide brief insights into different types of challenges, opportunities, recent developments, and future prospects in the field of functional foods. A comprehensive understanding of the abovementioned aspects would instil a scientific approach to the development of functional foods.

1.2 DEFINITION OF FUNCTIONAL FOODS

The definition of functional foods has remained as interesting a subject as the functional foods themselves. Earlier, the definition of functional foods often remained misunderstood, as it was not formalised or legally recognised in most countries (Ye et al., 2018). However, over the years, the definition of functional foods has evolved by considering and imbibing in it the clarity achieved in this field. In this context, it is vital to know about the Functional Foods Center (FFC), USA, which is a leading global organisation for assessing functional food products and educating the public on functional food science. According to the FFC, the need for a formal definition is to enable scientific organisations to work on the future of functional foods (Martirosyan & Singh, 2015). Since 2004, the FFC has held an annual international conference on the theme "Functional Foods for the Prevention and Treatment of Chronic Disease." The conference held in 2009 proposed the objective of devising a standard definition for functional foods. Accordingly, the 2009 FFC definition for functional foods was put forth as "Natural or processed foods that contain known or unknown biologically-active compounds; which, in defined, effective, and non-toxic amounts, provide a clinically proven and documented health benefit for the prevention, management, or treatment of chronic diseases" (Martirosyan, Zolaedov & Miroshnichenko, 2009). In the above definition, the term "biologically-active compounds" refer to extra-nutritional bioactive or chemical compounds that occur in trace amounts in foods. These bioactive compounds function as antioxidants and alleviate the risk or prevent the onset of certain diseases (Martirosyan & Pisarki, 2015). Due to the enhanced understanding of food bioactive compounds, the above definition has been recently revised by the FFC (2018) to be more specific and precise, which is stated as follows:

> Natural or processed foods that contain biologically-active compounds; which, in defined, effective, non-toxic amounts, provide a clinically proven and documented health benefit utilizing specific biomarkers, to promote optimal health and reduce the risk of chronic/viral diseases and manage their symptoms.

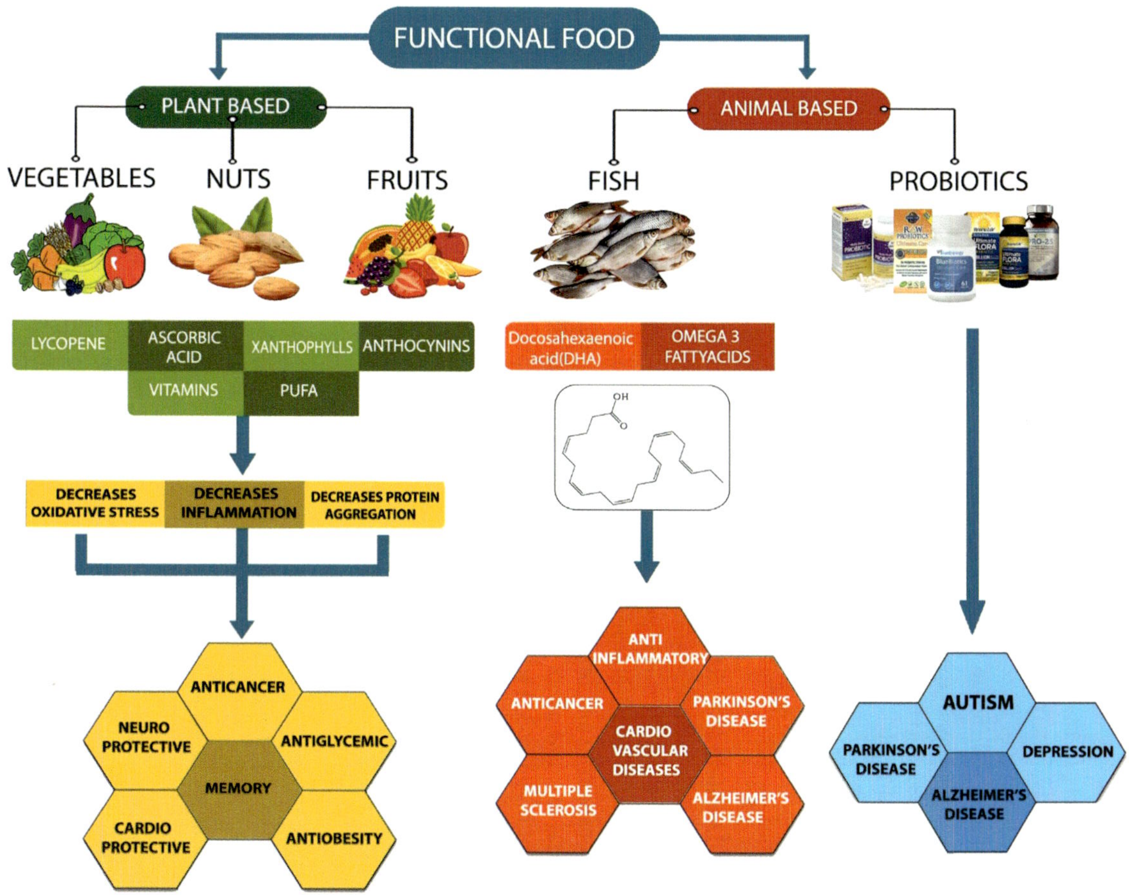

Figure 1.1 Health benefits of functional foods (reproduced with permission from Essa et al., 2021)

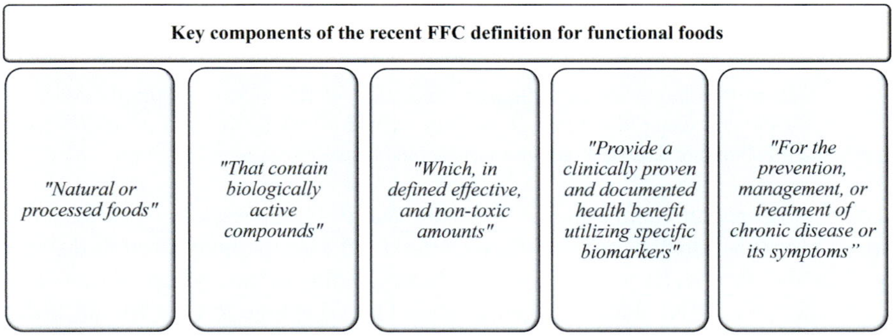

Figure 1.2 Important clauses in the new FFC definition for functional foods (based on Gur, Mawuntu & Martirosyan, 2018)

The vital components of the latest FFC definition of functional foods are depicted in Figure 1.2. The phrase "natural or processed foods" clarifies the food categories that can be included under functional foods. Then, the phrase "that contain biologically active compounds" narrows down the components in the above food types that influence health and bodily functions. Further, the part of the definition that mentions, "which, in defined, effective, and non-toxic amounts" emphasises the quantity, efficiency and safety level that the functional foods must possess. Similarly, the remark in the definition that functional foods should "provide a clinically proven and documented health benefit utilizing specific biomarkers" clarifies that in order for the claim of being a functional food to be substantiated, the food must undergo a rigorous evaluation and present clinically demonstrated evidence of its effects on consumer health. Finally, the phrase "reduce the risk of chronic/viral diseases and manage their symptoms" considers the three vital elements that the functional foods attempt to achieve, i.e., to prevent, manage and treat the disease and its symptoms (Functional Food Center, 2018; Gur, Mawuntu & Martirosyan, 2018).

Moreover, the new FFC definition includes three major revisions compared to the earlier definition, so that a widespread group of individuals obtain the benefits of functional foods. The rationale for each of the amendments in the definition is explained as follows:

1) Removal of the term "unknown" that preceded the word "biologically-active compounds" in the old definition. Without knowing the number of bioactive compounds, it is difficult to quantify their effects in a functional food.
2) Addition of the term, "specific biomarkers" to confirm the health benefit claim of functional foods. Biomarkers are indicators in the body that send out signals in tissues, organs, or systems and point to changes that occur at the cellular level. By measuring the biomarkers, the health benefits of bioactive compounds in a functional food can be ascertained. Some examples of biomarkers for obesity, cancer, cardiovascular disease and diabetes are circulating hormones and cytokines (e.g. adipokines), proteins found in body fluids or tissue (e.g. prostate-specific antigen),

myoglobin and HbA1c (which forms when glucose attaches to the amino-terminal group of the β subunit of haemoglobin).

3) Addition of the phrase "or its symptoms" to further elaborate the role of functional foods in preventing chronic diseases. Inclusion of this phrase in the definition broadens the scope of functional foods to include all the factors that contribute to the final state of a disease and not just its final state.

Overall, the recent definition put forth by the FFC suggests the possibility of prescribing functional foods as a potential relief for the symptoms of lifestyle diseases or disorders. It provides a comprehensive understanding of the concept of functional foods and their expected role and deliverables (Gur, Mawuntu & Martirosyan, 2018).

In addition to the FFC, other national and international organisations and researchers have established their own definitions for functional foods, which place emphasis on different aspects and differ in detail, specificity and focus. For instance, Granato et al. (2017) defined functional foods as "industrially processed or natural foods that when regularly consumed within a diverse diet at efficacious levels have potentially positive effects on health beyond basic nutrition." This definition confines the scope of the term "functional," such that any fresh, unprocessed or processed food can be considered as a functional food only based on sound evidence from appropriate clinical trials (randomised, double-blind, placebo-controlled clinical trials) and significant experimental proof in terms of its toxicological (safety) and functionality aspects (Assmann et al., 2014; Granato et al., 2020). Definitions of functional foods put forth by other global organisations are compiled in Table 1.2.

1.3 CLASSIFICATION OF FUNCTIONAL FOODS

Functional foods have been classified based on different criteria as listed in Figure 1.3. The earlier methods of classification have been based mainly on the food type involving its origin, degree of modification and nutritional composition. With the evolution of both consumers and functional food science, the recent mode of classification considers the extent of research outcomes and experimental evidence gathered on the claims as the criteria to classify functional foods. This section is intended to provide an overview of both the current and conventional approaches to classifying functional foods.

1.3.1 Research on Epidemiological and After-Market Studies

Similar to the definition, the FFC has proposed a new approach to classify functional foods based on epidemiological research and after-market studies (Martirosyan, Lampert, & Ekblad, 2022). Along with the Academic Society of Functional Foods and Bioactive Compounds (ASFFBC), the FFC is in communication with the Food and Drug Administration (FDA) to formalise the classification of functional foods. Accordingly, the quality of evidence for a functional food product is hierarchised as A, B and C, based on multiple integral factors. The approved functional food products are graded following an alphanumeric system. Accordingly, class A signifies the

Table 1.2 Definition of functional food given by different organisations (adapted from Martirosyan & Singh, 2015; Nazir et al., 2019)

Organisation	Definition	Reference
The Institute of Medicine of the U.S. National Academy of Sciences	"Foods that encompass potentially healthful products, including any modified foods or food ingredients that may provide a health benefit beyond the nutrients it contains."	Thomas & Earl (1994)
FOSHU ("Food for Specific Health Use"), Ministry of Health and Welfare, Japan	"Foods that are expected to have a specific health effect due to relevant constituents, or foods from which allergens have been removed, and foods where the effect of such an addition or removal has been scientifically evaluated, and permission has been granted to make claims regarding the specific beneficial effects on health expected from their consumption."	Roberfroid (2000)
	"Processed foods containing ingredients that aid specific bodily functions in addition to being nutritious."	Shimizu (2003)
State Food and Drugs Administration (SFDA), China	"Health (functional) food means that a food has special health functions or is able to supply vitamins or minerals. It is suitable for consumption by special groups of people and has the function of regulating human body functions but is not used for therapeutic purposes. And it will not cause any harm whether acute or sub-acute or chronic."	Yang (2008)
The Institute of Food Technologists (IFT)	"[S]ubstances [that] provide essential nutrients often beyond quantities necessary for normal maintenance, growth, and development, and/or other biologically active components that impart health benefits or desirable physiological effects."	Verbeke, Scholderer, & Lähteenmäki (2009)
American Dietetic Association, USA	Describes functional foods as "whole, fortified, enriched or enhanced" that should be consumed regularly and at effective amounts in order to derive health benefits.	Thomas and Earl (1994); Verhagen et al. (2010)
	"Whole, fortified, enriched or enhanced" that should be consumed regularly and at effective amounts in order to derive health benefits.	Martirosyan and Singh (2015)
International Life Sciences Institute (ILSI), Europe	"A food can be regarded as functional if it is satisfactorily demonstrated to beneficially affect one or more target functions in the body, beyond adequate nutritional effects, in a way which is relevant to either an improved state of health and well-being, or reduction of risk of disease."	Serafini, Stanzione, and Foddai (2012)

(Continued)

Table 1.2 (Continued)

Organisation	Definition	Reference
Foods for Special Dietary Use (As defined by the Federal Food, Drug, and Cosmetic Act)	"A particular use for which a food purports or is represented to be used, including to the following: • Supplying a special dietary need that exists by reason of a physiological, pathological, or physical condition. • Supplying a vitamin, mineral, or other ingredients for use by humans to supplement the diet by increasing the total dietary intake. • Supplying a special dietary need by reason of being a food for us as the sole item of diet."	Martirosyan and Singh (2015)
National Academy of Sciences Food and Nutrition Board, USA	"Any modified food or food ingredient that may provide a health benefit beyond the traditional nutrients it contains."	Martirosyan and Singh (2015)
Food Industry Division, Agriculture and Agri-Food, Canada	"Functional foods are foods enhanced with bioactive ingredients and which have demonstrated health benefits, such as probiotic yogurt, or breads and pasta with added pea fiber."	Canadian Food Business Spring (2017)

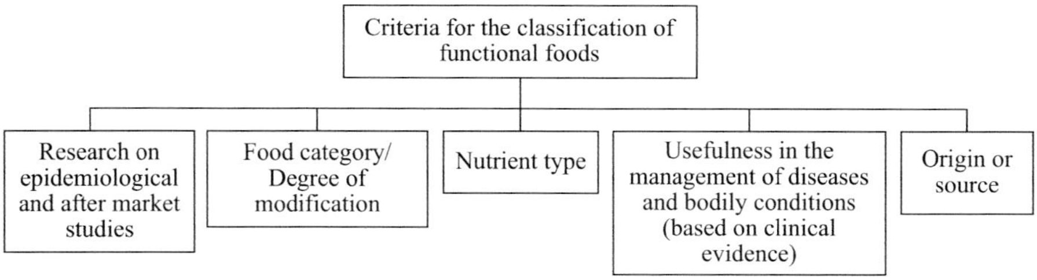

Figure 1.3 Criteria for the classification of functional foods

completion of aftermarket research, epidemiological investigations and certification of functional food status. The above process comprises a 16-step protocol put forth by the FFC for the introduction of foods into the functional food category and the means of launching them in the market (Figure 1.4). Hence, "class A" is the premium rating for the approval of functional food products (FFP) and the gold standard for

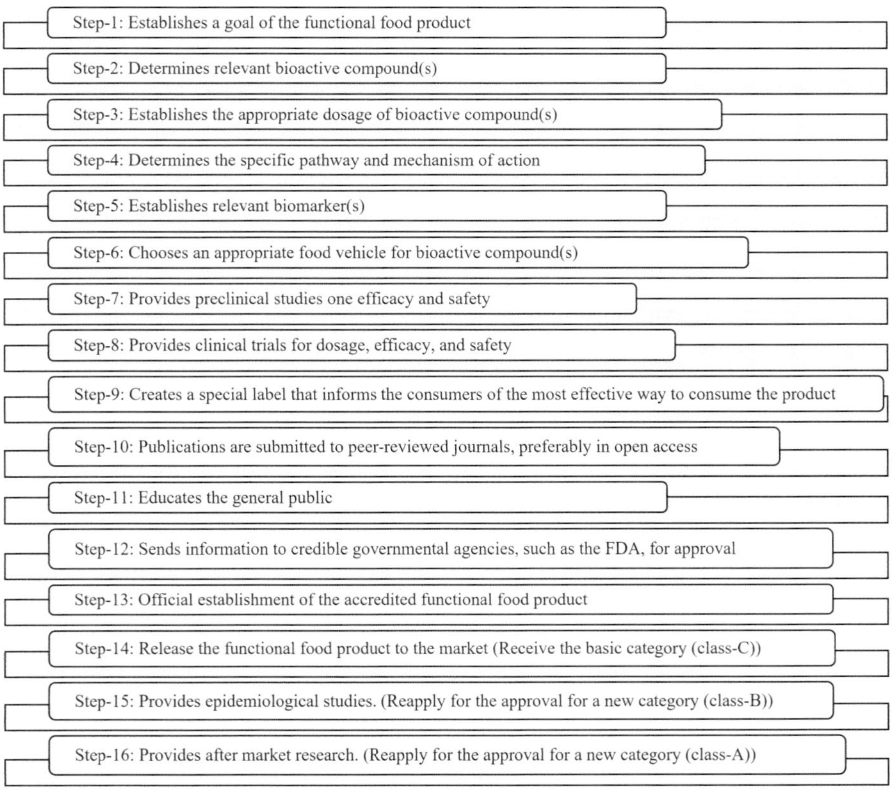

Figure 1.4 The 16-step protocol proposed by the FFC to streamline the development and market launch of functional food products (based on Martirosyan, Lampert, & Ekblad, 2022)

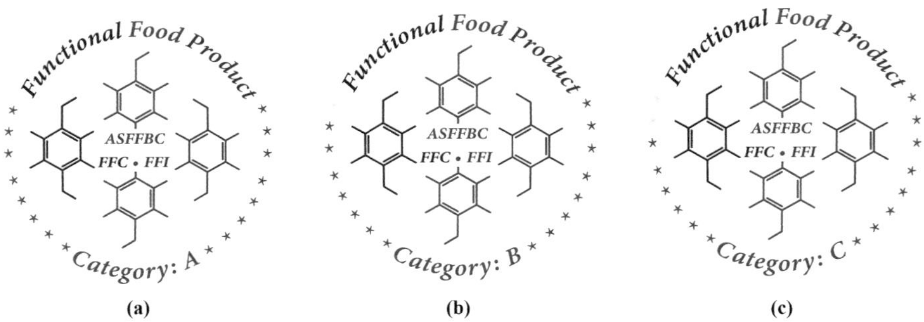

Figure 1.5 Logos for the certification of functional food products: (a) Category A; (b) Category B; (c) Category C (reproduced with permission from Martirosyan, Lampert, & Ekblad, 2022)

manufacturers to present experimental evidence on the efficacy, legitimacy and safety of the FFP.

On the other hand, class B comprises those foods for which the epidemiological studies and certification of functional food status are complete (steps 1–15 in Figure 1.4), but the after-market research is due. The products that fall under this group would be of a high standard and would have benefits for consumers. Finally, class C is limited to products that have been just granted the certification as functional foods (steps 1-14 in Figure 1.4). Thus, class C represents the basic classification or the minimum requirement for approval as a FFP according to the protocol of FFC, before the food receives its logo of certification. Thus, the rank for functional foods ranges between A and C, wherein A is the highest and C is the lowest, and foods with class A labels hold a superior factor of support than the foods with class B labels. The FFC logos of certifications for class A, class B and class C functional foods are depicted in Figure 1.5[a–c]. The aim of this new mode of classification proposed by the FFC is to synchronise with the other commonly observed grading systems in the world (Martirosyan, Lampert & Ekblad, 2022).

Having understood the FFC's 16-step process for the development and establishment of functional foods, it is also important to become familiar with the mandatory conditions for FOSHU recognition, as given below (Maeda-Yamamoto, 2017):

1. Enhanced dietary routine and health experienced by the consumers of the FFP.
2. Availability of scientific evidence to demonstrate the health claims associated with the product.
3. Proven clinical and nutritional evidence and/or functional ingredients.
4. Established safety of the product and/or functional ingredients for human consumption.
5. The functional ingredients are described based on their physical, chemical and biological characteristics, in addition to the methods for their qualitative and quantitative analytical determination.
6. Insignificant changes in the nutritional components of the food.
7. Consumption of foods on a daily basis and not occasionally.
8. The product or its functional ingredient is not deemed as a medicine (pharmaceutical).

1.3.2 Nutrient Type

Based on a nutritional perspective, functional foods can be classified as:

1. Nutrients: This category includes but is not limited to lipids (ω-3 fatty acids), vitamins (vitamin E, vitamin C and folates) and minerals (selenium).
2. Non-nutrients: This group of functional foods comprise fibre (insoluble and soluble fibre), phenolic compounds (phenolic acids, flavonoids, isoflavones, tannins), non-digestible oligosaccharides (fructans, galacto-oligosaccharides, isomalto-oligosaccharides, xylooligosaccharides, soybean oligosaccharides), phytosterols, glucosinolates, carotenoids (lutein, cryptoxanthin, lycopene) and organosulphur compounds.

1.3.3 Target Systems/Organs Benefited

Classification of functional foods based on their specified health benefits is presented in Table 1.3.

Table 1.3 Classification of functional foods based on intended functionality or beneficial action on target organs/systems (based on Maeda-Yamamoto, 2017)

Specified health use/target systems or organs	Example
Gastrointestinal tract	Probiotics (Bifidobacterium, Lactic acid bacteria), prebiotics (dietary fibre, ingestible dextrin, polydextrose, guar gum, psyllium seed coat, non-digestible oligosaccharides, lactose), polyphenols, phytates, ω-3 fatty acids
Cardiovascular system	ω-3 fatty acids, polyphenols, micronutrients, soluble fibres
Immune system	Probiotics, ω-3 fatty acids, polyphenols
Skeletal system	Fructans
Kidney	Fructans
Blood cholesterol levels	Chitosan, soybean protein, degraded sodium alginate
Blood glucose levels	Indigestible dextrin, wheat albumin, guava tea polyphenol and L-arabinose
Blood pressure	Lactotripeptide, casein dodecapeptide, tochu leaf glycoside (geniposidic acid) and sardine peptide
Dental hygiene	Palatinose, maltitol and erythritol
Simultaneous action on cholesterol and gastrointestinal conditions	Degraded sodium alginate and dietary fibre from psyllium seed husk
Mineral absorption	Calcium-citrated malate, casein phosphopeptide, heme iron and fructooligosaccharide
Promotion of osteogenesis	Soybean isoflavone and milk basic protein
Foods related to triacylglycerol	Middle-chain fatty acids, green tea catechin, oolong tea polyphenol and cholorogenic acid
Alleviate the risk of diseases such as osteoporosis	Calcium
Prevention of neural tube defects (spondyloschisis)	Folic acid

1.3.4 Origin or Source

Depending on their source, functional foods are classified as plant-based, animal-based or microbial-based. Some examples under each category are listed below:

- Plant-based FFPs: Fibre, polyphenols, fructans, ω-3 fatty acids, phytates, carotenoids, non-glycerides in edible oils,
- Animal-based FFPs: ω-3 and ω-6 fatty acids present in fish oils, chitosan, conjugated linolenic acid (dairy products).
- Microbial-based FFPs: Dairy products such as curd and yoghurt containing probiotics.

1.3.5 Degree of Modification

A functional food can fall under either one of the following categories (Figure 1.6) or a combination thereof based on its degree of modification (Henry, 2010):

1. A natural food.
2. A food enhanced with a new/additional macronutrient or micronutrient that is not available in it naturally (e.g. golden rice, which is a genetically modified crop to include β-carotene in the grains to address vitamin A deficiency).
3. A food enhanced with a micronutrient that is inherently present in it to a level that will lead to beneficial effects (e.g. rice and wheat fortified with iron, eggs enriched with omega-3 fatty acids resultant from the supplementation of chicken feed with dietary flaxseeds).
4. A food from which a constituent with potentially harmful effects has been removed (e.g. allergenic protein).
5. A food in which one or more of its nutrients have been modified or replaced (e.g. replacing fat with modified starch).
6. A food with modified bioavailability or stability of an ingredient that is known to have the ability to reduce the risk of incidence of a disease.

1.4 FUNCTIONAL FOODS: CONCERNS, CHALLENGES AND OPPORTUNITIES

1.4.1 Concerns and Challenges

The demand for functional foods is on the rise in the light of COVID-19 pandemic and increasing consumer preference for foods that provide more than nutrition. Nevertheless, many challenges exist in the future of the functional foods sector, which can be classified under three headings: "scientific," "regulatory" and "commercialisation/marketing." The scientific challenges are often related to establishing the efficacy of functional foods and elucidating the complex interactions and factors involved in the relationship between food and health (Singletary & Morganosky, 2004). Consequently, the development of functional food demands extensive

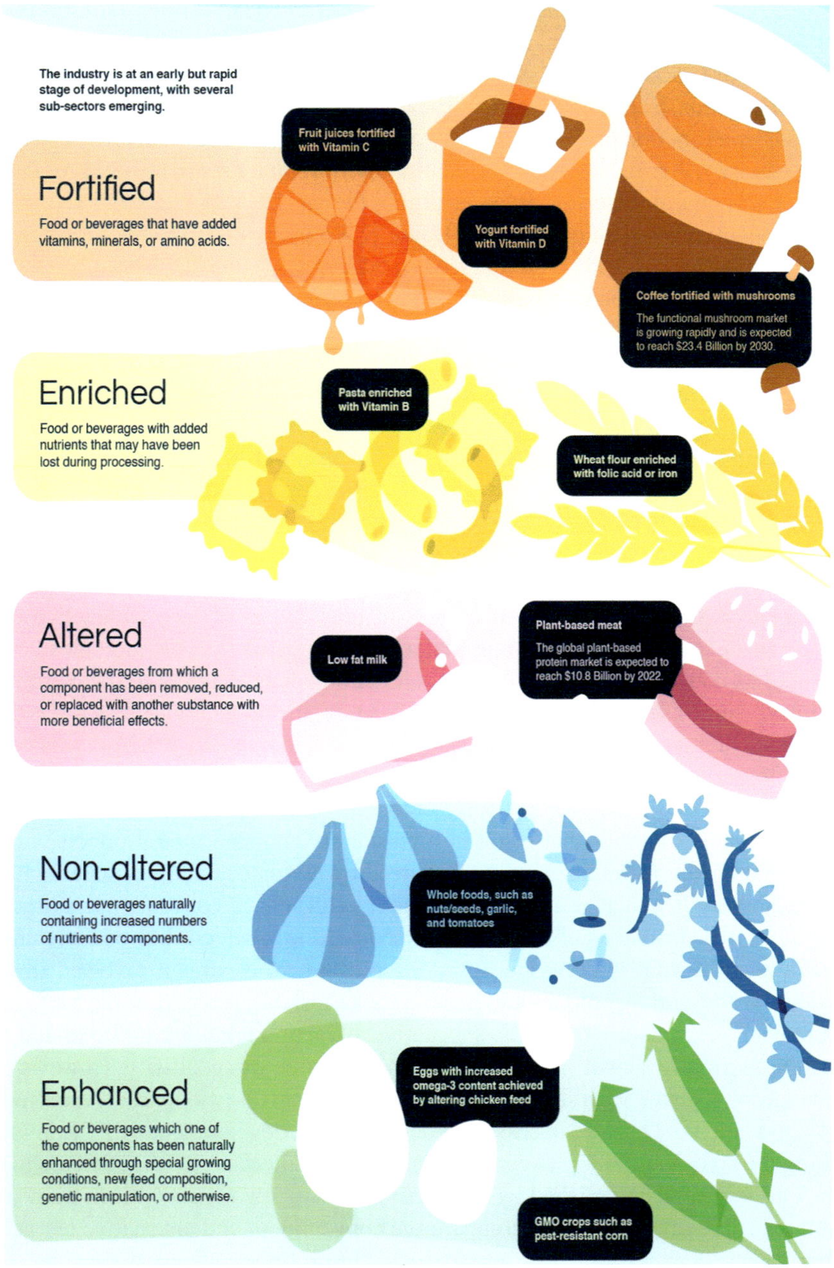

Figure 1.6 Categorization of functional foods based on the degree of modification (source: Jones, 2022; www.visualcapitalist.com/sp/the-rise-of-functional-food-what-investors-need-to-know/)

research and development efforts, which impose a financial burden on food industries due to budget limitations. In addition, differences are observed in the consumer acceptance of functional food products (Nazir et al., 2019). Moreover, variations in the health responses of individuals to the proposed benefits of functional foods are observed as a function of genetic, gender and metabolism-related factors (Singletary & Morganosky, 2004). Nevertheless, continuous developments in the fields of nutrigenomics and neuroscience aid in ascertaining the mechanism of disease risk reduction* by functional food constituents (Kaput & Rodriguez, 2004) and gain insights into consumers' motivations, preferences and decisions (Harrell, 2019), respectively. Further, certain functional ingredients upon consumption may interfere with digestion and/or metabolism and thereby affect the bioavailability of other ingredients. However, they might remain stable in the product. Yet another scientific challenge is associated with arriving at an *optimal dosage* of functional ingredients to be added to a product. Optimal dosage is the dose/quantity of any functional ingredient that is high enough to result in a beneficial effect without any negative effects on other functional constituents (Jackson & Paliyath, 2011). Moreover, it is imperative to maintain the integrity, bioactivity and bioavailability of the functional ingredients after processing and storage (Day et al., 2009).

Yet another scientific challenge in functional food production is the choice of appropriate extraction methods to obtain different types of target bioactive compounds from chemically-varied raw materials. The extraction technologies can be classified as conventional or alternative, wherein the former refers to the commonly employed solid-liquid extraction and the latter includes high-pressure-assisted extraction, high-voltage electric discharge-assisted extraction, microwave-assisted extraction, pulsed electric field-assisted extraction, supercritical fluid extraction and ultrasound-assisted extraction (Poojary et al., 2017). Irrespective of the technology, green extraction methods that result in high-quality and safe food products without imposing environmental hazards are preferred by regulatory bodies, manufacturers and consumers (Chemat et al., 2015). This is relevant, as conventional extraction techniques pose problems of toxic waste disposal, due to which they become unsustainable despite their wide usage. In addition, the heat treatment involved in conventional extraction techniques can denature and destabilise a part of the high-value functional constituents. Contrarily, alternative extraction methods are rapid and sustainable, thermally sensitive and exhibit a high degree of specificity. However, the challenges associated with scalability and higher capital costs jeopardise their industrial applications. Thus, a successful process for the development of FFP involves making the right trade-off between conventional and alternative extraction technologies. This can be accomplished by conducting statistically robust optimisation studies on the same raw material to compare the conventional and alternative extraction technologies for factors such as energy expenditure, mode of processing (batch or continuous), extraction time, extraction rate or velocity of extraction (milligrams of a target compound in a gram of dry matter, which can be extracted in 1 min;

* The dietary signals generated by food constituents control several gene-expression patterns in cells and tissues, and eventually reduce disease risk.

Putnik et al., 2018), extraction yield and other engineering aspects. Based on the comparative studies, the best extraction method must be selected based on statistical standardisation of extraction parameters such as solvent type, particle size, plant material-to-solvent ratio, time and temperature of extraction, and pressure. The above protocol would indicate the most cost-effective extraction technology that produces the best-quality extracts (Granato et al., 2020).

With respect to the regulatory aspects, validating the health claims attached to functional food products, compliance with the law of the land, and appropriate labelling of new products are the main challenges. This is because the functional food concept is perceived differently in different countries. For instance, Japan identifies functional foods as those that include natural ingredients. However, in the United States, functional foods are permitted to contain ingredients derived from biotechnological techniques (Bagchi, 2008). In India, functional foods can comprise herbal extracts, spices, fruits and nutritionally enriched foods or food products with additional functional components (Wani et al., 2022b; Keservani et al., 2014). Furthermore, obtaining approval for health claims based on scientific evidence involves tedious procedures. For example, the validation of biomarkers and dose-response curves to define the effective dose involve clinical studies that are time-consuming and may, under certain circumstances, demand additional studies to provide robust evidence (Nazir et al., 2019). Though FFP with novel ingredients are patentable, many of them contain non/patented ingredients and hence can be legally imitated. This reduces the competitive advantage of the parent company that initially filed the petition and expended resources to obtain approval for the health claim (Daliri & Lee, 2015). While many countries are coming up with regulations on functional foods, ambiguities prevail regarding the label claims on functionality (nutritional and health claims) of a product. Due to differences in the regulations between countries, the claims permitted in one nation are not permitted in another. For instance, certain health claims legalised in the United States are not accepted in Europe and vice versa (Nazir et al., 2019).

Apart from regulatory concerns, labelling-related ambiguities present challenges in the marketing of functional foods due to the missing information on health benefits. A recent article by Nilova and Malyutenkova (2020) discuss the challenges in promoting functional foods in e-commerce platforms. The study provides various insights into the limitations involved in the purchase of functional foods. Generally, consumers are alienated from functional foods due to the lack of information on their health benefits. In addition, they hesitate to buy FFP due to their high prices. For example, the inclusion of vitamin-mineral complexes and plant-based natural antioxidants and ingredients increases the cost of functional bread and bakery products. One main reason for the missing label information on the health benefits of functional foods is the difficulty of including this within the limited space offered by the small size of the packaging. Though e-commerce websites offer space to include the required information, challenges exist in selling functional food products online. Consumers lack confidence in the quality and freshness of delivered products and face problems in returning substandard products. A survey showed that despite the consumers recognising sour-milk drinks (yoghurt) as functional food products based

Table 1.4 Scientific challenges in the development of functional beverages (based on Nazir et al., 2019)

Challenges	Examples of challenges encountered	Reference
• Incorporation of multiple ingredients and possibility of interactions between the ingredients. • Interactions or chemical reactions might lead to insolubility, oxidation, precipitation or degradation of the ingredients. Consequently, the functionality of the beverage could be affected. • Technical know-how should be developed on the possible interactions that can occur between ingredients mixed in a product.	**Dairy-based beverages:** • Reports show that milk exerts a negative effect on the metabolic pathways and bioavailability of flavonoids added to it. Hence, fortification of milk-based beverages with flavonoids should be avoided. **Fruit and vegetable-based beverages:** • Sensitivity of fruits, vegetables, and herbal-based beverages to oxygen during manufacturing and storage, due to the inherent antioxidant properties.	Sun-Waterhouse (2011); Zhao (2012); Rodríguez-Roque et al. (2014); Paquin (2009)

on advertisements, they were hesitant to buy these products online due to uncertainties in delivery (Nilova & Malyutenkova, 2020).

Table 1.4 presents details on the challenges that are encountered in the development of different categories of functional beverages. Table 1.5 lists the corresponding solutions for the various challenges discussed above.

1.4.2 Opportunities in the Functional Foods Sector

Regardless of the multiple challenges and concerns related to their scientific claims, regulations and legal approval, novel functional food products with new ingredients and proven health benefits against particular diseases are being produced effectively (Nazir et al., 2019). To move forward, various opportunities are aligned to open up exciting avenues in the field of functional foods. Especially in the post-pandemic scenario, consumers have started relating their health to diet and prefer foods that boost nutrition and immunity. The outcomes of a global survey revealed that 42% of respondents increased their purchase of functional foods and beverages during the COVID-19 pandemic. While a majority of consumers preferred to consume health-promoting ingredients through foods (85%) and beverages (57%), only 23% of them favoured the intake of conventional oral supplements such as pills and tablets (NFRA, 2022). This shift in consumer behaviour signifies an important opportunity for the functional foods industry. With the development of innovative food technologies, functional foods would evolve as the "food of the future" (Burton, 2022). Nevertheless, it is necessary that nutrient-dense and high-quality functional foods should also have optimal sensory characteristics. Hence, developing flavourful functional foods is an interesting opportunity for food technologists. Similarly, the preference of the younger generation for nutritional drinks has created

Table 1.5 Challenges encountered during the clinical trials on functional foods and their corresponding solutions (Brown et al., 2018; adapted based on the information from Nilova & Malyutenkova, 2020; Lestari, 2021)

Challenge	Solution
Difficulty in obtaining industrial funding	Collaboration with university research centres could lead to funding proposals
Shortfalls in technical support	Use of university's expertise
Challenges in the identification of a control group (placebo) for the clinical trials	Detailed development of alternative products without active ingredients
Difficulty in delivering fresh products to consumers	Launch of home delivery service
Compliance by participants	Trust – develop a range of products to maintain interest
Identification of suitable biomarkers	Plasma biomarkers must be unique to functional food
Accidental intake of bioactive ingredients by the placebo group	Identify sensitive biomarkers in plasma
Statistical analysis	Careful planning of study design in the early stages
Response of the general public	Publication of study results in learned journals to increase trust in results
Response of the medical community	Well-designed studies provide credibility
Difficulties in the e-commerce sales of functional bakery products	"Baking-to-order" option will offer the consumers guaranteed delivery of fresh bakery products. Calculation of delivery time after order should account for the production time of the product.
Lack of regulations	Development of institutional capacity is essential to frame the required regulations. This can include food research centres, advisory services for producers, mentors in food sector marketing and management, and officials approving health claims for functional foods.

new opportunities for beverage manufacturers to develop non-alcoholic, sugar-free products that deliver functional benefits such as hydration and cognitive support. Consequently, smoothies, hydrating drinks and cold beverages such as tea and coffee with added nutrients, phytochemicals and probiotics are dominating the functional food market (NFRA, 2022).

1.5 FUTURE TRENDS IN FUNCTIONAL FOODS

An evolving concept in the field of functional foods is "postbiotics," which refer to inactivated probiotics or non-viable bacterial cells with substantial immunomodulatory effects. Postbiotics can be whole-cell or cell-free extracts, purified cell wall components or culture supernatant. They can also include soluble factors, i.e., products or metabolic byproducts secreted by live bacteria or released after bacterial lysis. These encompass cell surface proteins, enzymes, organic acids, peptides, peptidoglycan-derived muropeptides, polysaccharides and teichoic acids (Aguilar-Toala et al., 2018). Based on their immunomodulatory activity, postbiotics exhibit functional properties comprising systemic effects such as anti-hypertensive, anti-inflammatory, anti-obesogenic, antioxidant, anti-proliferative, hepato-protective and hypocholesterolemic properties (de Almada et al., 2016; Korpela, 2018; Pojic et al., 2018; Rodiño-Janeiro et al., 2018). Compared to probiotics, the benefit of including postbiotics in functional foods is their positive safety characteristics. With postbiotics, there is a lesser probability of the passing of antibiotic resistance genes by horizontal transfer mechanism, which can happen with certain probiotic strains (Taverniti & Guglielmetti, 2011; Imperial & Ibana, 2016). The scope for future research on postbiotics has been realised in the following two aspects (Aguilar-Toalá et al., 2018; Birch & Bonwick, 2019):

1. Elucidating the mechanism of action by which postbiotics may improve specific physiological functions and thereby promote host health.
2. Investigation of whether the probiotic strains that lost viability during their transit through the high-acid milieu of the stomach would transform as postbiotics in the intestine and still be beneficial to the gastrointestinal tract.

The future of functional foods demands a complete elucidation of the relationship between the functionality of chemical constituents and their prescribed dose, suitable delivery vehicles, safety, stability and price. An interdisciplinary research approach is essential to understand the above relationship. Collaboration between food scientists and experts in the fields of clinical studies and health would aid in establishing the effects of functional foods on human metabolism. Not limited to the above, investigations should be focused on the incorporation of natural extracts in different food matrices and their effects on the organoleptic, textural, physicochemical, rheological and functional characteristics. Indeed, taste and texture are the vital elements of a functional food. In this context, a recent innovation is "3D food printing," which enhances the appearance, taste and texture of traditional foods whilst being rich in nutrients. "3D food printing" is an "additive" manufacturing (AM) process that involves precise layering of small semi-liquefied food particles atop each other. It is an advanced food structuring technique to develop personalised processed foods with unique shapes, structures, textures and flavours (Anandharamakrishnan & Ishwarya, 2019). Moreover, 3D-printed functional foods can be personalised to accommodate the specific needs and expectations of an individual consumer (Jędrusek-Golińska et al., 2020), especially older adults or people with swallowing difficulties. A syndicate of ten companies from European Union (EU) countries conducted research on 3D-printed foods under the "Performance" project to develop innovative food products that are customised to cater to the definite dietary needs of individuals or groups (Grunewald, 2015; Kouzani et al., 2017). Another suggested approach to improve the organoleptic characteristics of functional foods is to explore a blended formulation of bioactive compounds, which conceal their undesirable effects in the food matrices and improve consumer acceptance (Birch & Bonwick, 2019). Further, in view of the recent scientific advancements in human nutrition and food science, the functional food market must also be examined from technological and marketing standpoints (Granato et al., 2020). Regarding the market for FFP, the potential areas of growth include food products derived from plant-based proteins, meat alternatives, whole-fat dairy, almonds, blueberry, green juices, seaweed snacks and other tree nuts-based foods (Puhakka et al., 2018; Tanna & Mishra, 2018).

Further, the 21st century is the "era of nanotechnology," and the concept of functional foods is no exception to this. According to the British Standards Institution (BSI, 2005), nanotechnology is defined as "the design, production, and application of structures, devices, and systems through control of the size and shape of the material at the nanoscale." Nanoscale refers to "structures having one or more dimensions of the order of 100 nm or less" (European Commission, 2006). The nanotechnology interventions in food processing have seen enormous developments in the last decade, as the nanoscale dimensions impart unique physical, chemical and biological properties to a food ingredient (Weiss, Takhistov & McClements, 2006). Nanotechnology has the capability to bring a new dimension to functional food product development, specifically in enhancing the bioavailability of bioactive ingredients and masking their negative impact on the organoleptic quality of food. The recent focus is on the application of nanotechnology approaches to develop delivery vehicles (Figure 1.7) for the encapsulation of food bioactives to improve their stability during food

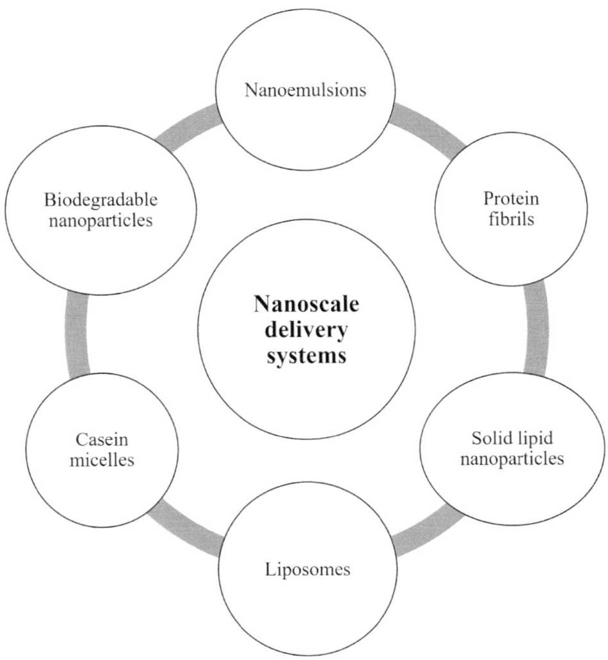

Figure 1.7 Nanoscale systems for the delivery of bioactive compounds

processing, solubility in water (mainly, of lipophilic bioactives) and gastric stability and uptake (enhanced bioavailability and circulation time) in the body. The higher surface-to-volume ratio of nanoscale delivery systems improves the solubility, controlled release pattern and absorption of the encapsulated core active ingredients. Moreover, nanotechnology offers innumerable opportunities to produce tailored nanoparticles with unique surface properties to deliver active compounds directly at the target sites within the body (e.g. colon-targeted delivery systems). Hence, the tools of nanotechnology are expected to boost the efficacy of bioactive compounds in functional foods and augment consumer health. Some examples of nanoscale delivery systems include nanoemulsions and nanoparticles of bioactive constituents such as catechins, vitamins, minerals and carotenoids. However, nano food products are still at the laboratory scale and have yet to achieve the "lab-to-land" transformation. Furthermore, extensive research is underway on clarifying the ambiguities related to the safety of nanoparticles related to their bioaccumulation, hindered excretion, toxicity, and permeation through biological barriers. Moreover, the daily intake limits for nanoparticles have to be established (Singh, 2016). Thus, the applications of nanotechnology have a multitude of future prospects in the functional food industry.

1.6 CONCLUSIONS

The information presented in this introductory chapter emphasises that the field of functional foods has come a long way over a period of more than four decades to be established in terms of its definition, classification, scientific foundation, regulations and consumer acceptability. The functional food sector relies upon the association between food scientists and technologists, health experts, regulatory bodies, the food industry and consumers. Developing countries are expected to be the future marketing hub for functional foods by promoting the above collaboration. Nevertheless, the success of a functional food product in higher markets depends on establishing its health claims with scientific evidence and showcasing its capability to consumers. This is because consumers of today of all age groups and health/nutritional requirements finalise their purchase decision of a food product based on sound technical evidence of its health claims. Therefore, technologists and manufacturers striving to launch novel functional foods in the market must gather scientific proof to demonstrate their efficacy and intended health claims. The above stakeholders are responsible for ensuring that any health claim declared on the packaging label of a functional food product is derived based on detailed clinical research. Science-based product innovations, standardisation of national and international regulations for functional foods, exploration of export market potential and improving consumer demand are the factors that are expected to govern the future of this sector. A multidisciplinary research approach is vital to identify and resolve the critical bottlenecks in the development of functional foods and transform them into "future foods."

REFERENCES

Aguilar-Toalá, J. E., Garcia-Varela, R., Garcia, H. S., Mata-Haro, V., González-Córdova, A. F., Vallejo-Cordoba, B., & Hernández-Mendoza, A. 2018. Postbiotics: An evolving term within the functional foods field. *Trends in Food Science and Technology*, 75, 105–114. https://doi.org/10.1016/j.tifs.2018.03.009.

Anandharamakrishnan, C., & Ishwarya, S. P. 2019. *Essentials and Applications of Food Engineering*. CRC Press, Boca Raton, FL.

Ares, G., Giménez, A., & Gámbaro, A. 2008. Influence of nutritional knowledge on perceived healthiness and willingness to try functional foods. *Appetite*, 51(3), 663–668. https://doi.org/10.1016/j.appet.2008.05.061.

Assmann, G., Buono, P., Daniele, A., Della Valle, E., Farinaro, E., Ferns, G., Krogh, V., Kromhout, D., Masana, L., Merino, J., Misciagna, G., Panico, S., Riccardi, G., Rivellese, A. A., Rozza, F., Salvatore, F., Salvatore, V., Stranges, S., Trevisan, M., & Vetrani, C. 2014. Functional foods and cardiometabolic diseases. *Nutrition, Metabolism, and Cardiovascular Diseases*, 24(12), 1272–1300. https://doi.org/10.1016/j.numecd.2014.10.010.

Bagchi, D. 2008. *Nutraceutical and Functional Food Regulations in the United States and Around the World*. Academic Press, Boston, MA, pp. 115–364.

Bashir, K. M. I., & Choi, J. S. 2017. Clinical and physiological perspectives of β-glucans: The past, present, and future. *International Journal of Molecular Sciences*, 18(9), 1906. https://doi.org/10.3390/ijms18091906; PMID: 28872611; PMCID: PMC5618555.

Birch, C. S., & Bonwick, G. A. 2019. Ensuring the future of functional foods. *International Journal of Food Science and Technology*, 54(5), 1467–1485. https://doi.org/10.1111/ijfs.14060.

Brown, L., Caligiuri, S. P. B., Brown, D., & Pierce, G. N. 2018. Clinical trials using functional foods provide unique challenges. *Journal of Functional Foods*, 45, 233–238. https://doi.org/10.1016/j.jff.2018.01.024.

BSI, British Standards Institution 2005. *Vocabulary – Nanoparticles, Publicly Available Specification*, PAS 71:2005. BSI, London.

Burton, J. 2022. *The Growth of Functional Foods Post Pandemic*. Available from: https://www.newfoodmagazine.com/article/163783/the-growth-of-functional-foods-post-pandemic/ (Last accessed on 25 July 2022).

Canadian Food Business Spring 2017. *Functional Foods*, p. 4. Available from: https://issuu.com/dovetailcommunications6/docs/cfb_spring_ebook (Last accessed on 24 July, 2022).

Chemat, F., Rombaut, N., Fabiano-Tixier, A.-S., Pierson, J. T., & Bily, A. 2015. Green extraction: From concepts to research, education, and economical opportunities. In: F. Chemat & J. Strube (Eds.), *Green Extraction of Natural Products*. Wiley-VCH, Weinheim, pp. 1–36. https://doi.org/10.1002/9783527676828.ch1.

Committee on Opportunities in the Nutrition and Food Sciences, Food and Nutrition Board, Institute of Medicine 1994. Enhancing the food supply. In: P.R. Thomas & R. Earl (Eds.), *Opportunities in the Nutrition and Food Sciences: Research Challenges and the Next Generation of Investigators*. National Academy Press, Washington, DC, pp. 98–142.

Daliri, E. B. M., & Lee, B. H. 2015. Current trends and future perspectives on functional foods and nutraceuticals. In: M. T. Liong (Ed.), *Beneficial Microorganisms in Food and Nutraceuticals*. Microbiology monographs: Vol. 27. Springer, Cham, pp. 221–244.

Day, L., Seymour, R. B., Pitts, K. F., Konczak, I., & Lundin, L. 2009. Incorporation of functional ingredients into foods. *Trends in Food Science and Technology*, 20(9), 388–395. https://doi.org/10.1016/j.tifs.2008.05.002.

de Almada, C. N., Almada, C. N., Martinez, R. C. R., & Sant'Ana, A. S. 2016. Paraprobiotics: Evidences on their ability to modify biological responses, inactivation methods and perspectives on their application in foods. *Trends in Food Science and Technology*, 58, 96–114. https://doi.org/10.1016/j.tifs.2016.09.011.

Ellis, E. 2022. *Functional Foods*. Available from: https://www.eatright.org/food/nutrition/healthy-eating/functional-foods (Last accessed on 24 July 2022).

Essa, M. M., Bishir, M., Bhat, A., Chidambaram, S. B., Al-Balushi, B., Hamdan, H., Govindarajan, N., Freidland, R. P., & Qoronfleh, M. W. 2021. Functional Foods and their impact on health. *Journal of Food Science and Technology*. https://doi.org/10.1007/s13197-021-05193-3.

European Commission. 2006. The appropriateness of existing methodologies to assess the potential risks associated with engineered and adventitious products of nanotechnologies. Available from: https://ec.europa.eu/health/ph_risk/committees/04_scenihr/docs/scenihr_o_003b.pdf (Last accessed on 25 July 2022).

Farid, M., Kodama, K., Arato, T., Okazaki, T., Oda, T., Ikeda, H., & Sengoku, S. 2019. Comparative study of functional food regulations in Japan and globally. *Global Journal of Health Science*, 11(6), 132. https://doi.org/10.5539/gjhs.v11n6p132.

Granato, D., Barba, F. J., Bursać Kovačević, D., Lorenzo, J. M., Cruz, A. G., & Putnik, P. 2020. Functional foods: Product development, technological trends, efficacy testing, and safety. *Annual Review of Food Science and Technology*, 11(1), 93–118. https://doi.org/10.1146/annurev-food-032519-051708.

Granato, D., Nunes, D. S., & Barba, F. J. 2017. An integrated strategy between food chemistry, biology, nutrition, pharmacology, and statistics in the development of functional foods: A proposal. *Trends in Food Science and Technology*, 62, 13–22. https://doi.org/10.1016/j.tifs.2016.12.010.

Grunewald, S. J. 2015. *Printed Foods to Help Patients Suffering with Dysphagia*. Available from: https://3dprint.com/102674/3d-printed-food-dysphagia/ (Last accessed on 25 July 2022).

Gur, J., Mawuntu, M., & Martirosyan, D. 2018. FFC's advancement of functional food definition. *Functional Foods in Health and Disease*, 8(7), 385. https://doi.org/10.31989/ffhd.v8i7.531.

Harrell, E. 2019. *Neuromarketing: What You Need to Know*. Available from: https://hbr.org/2019/01/neuromarketing-what-you-need-to-know (Last accessed on 25 July 2022).

Henry, C. J. 2010. Functional foods. *European Journal of Clinical Nutrition*, 64(7), 657–659. https://doi.org/10.1038/ejcn.2010.101.

Imperial, I. C., & Ibana, J. A. 2016. Addressing the antibiotic resistance problem with probiotics: Reducing the risk of its double-edged sword effect. *Frontiers in Microbiology*, 7, Article 1983. https://doi.org/10.3389/fmicb.2016.01983.

Jackson, C.C. & Paliyath, G. 2011.In: G. Paliyath, M. Bakovic & K. Shetty (Eds.), *Functional Foods, Nutraceuticals, and, Degenerative Disease Prevention*, John Wiley & Sons, Inc., West Sussex, UK, pp. 11–43.

Jędrusek-Golińska, A., Górecka, D., Buchowski, M., Wieczorowska-Tobis, K., Gramza-Michałowska, A., & Szymandera-Buszka, K. 2020. Recent progress in the use of functional foods for older adults: A narrative review. *Comprehensive Reviews in Food Science and Food Safety*, 19(2), 835–856. https://doi.org/10.1111/1541-4337.12530.

Kaput, J., & Rodriguez, R. L. 2004. Nutritional genomics: The next frontier in the postgenomic era. *Physiological Genomics*, 16(2), 166–177. https://doi.org/10.1152/physiolgenomics.00107.2003.

Keservani, R. K., Sharma, A. K., Ahmad, F., & Baig, M. E. 2014. Nutraceutical and functional food regulations in India. In: D. Bagchi (Ed.), *Nutraceutical and Functional Food Regulations in the United States and Around the World*. Academic Press, Elsevier Inc., London, pp. 327–342. https://doi.org/10.1016/b978-0-12-405870-5.00019-0.

Korpela, K. 2018. Diet, microbiota, and metabolic health: Trade-off between saccharolytic and proteolytic fermentation. *Annual Review of Food Science and Technology*, 9(1), 65–84. https://doi.org/10.1146/annurev-food-030117-012830.

Kouzani, A. Z., Adams, S., Whyte, D. J., Oliver, R., Hemsley, B., Palmer, S., & Balandin, S. 2017. 3D printing of food for people with swallowing difficulties. *KnE Engineering*, 2(2), 23. https://doi.org/10.18502/keg.v2i2.591.

Lestari, L. A. 2021. Opportunities and challenges of functional foods and nutraceuticals development during Covid-19 pandemic. *BIO Web of Conferences*, 41, 02009. https://doi.org/10.1051/bioconf/20214102009.

Maeda-Yamamoto, M. 2017. Development of functional agricultural products and use of a new health claim system in Japan. *Trends in Food Science and Technology*, 69, 324–332. https://doi.org/10.1016/j.tifs.2017.08.011.

MarketsandMarkets Research Private Ltd. 2020. *Functional Food Ingredients Market by Type (Probiotics, Prebiotics, Proteins & Amino Acids, Phytochemicals & Plant Extracts, Omega-3 Fatty Acids, Carotenoids, and Fibers & Specialty Carbohydrates), Source, Application, and Region - Global Forecast to 2026*. Available from: https://www.marketsandmarkets.com/Market-Reports/functional-food-ingredients-market-9242020.html (Last accessed on 24 July 2022).

Martirosyan, D., Lampert, T., & Ekblad, M. 2022. Classification and regulation of functional food proposed by the functional food center. *Functional Food Science*, 2(2), 25. https://doi.org/10.31989/ffs.v2i2.890.

Martirosyan, D., & Pisarski, K. 2015. Bioactive compounds: Their role in functional food and human health, classifications, and definitions. In: D. Martirosyan & J. Zhou (Eds.), *Bioactive Compounds and Cancer*. Food Science Publisher, San Diego, CA, pp. 238–277.

Martirosyan, D., Zolaedov, V., & Miroshnichenko, L. 2009. *Functional Foods for Chronic Diseases: Diabetes and Related Diseases (Volume 6)*. In: D. Martirosyan & C. Prasad, The 6th International Conference Proceedings, Functional Food Center Inc., TX, USA, pp. 163–170.

Martirosyan, D. M., & Singh, J. 2015. A new definition of functional food by FFC: What makes a new definition unique? *Functional Foods in Health and Disease*, 5(6), 209. https://doi.org/10.31989/ffhd.v5i6.183.

Nazir, M., Arif, S., Khan, R. S., Nazir, W., Khalid, N., & Maqsood, S. 2019. Opportunities and challenges for functional and medicinal beverages: Current and future trends. *Trends in Food Science and Technology*, 88, 513–526. https://doi.org/10.1016/j.tifs.2019.04.011.

NFRA. 2022. *Functional Foods: Fad or Opportunity?* Available from: https://nfraweb.org/functional-foods-fad-or-opportunity/ (Last accessed on 25 July 2022).

Nilova, L., & Malyutenkova, S. 2020. Challenges and opportunities for promoting functional foods in e-commerce. In: *Proceedings of the E3S Web of Conferences*, TPACEE-2019, 164, 09012. https://doi.org/10.1051/e3sconf /202016409012.

Paquin, P. 2009. *Functional and Speciality Beverage Technology*. CRC Press, Woodhead Publishing Limited, Cambridge.

Pem, D., & Jeewon, R. 2015. Fruit and vegetable intake: Benefits and progress of nutrition education interventions – Narrative review article. *Iranian Journal of Public Health*, 44(10), 1309–1321. PMID: 26576343; PMCID: PMC4644575.

Pojić, M., Mišan, A., & Tiwari, B. 2018. Eco-innovative technologies for extraction of proteins for human consumption from renewable protein sources of plant origin. *Trends in Food Science and Technology*, 75, 93–104. https://doi.org/10.1016/j.tifs.2018.03.010.

Poojary, M. M., Putnik, P., Bursać Kovačević, D., Barba, F. J., Lorenzo, J. M., Dias, D. A., & Shpigelman, A. 2017. Stability and extraction of bioactive sulfur compounds from allium genus processed by traditional and innovative technologies. *Journal of Food Composition and Analysis*, 61, 28–39. https://doi.org/10.1016/j.jfca.2017.04.007.

Puhakka, R., Valve, R., & Sinkkonen, A. 2018. Older consumers' perceptions of functional foods and non-edible health-enhancing innovations. *International Journal of Consumer Studies*, 42(1), 111–119. https://doi.org/10.1111/ijcs.12400.

Putnik, P., Bursać Kovačević, D., Ježek, D., Šustić, I., Zorić, Z., & Dragović-Uzelac, V. 2018. High-pressure recovery of anthocyanins from grape skin pomace (*Vitis vinifera* cv. Teran) at moderate temperature. *Journal of Food Processing and Preservation*, 42(1). https://doi.org/10.1111/jfpp.13342.

Reportlinker 2022. *Functional Food Global Market Opportunities and Strategies to 2030: COVID-19 Growth and Change*. Available from: https://www.reportlinker.com/p06039376/Functional-Food-Global-Market-Opportunities-And-Strategies-To-COVID-19-Growth-and-Change.html (Last accessed on 24 July 2022).

Roberfroid, M. B. 2000. Defining functional foods. In: G. Gibson & C. Williams (Eds.), *Functional Foods*. Woodhead, Cambridge, pp. 9–27.

Roberfroid, M. B. 2002. Global view on functional foods: European perspectives. *British Journal of Nutrition*, 88(S2). https://doi.org/10.1079/bjn2002677.

Rodiño-Janeiro, B. K., Vicario, M., Alonso-Cotoner, C., Pascua-García, R., & Santos, J. 2018. A review of microbiota and irritable bowel syndrome: Future in therapies. *Advances in Therapy*, 35(3), 289–310. https://doi.org/10.1007/s12325-018-0673-5.

Rodríguez-Roque, M. J., Rojas-Graü, M. A., Elez-Martínez, P., & Martín-Belloso, O. 2014. In vitro bioaccessibility of health-related compounds as affected by the formulation of fruit juice-and milk-based beverages. *Food Research International*, 62, 771–778. https://doi.org/10.1016/j.foodres.2014.04.037.

Serafini, M., Stanzione, A., & Foddai, S. 2012. Functional foods: Traditional use and European legislation. *International Journal of Food Sciences and Nutrition*, 63(sup1), 7–9. https://doi.org/10.3109/09637486.2011.637488.

Shimizu, T. 2003. Health claims on functional foods: The Japanese regulations and an international comparison. *Nutrition Research Reviews*, 16(2), 241–252. https://doi.org/10.1079/nrr200363.

Singh, H. 2016. Nanotechnology applications in functional foods; opportunities and challenges. *Preventive Nutrition and Food Science*, 21(1), 1–8. https://doi.org/10.3746/pnf.2016.21.1.1.

Singletary, K. W., & Morganosky, M. A. 2004. Functional foods: Consumer issues and future challenges. *Journal of Food Distribution Research*, 35, 1–5.

Sun-Waterhouse, D. 2011. The development of fruit-based functional foods targeting the health and wellness market: A review. *International Journal of Food Science and Technology*, 46(5), 899–920. https://doi.org/10.1111/j.1365-2621.2010.02499.x.

Swinbanks, D., & O'Brien, J. 1993. Japan explores the boundary between food and medicine. *Nature*, 364(6434), 180–180. https://doi.org/10.1038/364180a0.

Tanna, B., & Mishra, A. 2018. Metabolites unravel nutraceutical potential of edible seaweeds: An emerging source of functional food. *Comprehensive Reviews in Food Science and Food Safety*, 17(6), 1613–1624. https://doi.org/10.1111/1541-4337.12396.

Taverniti, V., & Guglielmetti, S. 2011. The immunomodulatory properties of probiotic microorganisms beyond their viability (ghost probiotics: proposal of paraprobiotic concept). *Genes and Nutrition*, 6(3), 261–274. https://doi.org/10.1007/s12263-011-0218-x.

Verbeke, W., Scholderer, J., & Lähteenmäki, L. 2009. Consumer appeal of nutrition and health claims in three existing product concepts. *Appetite*, 52(3), 684–692. https://doi.org/10.1016/j.appet.2009.03.007.

Verhagen, H., Vos, E., Francl, S., Heinonen, M., & van Loveren, H. 2010. Status of Nutrition and health claims in Europe. *Archives of Biochemistry and Biophysics*, 501(1), 6–15. https://doi.org/10.1016/j.abb.2010.04.012.

Wani, S. A., Naik, H. R., Wagay, J., Ganie, N. A., Mulla, M. Z., & Dar, B. N. 2022b. Mentha: A review on its bioactive compounds and potential health benefits. *Quality Assurance and Safety of Crops and Foods*, 14(4), 154–168.

Wani, S. A., Singh, A., & Kumar, P. (Eds.). 2022a. *Spice Bioactive Compounds: Properties, Applications, and Health Benefits* (1st ed.). CRC Press. https://doi.org/10.1201/9781003215387.

Weiss, J., Takhistov, P., & McClements, D. J. 2006. Functional materials in food nanotechnology. *Journal of Food Science*, 71(9). https://doi.org/10.1111/j.1750-3841.2006.00195.x.

World Health Organization. 2003. *Diet, nutrition and the prevention of chronic diseases.* WHO Technical Report Series 916. Geneva. Available from: http://www.fao.org/docrep/005/AC911E/AC911E00.HTM#Contents (Last accessed on 25 July 2022).

Yang, Y. 2008. Scientific substantiation of functional food health claims in China. *The Journal of Nutrition*, 138(6). https://doi.org/10.1093/jn/138.6.1199s.

Ye, Q., Georges, N., & Selomulya, C. 2018. Microencapsulation of active ingredients in functional foods: From research stage to commercial food products. *Trends in Food Science and Technology*, 78, 167–179. https://doi.org/10.1016/j.tifs.2018.05.025.

Zhao, Y. 2012. *Specialty Foods: Processing Technology, Quality, and Safety.* CRC Press, Boca Raton, FL.

WEB REFERENCES

Functional Food Center. 2018. Available from: https://www.functionalfoodscenter.net/ (Last accessed on 24 July 2022).

Jones, K. 2022. *The Rise of Functional Food: What Investors Need to Know.* Available from: https://www.visualcapitalist.com/sp/the-rise-of-functional-food-what-investors-need-to-know/; © 2022 Visual Capitalist (Last accessed on 25 July 2022).

CHAPTER 2

Technological Advancement in the Development of Functional Food

Supriya Kumari*, **Shyam Sundar**†, **Sarvesh Rustagi**‡ and **Ayman Younes Allam**§

2.1 INTRODUCTION

Functional foods are those which are designed to provide tremendous health benefits (Mollet & Rowland, 2002), provide nutrition according to recommended daily allowance (RDA) of humans, be low cost and improve mental and physical ailments (Nöthlings et al., 2007, Takachi et al., 2008). Attributable to rising consumer buying power and dietary demand, there has been great advancement in the development of functional foods (Shah, 2007). Beyond their nutritional content, functional foods have potential health advantages. These foods could include extra vitamins, minerals, herbs or other compounds with potential health advantages (Huggett and Schliter, 1996; Charalampopoulos et al., 2002).

New scientific and technical approaches to food processing and the introduction of innovative foods are the main topics of current innovations in the food business (Bigliardi & Galati, 2013). Functional foods can be fortified, incorporated or enriched with one or more functional ingredients. As an illustration, consider a functional food or beverage with additional minerals, vitamins or phytonutrients to improve its nutritional value. With health advantages, functional food should be affordable and have good customer acceptance in terms of sensory qualities. Functional foods that lack desirable sensory qualities won't be accepted. Cost is yet another crucial factor in creating nutritious meals. It's crucial to keep in mind that functional foods might not provide as many health advantages as supplements. However, frequent use of

* Corresponding author E-mail: supriyakumari33@gmail.com
† Corresponding author E-mail: hishyam@outlook.com
‡ Corresponding author E-mail: sarveshrustagi@uttaranchaluniversity.ac.in
§ Corresponding author E-mail: ayman.younis21@agr.menofia.edu.eg

DOI: 10.1201/9781003315100-2

functional foods may have a major long-term influence on human health. The development of functional foods entails the addition of certain bioactive substances with advantageous physiological effects (Hamer, Owen & Kloek, 2005).

The Department of Health and Human Services has suggested the crucial role of food and the selection of particular food in the diet to affect health in a significant way. Improper selection of food may lead to various diseases and can even be a cause of death. Today, several new technologies are being developed for the creation of functional foods. Therefore, it is clear that the market will soon present a wider variety of cutting-edge health-food goods. However, in order to support claims that the consumption of functional foods or food components has an impact on health outcomes, solid scientific data and broad scientific consensus are required. To promote health, several technological techniques can modify foods to develop functional food (Duchateau & Klaffke, 2008). Before the development of a functional food, it is crucial to understand the physicochemical properties of food, microstructure, bioavailability and physiological processes that take place during food ingestion, digestion and absorption (McClements, Decker, Park & Weiss, 2008; Sanguansri & Augustin, 2006). Before introducing any product to the market, development calls for an in-depth understanding of the products and the target consumers (Beardsworth & Keil, 1992). The main purpose of presenting this chapter is to overview more advanced technical processes that meet the requirements of the consumer without any loss of nutrients during processing. To meet demands and expectations in the field of functional meals, a wide range of technical strategies are needed that can provide a wide perspective to save functional foods from deterioration and also improve the functionality of food by enhancing the bioactive compounds, vitamins, and dietary fibres. This chapter also focuses on recent technologies that contribute to the design of personalised functional food without any kind of deterioration and providing nutrition. Figure 2.1 shows market research on functional foods.

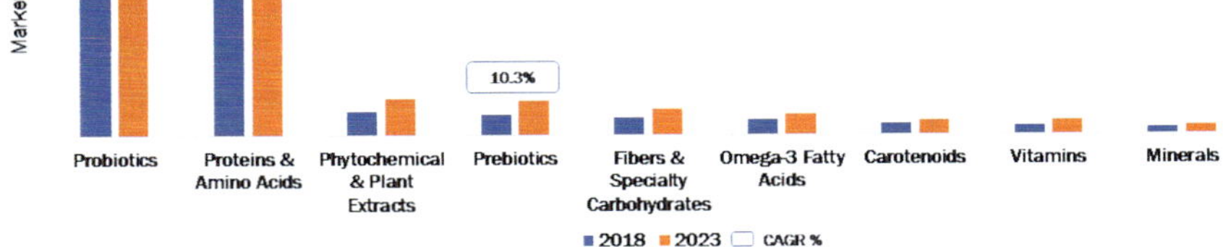

Figure 2.1 Showing market research on functional foods (source: Market and Markets Blog)

2.2 ADVANCEMENTS IN FUNCTIONAL FOOD INGREDIENTS

The food sector industry has been impacted by technological and economic changes in society, seen in the manufacturing of different functional foods and food processing. This has had a substantial effect on the whole food chain up until the point at which food is distributed to final consumers and has forced businesses to provide food products that satiate consumers' demand for a healthy lifestyle, being a high priority. Recent advancements in the food sector have primarily involved the introduction of innovative foods as well as new scientific and technical methods for preparing food. The selection of functional food ingredients and their preparation play a crucial role in the development of the functional food. The most well-known functional food ingredients, of both animal and plant origin, aim to lower blood pressure, cholesterol, blood sugar and osteoporosis (Kaur & Das, 2011). There are various novel functional food ingredients from various sources such as carotenoids, bioactive components, omega-3 and 6 fatty acids. They have a good impact on vision and cognitive function, as well as preventing diabetes, cancer, heart disease and other ailments that cause premature ageing (Shahidi & Ambigaipalan, 2015; Silva, Albuquerque, Alves, Oliveira & Costa, 2020). Table 2.1 shows various techniques for the development of functional foods.

2.2.1 Recent Advancements in Functional Food Ingredients

1. **Plant-based ingredients:** The use of plant-based ingredients including fruits, vegetables and herbs to deliver health advantages has gained popularity. Antioxidants, phytochemicals and other useful substances abound in these foods. Natural substances like vitamin E, rosemary extract and other antioxidants can preserve food in response to the rising demand for natural food preservatives. Herbs are the origins of a variety of useful substances that are advantageous when added to food, either whole or extracted. They are helpful in treating several diseases, including cardiovascular disease (CVD), obesity, fatigue Alzheimer's, multiple sclerosis, Parkinson's, cerebral injury, cancer, allergy, bronchitis, rheumatoid arthritis, psoriasis and diabetes (Aggarwal & Harikumar, 2009; Ernst, 2002; Ramírez et al., 2006). Fibre is found in vegetables, fruits, whole grains and other natural foods, and can help lower cholesterol and regulate blood sugar.
2. **Microalgae:** Microalgae are a class of photosynthetic microorganisms that can produce biomass from solar energy. They provide an abundance of protein, minerals and vitamins as well as omega-3 fatty acids and other beneficial bioactive compounds. They can be used to produce functional ingredients (Plaza, Herrero, Cifuentes & Ibanez, 2009; Matos, Cardoso, Bandarra & Afonso, 2017).
3. **Synbiotic:** Probiotics and prebiotics are combined to create synbiotics, which help the development of good bacteria in the stomach. Live microorganisms known as probiotics are comparable to the helpful microbes present in the human stomach. Yoghurt, kefir and other fermented foods can all benefit from their addition. Non-digestible carbohydrates known as prebiotics serve as food for probiotics and other good bacteria in the stomach. Oligofructose and inulin are two examples (Barros et al., 2022; Kareb & Aïder, 2019; Ziemer & Gibson, 1998).
4. **Bioactive peptides:** Peptides are short chains of amino acids that can have a variety of health benefits. They can be derived from food proteins such as casein and

Table 2.1 Various techniques for the development of functional foods

	Techniques for functional food development	Techniques	Compound of interest	References
1	Traditional methods/ techniques	Blending	Bioactive compound	Rivas et al., 2007; Zhang et al., 2010
		Fermentation	Anti-nutritional compound, probiotics	Yadav et al., 2011, Yadav et al., 2007
		Biotechnological	Bioactive compound, fatty acids, vitamin E	Zhao et al., 2011; Woods et al., 2009
2	Technological advancements	Micro-encapsulation/ nano-encapsulation	Bioactive compound, probiotics	Thies, 2005; Gibbs et al., 1999
		Emulsification	Bioactive compound, probiotics, vitamins	Chen & Chen, 2007; Kailasapathy, 2002
		Spray-drying	Bioactive compound, probiotics	Morgan et al., 2006
		Freeze-drying	Bioactive compound, probiotics, phenolics	Maltesen et al., 2008; Fowler et al., 2006
		Extrusion	Bioactive compound, proteins	Krasaekoopt et al., 2008; Vaziri et al., 2018
		Coacervation	Bioactive compound, essential oils, probiotics	Gouin, 2004; Yuan et al., 2017
		Electrospinning	Bioactive compound, probiotics	López-Rubio et al., 2012; Sobel et al., 2014
3	Novel techniques	Reverse engineering	Bioactive compounds, probiotics, Vitamins	Dima et al., 2020
		3D printing	Bioactive compounds, probiotics, Vitamins	Wilms et al., 2021
		Nutrigenomics	Bioactive compounds, probiotics, vitamins	García-Cañas et al., 2010

whey, or they can be produced through biotechnology. Based on their content and sequencing, bioactive peptides are protein-specific fragments of many amino acid residues that have desired biological properties (Bersi, Barberis, Origone & Adaro, 2018; Dullius, Goettert & de Souza, 2018).

5. **Bioactive compounds:** Scientists have isolated and identified bioactive compounds from food and food by-products for their health benefits. These include compounds such as anthocyanins, omega-3, omega-6, vitamin E, flavonoids and polyphenols (Li et al., 2021; Marsanasco, Márquez, Wagner, Chiaramoni & Alonso, 2015; Murray, Dordevic, Ryan & Bonham, 2018). Omega-3 fatty acids are found in fish and various oilseeds which are believed to be beneficial for heart health. Phytosterols are plant-based compounds that are similar to cholesterol and can help lower cholesterol levels in the blood. This is rich with antioxidants and also provides protection to the body, which helps prevent free radical damage. Various naturally existing antioxidants including vitamin C, vitamin E, beta-carotene etc. serve as functional ingredients in the preparation of functional food (Diplock et al., 1998; Misra, Pandey & Mishra, 2021).

2.3 RECENT TECHNOLOGICAL DEVELOPMENT IN FUNCTIONAL FOODS

Recently, a number of novel approaches have been developed and are being developed to include alternate sources of functional ingredients in mainstream food products including meat substitutes, protein and energy bars and beverages. Other bioactive ingredients have been added to a number of probiotic meals and drinks. To create functional food, various fruits, vegetables and herbs that are recognised for their mineral, vitamin, dietary fibre and phytochemical content and that offer a variety of health advantages have been combined. Many technical advancements have been made recently to improve food quality, sustainability and the advancement of functional foods, including the utilisation of microwaves, high hydrostatic pressure, pulsed electric fields and ultrasounds. Recent technological developments in functional foods have focused on several areas, as will be discussed in the following sections.

2.3.1 Fermentation

Fermentation is a natural and oldest process that can be used to produce functional foods by introducing beneficial microorganisms into food products. This technology is used to produce probiotic-rich foods such as yoghurt and kefir. The advancement in the fermentation process involving the microbial generation of bioactive components, such as specific vitamins, organic acids, bioactive peptides and fatty acids, is what gives fermented functional foods their biogenic qualities (Shiby & Mishra, 2013; Stanton, Ross, Fitzgerald & Van Sinderen, 2005).

2.3.2 Biotechnology and Nanotechnology

Biotechnology can be used to produce functional components such as omega-3 fatty acids, phytosterols and other compounds that have potential health benefits. In order to give particular nutritional and health benefits, GM (genetic modification) has also been applied in the advancement of functional foods (Bagchi, Lau & Ghosh, 2010). The use of nanotechnology on the other hand in functional food development has the potential to improve the stability, bioavailability and delivery of functional ingredients in foods. Nanotechnology advancement has several advantages for the food sector, including improved nutrient benefits, safety, food fabrication, smart packaging, food preservation from physical and chemical deterioration, and food characterisation. The solubility and bioavailability of the ingredients have been improved by the use of specific novel nanoparticles and nanomaterials (Kanugo, Goyal, Sharma & Gautam, 2022).

2.3.3 Encapsulation

Micro- and nano-encapsulation technology allows functional ingredients to be encapsulated in a protective coating, which can improve their stability and bioavailability. This technology can also be used to deliver functional ingredients in a controlled-release manner. Functional ingredients including antioxidants, vitamins, flavours, minerals, natural colours, nutraceuticals, probiotics, bioactive lipids, antimicrobials, proteins and carbohydrates are encapsulated and utilised as functional food or incorporated to form functional food (Abdel-Aty, Elsayed, Salah, Bassuiny & Mohamed, 2021; Delshadi, Bahrami, Tafti, Barba & Williams, 2020; Dias, Botrel, Fernandes & Borges, 2017; McClements, 2018).

2.3.4 Protein Fortification and Meat Analogues

A growing interest in adding plant-based proteins to meals to serve as a source of complete protein has coincided with the focus on plant-based diets. Microalgae and seaweeds also produce good-quality protein, which plays a role in the development of functional food. The protein fraction in various kinds of seaweed is comparable to a plant source such as soybeans (Holdt & Kraan, 2011; Plaza et al., 2009). Meat analogue is another sustainable approach to producing meat substitutes. These plant or algal protein-based analogues can be enhanced with a variety of necessary nutrients or phytonutrients to develop functional foods (Asgar, Fazilah, Huda, Bhat & Karim, 2010; Singh et al., 2021).

2.3.5 Precision Nutrition and Nutraceutical Utilisation

With the integration of technology and data analytics, industries are now able to personalise food products to meet the specific nutritional needs of the individual (Boye, 2015).

2.4 SPECIFIC TECHNOLOGIES FOR FUNCTIONAL FOOD DEVELOPMENT THAT PROTECT THE PHYSIOLOGICALLY ACTIVE COMPOUNDS FROM DETERIORATION

The purpose of food nowadays is to avoid diseases connected to malnutrition, increase physical and mental well-being and satisfy hunger, as well as provide individuals with the nutrients they need (Takachi et al., 2008; Nöthlings et al., 2007). Major modifiable risk factors for the development of coronary heart disease, cancer, type 2 diabetes, obesity, osteoporosis and periodontal disease include a range of dietary patterns and lifestyle choices, according to the World Health Organization and the Food and Agriculture Organization. Functional foods are quite important in this aspect. The increasing demand for such foods can be explained by the rising cost of healthcare, the constantly increasing life expectancy and the desire for a high quality of life in later years (Roberfroid, 2000). The development of new functional food products becomes more difficult in this context since it must satisfy consumer expectations for products that are both delicious and healthy (Shah, 2007). The inclusion of bioactive ingredients including flavours, essential oils, antioxidants, vitamins, natural extracts and probiotics has received a lot of attention as people become more aware of the potential health advantages of functional food items (Chawda et al., 2017). The incorporation of these important ingredients for the development and advancement of functional foods is being done in such a way that their bioactive compounds, important nutrients, and probiotic cells can remain intact or preserved, and so that their organoleptic properties remain.

2.4.1 Functional Food Components

Developing a sustainable edible delivery system for a few functional components is very difficult because these components can easily become destroyed during processing. These functional components include bioactive compounds, antioxidants, fatty acids, carotenoids, minerals, protein and vitamins, which are essential for the growth of individuals (Ubbink, 2002; Shefer and Shefer, 2003; Chen et al., 2006; Ubbink and Kruger, 2006). These nutritional substances have different molecular weights, polarities and charges, which may become altered after processing, leading to physiological changes. As a result, various delivery mechanisms are frequently required to state particular molecular and physicochemical issues related to each functional or nutraceutical component. Various forms of delivery systems, such as association colloids, simple solutions, suspensions, emulsions, gels, solid matrices, etc., have been designed to encapsulate functional chemicals. For the encapsulation, preservation and distribution of functional agents, each type of delivery method has particular benefits and drawbacks. Figure 2.2 shows the concept of the functional food development chart

2.4.2 Traditional Techniques for the Development of Functional Food

2.4.2.1 Blending and Fortification

Formulating a functional food product using blending techniques to boost the food product nutritionally is one of the cheapest technologies but is quite old nowadays (Betoret, Betoret, Vidal & Fito, 2011). Its successful management of deficits of vitamins A and D, various B vitamins (thiamine, riboflavin and niacin), iodine and iron has a long history of usage in the development of functional foods. Iodised salt was first made available in Switzerland and the United States in the early 1920s, and

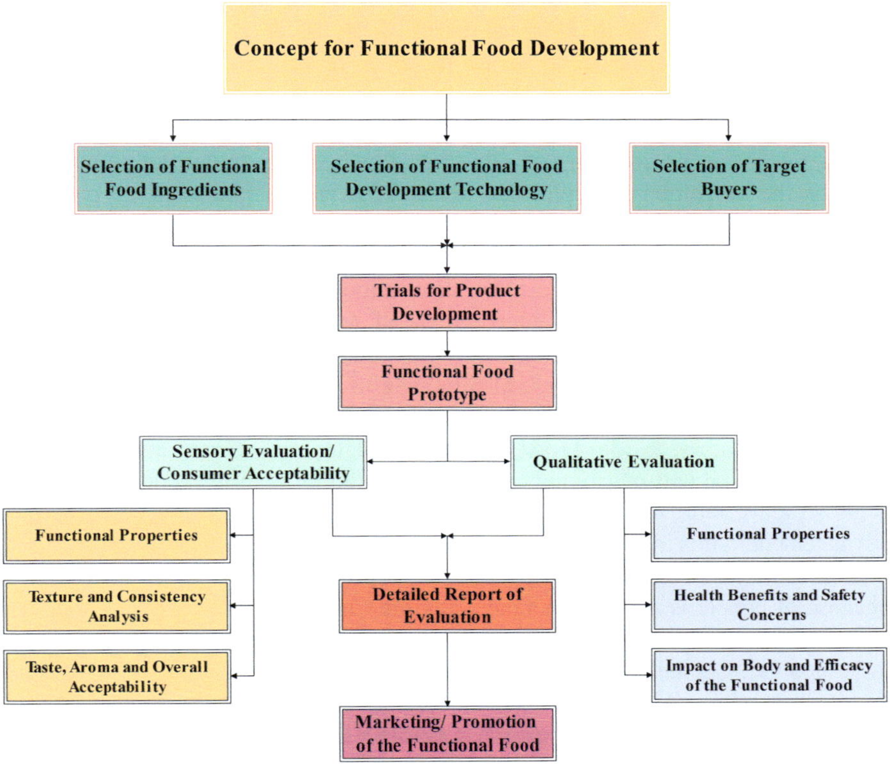

Figure 2.2 Concept of functional food development chart (source: Burdock & Carabin, 2019; Jones & Jew, 2007)

it has since spread steadily over the world to the point that it is currently used in the majority of nations (Bürgi, Supersaxo & Selz, 1990; Marine & Kimball, 1920). In Denmark, milk was incorporated with vitamin D (FAO & WHO, 2006), and margarine, which is a fat, was augmented with vitamin A (FAO & WHO, 2006). In more recent times, the inclusion of dietary fibres with health advantages presented a better chance to enhance people's health; as a result, the scientific community, consumers and food manufacturers paid close attention to this category of compounds. There are countless dietary active substances (vitamins, probiotics, bioactive peptides, antioxidants, etc.), and the range of obtained final products is continually expanding. Juices are often enhanced with flavonoids and vitamins to give children the necessary nutrients for optimum growth and development (Rivas et al., 2007; González-Molina, Moreno & Garcıa-Viguera, 2009), and meats that are often augmented with a large number of bioactive components (Jimenez-Colmenero,

Sanchez-Muniz & Olmedilla-Alonso, 2010; Zhang et al., 2010) are developed using formulation techniques.

2.4.2.2 Fermentation

The fermentation process involves the breakdown of big molecules into smaller ones or the activation of molecular oxidation/reduction pathways by particular microorganisms. This process helps in producing various bioactive compounds (Yadav et al., 2011). One of the crucial subgroups of these microorganisms that takes part in the fermentation process of diverse food products is probiotic microbes, which are responsible for the improvement of gut health. Live microorganisms are referred to as probiotics because they can benefit a host's health when given in adequate quantities, probiotics are also thought to be usually safe for consumption by humans and have a range of beneficial health effects, including those that are immunomodulatory, cardio-protective, anti-oxidant, anti-cancer and anti-obese. They also reduce anti-nutritional compounds including tannin, saponin and phytic acids (Yadav, Jain & Sinha, 2007). Fermentation is a traditional process, and advancement in this traditional process is necessary because there are a lot of lactose intolerance issues beginning to appear. There is a need for vegetarian alternatives because of concerns about intolerance, allergies and other factors, and there is an increasing need for dairy substitutes. As a result, soy-based meals are becoming more popular. The newest kinds of food products available nowadays are probiotic goods made with soy protein blended fruit juices, an easy method to incorporate soy protein into a normal diet (Champagne & Gardner 2008). Sales of probiotic-rich soy products have increased over the past ten years. These products include yoghurt-like blends, fruit juice blends and fermented beverages. This trend indicates that new products are being developed with appropriate sensory and nutritional appeal as well as beneficial properties when regularly consumed.

2.4.2.3 Biotechnological Techniques for Breeding

Since the dawn of human history, biotechnology has been used in the breeding of crops and animals. Around 12,000 years ago, people began domesticating various animal and plant species with the goal of obtaining desired traits, which were determined by environmental, nutritional and social needs without any knowledge of the underlying molecular-level processes. This was the start of the selection and assessment of various breeds (National Research Council, 1989, p. 16). Agriculture and livestock have always been marketed as sources of very nutrient-dense commodities. When breeding and agronomic techniques fall short of considerably enhancing food products, biotechnology offers a good alternative. (Zhao & Shewry, 2011). The application of molecular biology tools and the creation of genetically modified seeds transformed biotechnology into a contemporary method that provides an extra approach to changing the composition of crops. Other food crops, such as potatoes, have also been fortified with vitamin A (Tanumihardjo et al., 2008); and maize (Zhu et al., 2008), tomato (de La Garza, Gregory & Hanson,

2007) and rice grain (Storozhenko et al., 2007) have been fortified with folates; and soybeans and corn seeds are incorporated with vitamin E (Karunanandaa et al., 2005). Additionally, animal breeding provides the opportunity to produce better food products. Considerable research has been conducted to look at the sources of nutrients that may be added to animal diets and how they are then transferred to the products that are produced. In a review, Woods and Fearon (2009) evaluated the sources of fatty acids that can be added to animal diets and then transferred to meat, eggs or milk.

2.4.3 Specific and Advanced Technologies for Functional Food Development That Prevent the Deterioration of Physiologically Active Compounds

2.4.3.1 Encapsulation

Small particles such as solids, liquid droplets and gases are encased in a coating by a process known as microencapsulation. Many different biologically relevant materials can be protected by microencapsulation, including cells derived from bacteria, yeast and animals, as well as tiny compounds and proteins (enzymes, hormones) (Thies, 2005). Applications ranging from cell treatment to drug delivery as well as applications in the food industry for the production of neutraceuticals, such as adaptable technology, are therefore extensively researched and used in the advanced domains of biomedicine and pharmaceutics (Smidsrød & Skjak-Braek, 1990). The biological activity of the newly introduced functional products enriched with bioactive substances may be hampered by other dietary components. Therefore, it is essential that the encapsulation technique protects bioactivity throughout all stages of processing, storage and delivery, as well as has appropriate packaging (Gibbs et al., 1999; McClements, Decker & Park, 2008). Studies have demonstrated the effectiveness of some microencapsulation materials. Some biopolymers have already been tested by using spray-drying to see if they can encapsulate certain substances, including starch, hydrocolloids, whey proteins, gelatins and maltodextrins (Gharsallaoui et al., 2007). There are some successful microencapsulation techniques that are efficiently utilised for the encapsulation of bioactive compounds.

2.4.3.2 Emulsification

This process comprises two phases process including a dispersed phase, which holds a suspension of the cell polymer, and a continuous phase, which is either an oil (plant or mineral oil) or an organic solution. The mixture is homogenised with the help of surfactants to create the emulsion. When a cross-linking agent or chilling technique is used to insolubilise the water-soluble polymer within the oil phase, the resultant particles are generated. After being produced, the microbeads are centrifuged or filtered. The rate of agitation, the rate of cross-linking agent addition, the concentration of surfactants and the water-to-oil ratio all affect the size of the beads which may vary from 25 μm to 2 mm (Kailasapathy, 2009; Shah & Ravula, 2000).

Emulsifiers reduce surface tension to increase emulsion stability, and as a result, the microsphere size is reduced (Krasaekoopt et al., 2004). With this approach, probiotics survive longer, and the microcapsules have a smaller diameter, making them simpler to expand (Chen & Chen, 2007). The primary drawbacks are the creation of micro and macro capsules in a variety of sizes and shapes, as well as the need for a second polymer solution for the cell's extra coating (Kailasapathy, 2009). Milk protein is used in emulsification with gelification to encapsulate probiotics since it has gelation capabilities and serves as a natural carrier for the probiotic substances (Livney, 2010). Casein micelles created by the release of k-casein have been applied for the gelation process using rennet, a proteolytic enzyme complex (Heidebach, Först & Kulozik, 2009). The flocculating micelles can guard the cells under a low pH environment when non-covalent cross-linkages are present at a temperature above 18 °C (Bansal et al., 2007). The adaptability of the microcapsule size, which affects the organoleptic acceptability of the final food product, has demonstrated that this technology is acceptable for usage in food applications. Emulsification with interfacial polymerisation is applied as the continuous phase, where the aqueous solution containing the probiotic substances is dispersed in order to increase the productivity of the microorganisms during the fermentation process (Yáñez-Fernández et al., 2008). The polymerisation process is started with a biologically compatible material that is solubilised easily in the continuous phase, resulting in membrane-coated encapsulated droplets (Kailasapathy, 2002). The emulsification process is an effective technique used efficiently for the encapsulation of important nutrients such as bioactive substances, vitamins and probiotics. Gaudreau et al. (2016) reported that the emulsification procedure can be utilised to encapsulate probiotic microorganisms and green tea extract (GTE), and pectin and whey protein isolates were used as a coating material.

2.4.3.3 Spray-Drying

Since the late 1950s, the food industry has used this well-established technique of microencapsulation to stop the oxidative destruction of flavour oils by turning them from suspension to powder form. Because liquid suspension is needed for the spray-drying process, the variety of coat materials with acceptable solubility qualities is constrained. To coat various probiotics, ingredients such as modified starch maltodextrin and their combinations with some hydrocolloids such as gum acacia, carboxymethyl cellulose, guar gum and soy or whey protein isolates are frequently used. The widely utilised and affordable technology of spray-drying offers flexibility and yields high-quality goods. This procedure produces matrix-type microcapsules by allocating the main material into a polymeric solution and homogenising it, and then the atomisation process is carried out inside the drying chamber (Jackson & Lee, 1991). The process conditions, such as the droplet sizes and temperatures at the inlet and output, must be adequately regulated in the size of 10–150 μm and a temperature of 150–250°C in order to achieve the appropriate suitability of the encapsulated microbes in this continuous process. Comparing the freeze-drying technique to the spray-drying technique, less energy is used. The tolerance of the probiotic strain to osmotic, oxidative and thermal stress should be taken into consideration when spray-drying. Due to the use of high temperatures, some species, including Lactobacillus and Bifidobacterium, are particularly more prone to becoming destroyed in spray-drying conditions. Bacteria become resistant during osmosis activity as a result of the decrease in water activity, and they continue to exist by assembling suitable solutes like carbohydrates, sugar and amino acids (Woods 2009). Spray driers are largely used at an industrial level, where pneumatic systems help the probiotic bacteria to retain their viability. The spray-drying technique is an encapsulation process, which provides a shield to the organism by forming a coating film. Trehalose is one of the most efficient carbohydrates used for coating material due to its ability to stabilise membranes and proteins (Morgan et al., 2006). Apart from trehalose, other carbohydrates, such as mannitol, inulin, xylose, maltose, sorbitol, lactose, glucose, dextrose and maltodextrin, have also been employed as coating materials in a spray-drying process at a low cost (Perdana et al., 2014). Rodrigues Pereira et al. (2009) reported that when using protein isolates from green pea and cowpea, spray-drying can produce ascorbic acid microparticles. These bioactive compounds, protein isolates and probiotic strains have good emulsification properties, and using them as coating materials for the encapsulation of orange oil in the spray-drier technique, which uses more temperature, protects from deterioration and prevents oxidation (Kim & Morr, 1996).

2.4.3.4 Freeze-Drying

Lyophilisation, often known as freeze-drying, is a popular and practical approach for improving the storability of microencapsulated probiotics. The three main phenomena of freezing, primary drying and secondary drying can be used to summarise the process. Because of the formation of smaller ice crystals, compared to a slow freezing rate, a fast freezing rate guards against more extreme cellular damage (Fowler & Toner, 2006). It is possible to control process variables like temperature and freezing rate that can impact ice crystal development. Freezing results in chemical and osmotic damage as a result of the crystallisation of water and concentration of solutes in the still-unfrozen portion. Due to the sublimation process, frozen water is eliminated by vacuum conditions during the primary drying phase, and any remaining frozen water is eluted during the second drying cycle (Maltesen & van de Weert, 2008). The cellular membrane, cell wall and proteins can all be damaged during the drying process, which lowers the likelihood that the cells will survive after drying because water is so important for maintaining the stability and integrity of cells. The process of freeze-drying is quite an expensive and long-term process. An additional step is required for the production of powder particles (Maltesen & van de Weert, 2008). The durability of the probiotic strain after freeze-drying is influenced by the choice of acceptable wall and coating materials, adjustment of the process parameters and pre-stressing of the bacterial sample. During the freeze-drying process, unfrozen probiotics are subjected to compression due to the development of crystal ice; therefore cryoprotectants are used and the unfrozen percentage rises, creating more room for the probiotics and protecting the cells from osmotic or mechanical stress as well as cellular damage. Due to the lower temperature and pressure, the freeze-drying process results in greater retention of bioactive chemicals but has high operational costs.

2.4.3.5 Extrusion

The extrusion process is an efficient, simple and cost-effective technique, which deals with the projection of suspension at high pressure through needles that are prepared with the mixing of different functional ingredients. It is also most efficiently used for enhancing the viability of probiotic microorganisms and bioactive components (Krasaekoopt et al., 2008). Factors that affect the extrudate products are consistency and flow rate of the suspensions (Brun-Graeppi et al., 2011). The main limitations of direct extrusion procedures are the inability of this technique to manufacture microcapsules smaller than 500 µm (Burgain et al., 2011). To overcome these drawbacks, some techniques, including prilling and centrifugal extrusion. are performed. The technique of prilling involves the formation of droplets at controlled-environment conditions This is carried out by the pulsation or vibration of the jet nozzle. Another method is centrifugal extrusion, which uses a nozzle with concentric orifices located on the outside of a spinning cylinder and is based on the extrusion principle. Apart from these methods, there are a few advanced extrusion techniques that have multi-nozzle systems using acoustic or vibration energy, liquid jet cutting techniques and rotating disc atomisation (Rathore et al., 2013). Various applications of the extrusion process are involved in enhancing the viability of probiotics and improving the efficiency of DHA-rich oil and the quality of functional food (Vaziri et al., 2018). When developing food products, the bigger particles generated by the extrusion method should be taken into account.

2.4.3.6 Coacervation Technique

The mechanism of this procedure involves removing the hydrocolloid from the primary solution, which is followed by agglomeration into a different liquid phase known as "coacervate." (Napiórkowska & Kurek, 2022). Through the phase separation of biopolymers under controlled composition, temperature or pH conditions, coacervate is created using this technique. The coacervate is then deposited around the essential elements to create microspheres (Gouin, 2004). The coacervation technique is one of the advanced methods for the formulation of functional food development and protects food from deterioration. This technique has gained popularity in the past few years and has been applied in many fields, such as agriculture, pharmaceuticals and food technology (Gouin, 2004). The coacervation technique is further classified into the simple process and the complex process. Simple coacervation comprises altering the conditions that lead to the dehydration of macromolecules and adding non-solvents or microions to the solution, as well as changing the temperature or pH. Complex coacervation involves the partition of macromolecules ionically into two immiscible liquid phases. Complex coacervation involves the phase separation technique because of the two opposite charged colloids like protein and polysaccharides. To reduce the oxidation process during the storage of soy proteins and chitosan as wall material, algae oil is effectively used in the complex coacervation technique (Yuan et al. 2017). Another study reported the encapsulation of probiotic and tuna oil rich in omega-3 fatty acids by using a complex coacervation technique followed by freeze-drying and spray-drying process in gum Arabic and WPI (whey protein isolates) matrix (Eratte et al. 2015). These complex microspheres were strengthened by adding some enzymes and cross-linking agents, which helped them to stabilise the dry products, although this lowered the stability and shelf life of the probiotics because of the dispersed phase, and the handling costs were also higher, making the efficacy of this process questionable. These drawbacks can be overcome by using different dehydration techniques.

2.4.3.7 *Electrospinning*

The encapsulation process includes a number of methods such as emulsification, spray-drying, extrusion and coacervation, which utilise organic materials at high temperatures and cause heat-sensitive bioactive chemicals to degrade. Due to the association of organic agents, toxicity can be caused (Birnbaum et al., 2000). There are some advanced alternative methods that do not work at high temperatures and use organic substances instead, comprising the spinning process, which uses high voltages and is known as electrospinning or electro-spraying. Electrospinning is a simple, efficient and versatile electro-hydrodynamic atomisation technique and is used to encapsulate the bioactive compounds and probiotics in the nano to micro range. Without the need for temperature and organic substances, this procedure applies an external electric field between two electrodes before being applied to a polymer solution. By adjusting the process conditions and changing the solution's properties by adding the right additives, it is possible to electro-spin the biopolymers from watery solutions, showing the possibility for use in food applications (López-Rubio et al., 2012). Some naturally occurring biopolymers, such as proteins and polysaccharides, can be employed as encapsulation matrices because they are amphiphilic macromolecules with greater significance in stabilising food formulations and are surface-active materials. There are some synthetic polymers such as polyethylene oxide that cannot be utilised for the formulation of food but, instead, are widely used in pharmaceutical applications (Sobel et al., 2014). These polymers help in improving the vitality as well as the mechanical and thermal stability of probiotics. However, there have not been many investigations done on this technique with regard to probiotics and bioactive compounds.

2.5 RECENT TECHNOLOGIES THAT CONTRIBUTE TO CUSTOMISED FUNCTIONAL FOODS

Conventional methods have drawbacks in terms of maintaining nutritional advantages, environmental factor preservation and lack of stability because processing in high temperatures leads to a loss of bioactive compounds and other nutritional substances. Therefore, advancements in functional food development are very much required for preserving nutrients (Kanugo et al., 2022). The development of nanotechnology affects the food sector in a number of ways, including the characterisation, manufacture, preservation of goods (against physical and chemical deterioration), intelligent packaging, improvement of nutritional advantages, safety, etc. (Dima et al., 2020). Thus, advances have been made in the study of ecologically responsible ecosystem exploitation, the creation of novel, environmentally friendly food nanomaterials, the reduction of water and energy use, and, last but not least, the reduction of food waste. This chapter focuses on the current state of nanomaterials in the context of the need to create and design advanced sustainable functional food with superior nutritional and health qualities, which target the selection of particular nutritional aspects.

2.5.1 Reverse Engineering

Multi-criteria reverse engineering is a sophisticated approach centred on the consumer and is used to build functional foods based on the encapsulation technique (Thomopoulos et al., 2019). Performing a sensorial and nutritional evaluation of functional foods in accordance with customer needs; choosing the target nanocarrier elements, bioactive compounds and nutraceuticals delivery methods; selecting the encapsulation technique and food matrix; choosing the packaging method; evaluation cost analysis and the effect on the environment; analysing the perception of the food product; and finalising the design (Dima et al., 2020) are all important aspects. Milk proteins such as casein, whey proteins, b-lactoglobulin, lactoferrin, etc. and plant proteins are the most often utilised proteins to encapsulate materials (soy proteins, zein, gliadin, etc.). The encapsulation of both lipophilic and hydrophilic biocomponents is possible using protein-based nanocarriers. Consideration should be given to the type and characteristics of the nanocarrier components to ensure that it is suitable for the food matrix in which they are to be incorporated while choosing encapsulation techniques. As a result, certain methods are used to create solid nanocarriers, such as polymeric powders used to fortify dairy products, cereals and baked goods, while other methods are used to create liquid nanocarriers like nano emulsions used in beverages and animal meat products.

2.5.2 3D Printing

3D printing is a novel technique for creating three-dimensional items by continuously adding layers of material (Wilms et al., 2021). It uses a layer-by-layer printing design system based on a predefined specific slice of the preferred object, and the item can be manufactured instantly without design complexity, offering great potential for novel geometric designs with no variation, particularly in the fields of medicine, food production and aerospace (Feng et al., 2021). The 3D printing technique has a wide range of possibilities in food manufacturing and processing for customising and personalising food materials and shapes without moulds (Pulatsu & Lin, 2021). Currently, a variety of foods, including chocolates, cakes and meat substitutes, have been successfully made utilising 3D printing technology. The hot melt extrusion method is currently the most popular 3D printing technology used to produce food material using nozzles. For successful 3D-printed foods, printing parameters are important factors and play an important role (Wilms et al., 2021). Every material has different properties in the context of rheology, which majorly influence the printing parameters; therefore choosing parameters accordingly is important (Wilms et al., 2021). Future advancements in 3D printing technology systems in the functional food sector will fundamentally alter the conventional food processing sector and add new and technologically advanced driving techniques to the industry's growth, particularly in terms of precise customisation of nutrient-dense foods. In addition, it is a simple and cost-effective technique, with fast prototyping (Wilms et al., 2021).

2.5.3 Nutrigenomics

Nutrigenomics is also referred to as nutritional genomics, With the help of "omics" technologies like transcriptomics, proteomics and metabolomics, the field of nutrigenomics focuses on understanding how certain nutrients and diets affect human health by expressing genetic information. The prevention of chronic or acute diseases by nutrition, such as any kind of cancer, metabolic syndrome and cardiovascular diseases, is one of the key subjects of interest in nutrigenomics research. Several phases in the expression of genetic information may be controlled by nutrients and other bioactive substances found in food. Consequently, the evaluation of mRNA expression changes by one of the main objectives of nutrigenomics research is to understand the transfer of molecular information from the genome to the proteome and metabolome. This is generally done by studying nutrients and bioactive food components. Using northern blotting to quantify mRNA, scientists have been able to identify the expression of specific genes for years. More sensitive techniques, such as real-time PCR, have gradually taken the place of this traditional method. However, only a few potential genes may be studied at a time using either method to assess gene expression. This is a significant drawback for their use in nutrigenomics research because the study of fewer genes may not reveal information regarding the causal link between a bioactive food component and its biological function. On the other hand, global gene expression analysis may provide better opportunities to determine how bioactive food components affect metabolic pathways and homeostatic control and how this regulation may be altered in the course of the development of specific chronic diseases (García-Cañas et al., 2010).

2.6 CONCLUDING REMARKS AND FUTURE DIRECTIONS

Without a doubt, consumers, businesses, governments and academic institutions are very interested in the creation of functional foods. From the perspective of research and development, functional foods offer a chance to obtain cutting-edge goods, significantly meeting this market need. The most commonly employed technologies have evolved significantly over time. Thus, even if traditional methods are most frequently employed in the production of functional foods, methods aiming at individualised nutrition have seen significant growth over the last decade. There are some technologies that have evolved significantly and are intended to stop the degradation of physiologically active substances. The development of innovative technologies for food processing that improves the health potential and the appeal of food products is very important. This is fulfilled through functional foods. The development of an adequate food carrier that delivers the active molecular form to the physiological target within the organism and maintains it until the time of consumption, is a crucial stage in the success of a functional food. Though the ideas of 3D printing, reverse engineering and nutrigenomics make sense, we are still a long way from personalised nutrition. Healthcare and food product development will continue regardless, and it is the collaboration between science and practice that will allow for the seamless integration and application of new information as it is developed in the field.

REFERENCES

Abdel-Aty, A. M., Elsayed, A. M., Salah, H. A., Bassuiny, R. I., & Mohamed, S. A. (2021). Egyptian chia seeds (Salvia hispanica L.) during germination: Upgrading of phenolic profile, antioxidant, antibacterial properties and relevant enzymes activities. *Food Science and Biotechnology*, *30*(5), 723–734.

Aggarwal, B. B., & Harikumar, K. B. (2009). Potential therapeutic effects of curcumin, the anti-inflammatory agent, against neurodegenerative, cardiovascular, pulmonary, metabolic, autoimmune and neoplastic diseases. *The International Journal of Biochemistry and Cell Biology*, *41*(1), 40–59.

Asgar, M., Fazilah, A., Huda, N., Bhat, R., & Karim, A. (2010). Nonmeat protein alternatives as meat extenders and meat analogs. *Comprehensive Reviews in Food Science and Food Safety*, *9*(5), 513–529.

Bagchi, D., Lau, F. C., & Ghosh, D. K. (2010). *Biotechnology in functional foods and nutraceuticals*. CRC Press, Boca Raton, Florida.

Bansal, N., Fox, P. F., & McSweeney, P. L. (2007). Aggregation of rennet-altered casein micelles at low temperatures. *Journal of Agricultural and Food Chemistry*, *55*(8), 3120–3126.

Barros, C. P., Silva, R., Guimarães, J. T., Balhtazar, C. F., Verruck, S., Pimentel, T. C., … da Cruz, A. G. (2022). Prebiotics and Synbiotics in functional foods. In N. Chhikara, A. Panghal, G. Chaudhary (Eds.) *Functional Foods* (pp. 21–53). Hoboken, NJ: John Wiley & Sons and Beverly, MA: Scrivener Publishing LLC.

Beardsworth, A., & Keil, T. (1992). The vegetarian option: Varieties, conversions, motives and careers. *The Sociological Review*, *40*(2), 253–293.

Bersi, G., Barberis, S. E., Origone, A. L., & Adaro, M. O. (2018). Bioactive peptides as functional food ingredients. In A.M. Grumezescu and A.M. Holban (Eds.) *Role of Materials Science in Food Bioengineering* (pp. 147–186). Academic Press, Elsevier.

Betoret, E., Betoret, N., Vidal, D., & Fito, P. (2011). Functional foods development: Trends and technologies. *Trends in Food Science and Technology*, *22*(9), 498–508.

Bigliardi, B., & Galati, F. (2013). Innovation trends in the food industry: The case of functional foods. *Trends in Food Science and Technology*, *31*(2), 118–129.

Birnbaum, D. T., Kosmala, J. D., Henthorn, D. B., & Brannon-Peppas, L. (2000). Controlled release of β-estradiol from PLAGA microparticles: The effect of organic phase solvent on encapsulation and release. *Journal of Controlled Release*, *65*(3), 375–387.

Boye, I., & Nutraceutical, J. (2015). *Functional Food Processing Technology*. Wiley-Blackwell, Hoboken, NJ, USA.

Brun-Graeppi, A. K. A. S., Richard, C., Bessodes, M., Scherman, D., & Merten, O. W. (2011). Cell microcarriers and microcapsules of stimuli-responsive polymers. *Journal of Controlled Release*, *149*(3), 209–224.

Burdock, G. A., & Carabin, I. G. (2019). Breaking down the barriers to functional foods, supplements, and claims. In D. Bagchi (Ed.) *Nutraceutical and functional food regulations in the United States and around the world* (pp. 121–146). Academic Press.

Burgain, J., Gaiani, C., Linder, M., & Scher, J. (2011). Encapsulation of probiotic living cells: From laboratory scale to industrial applications. *Journal of Food Engineering*, *104*(4), 467–483.

Bürgi, H., Supersaxo, Z., & Selz, B. (1990). Iodine deficiency diseases in Switzerland one hundred years after Theodor Kocher's survey: A historical review with some new goitre prevalence data. *European Journal of Endocrinology*, *123*(6), 577–590.

Champagne, C. P., & Gardner, N. J. (2008). Effect of storage in a fruit drink on subsequent survival of probiotic lactobacilli to gastro-intestinal stresses. *Food Research International*, *41*(5), 539–543.

Charalampopoulos, D., Wang, R., Pandiella, S. S., & Webb, C. (2002). Application of cereals and cereal components in functional foods: A review. *International Journal of Food Microbiology*, *79*(1–2), 131–141.

Chawda, P. J., Shi, J., Xue, S., & Young Quek, S. (2017). Co-encapsulation of bioactives for food applications. *Food Quality and Safety*, *1*(4), 302–309.

Chen, L. Y., Remondetto, G. E., & Subirade, M. (2006). Food protein-based materials as nutraceutical delivery systems. *Trends in Food Science and Technology*, *17*(5), 272–283.

Chen, M. J., Chen, K. N., & Kuo, Y. T. (2007). Optimal thermotolerance of Bifidobacterium bifidum in gellan–alginate microparticles. *Biotechnology and Bioengineering*, *98*(2), 411–419.

Díaz de la Garza, R. I., Gregory III, J. F., & Hanson, A. D. (2007). Folate biofortification of tomato fruit. *Proceedings of the National Academy of Sciences*, *104*(10), 4218–4222.

Delshadi, R., Bahrami, A., Tafti, A. G., Barba, F. J., & Williams, L. L. (2020). Micro and nano-encapsulation of vegetable and essential oils to develop functional food products with improved nutritional profiles. *Trends in Food Science and Technology*, *104*, 72–83.

Dias, D. R., Botrel, D. A., Fernandes, R. V. D. B., & Borges, S. V. (2017). Encapsulation as a tool for bioprocessing of functional foods. *Current Opinion in Food Science*, *13*, 31–37.

Dima, C., Assadpour, E., Dima, S., & Jafari, S. M. (2020). Bioactive-loaded nanocarriers for functional foods: From designing to bioavailability. *Current Opinion in Food Science*, *33*, 21–29.

Diplock, A., Charuleux, J.-L., Crozier-Willi, G., Kok, F., Rice-Evans, C., Roberfroid, M., … Vina-Ribes, J. (1998). Functional food science and defence against reactive oxidative species. *British Journal of Nutrition*, *80*(S1), S77–S112.

Duchateau, G. S., & Klaffke, W. (2008). Product composition, structure, and bioavailability. *Food Biophysics*, *3*(2), 207–212.

Dullius, A., Goettert, M. I., & de Souza, C. F. V. (2018). Whey protein hydrolysates as a source of bioactive peptides for functional foods–Biotechnological facilitation of industrial scale-up. *Journal of Functional Foods*, *42*, 58–74.

Eratte, D., McKnight, S., Gengenbach, T. R., Dowling, K., Barrow, C. J., & Adhikari, B. P. (2015). Co-encapsulation and characterisation of omega-3 fatty acids and probiotic bacteria in whey protein isolate–gum arabic complex coacervates. *Journal of Functional Foods*, *19*, 882–892.

Ernst, E. (2002). The risk–benefit profile of commonly used herbal therapies: Ginkgo, St. John's Wort, ginseng, Echinacea, Saw Palmetto, and Kava. *Annals of Internal Medicine*, *136*(1), 42–53.

Gibbs, F., Kermasha, S., Alli, I., & Mulligan, C. N. (1999). Encapsulation in the food industry: A review. *International Journal of Food Sciences and Nutrition*, *50*(3), 213–224.

FAO/WHO. (2006). Probiotics in food health and nutritional properties and guidelines for evaluation. *FAO Food and Nutrition*; 2006 pp 85 ISBN 92-5-105513-0. Also available at ftp://ftp.fao.org/docrep/fao/ 009/a0512e/a0512e00.pdf

Feng, Y., Zhu, S., Mei, D., Li, J., Zhang, J., Yang, S., & Guan, S. (2021). Application of 3D printing technology in bone tissue engineering: A review. *Current Drug Delivery*, *18*(7), 847–861.

Fowler, A., & Toner, M. (2006). Cryo-injury and biopreservation. *Annals of the New York Academy of Sciences*, *1066*(1), 119–135.

García-Cañas, V., Simó, C., León, C., & Cifuentes, A. (2010). Advances in nutrigenomics research: Novel and future analytical approaches to investigate the biological activity of natural compounds and food functions. *Journal of Pharmaceutical and Biomedical Analysis*, *51*(2), 290–304.

Gaudreau, H., Champagne, C. P., Remondetto, G. E., Gomaa, A., & Subirade, M. (2016). Co-encapsulation of Lactobacillus helveticus cells and green tea extract: Influence on cell survival in simulated gastrointestinal conditions. *Journal of Functional Foods*, *26*, 451–459.

Gharsallaoui, A., Roudaut, G., Chambin, O., Voilley, A., & Saurel, R. (2007). Applications of spray-drying in microencapsulation of food ingredients: An overview. *Food Research International*, *40*(9), 1107–1121.

González-Molina, E., Moreno, D. A., & García-Viguera, C. (2009). A new drink rich in healthy bioactives combining lemon and pomegranate juices. *Food Chemistry*, *115*(4), 1364–1372.

Gouin, S. (2004). Microencapsulation: Industrial appraisal of existing technologies and trends. *Trends in Food Science and Technology*, *15*(7–8), 330–347.

Hamer, M., Owen, G., & Kloek, J. (2005). The role of functional foods in the psychobiology of health and disease. *Nutrition Research Reviews*, *18*(1), 77–88.

Heidebach, T., Först, P., & Kulozik, U. (2009). Transglutaminase-induced caseinate gelation for the microencapsulation of probiotic cells. *International Dairy Journal*, *19*(2), 77–84.

Holdt, S. L., & Kraan, S. (2011). Bioactive compounds in seaweed: Functional food applications and legislation. *Journal of Applied Phycology*, *23*(3), 543–597.

Huggett, A. C., & Schliter, B. (1996). Research needs for establishing the safety of functional foods. *Nutrition Reviews*, *54*(11 Pt 2), S143–S148.

Jackson, L. S., & Lee, K. (1991). Microencapsulation and the food industry. *Lebensmittel-Wissenschaft und --Technologie*, *24*(4), 289–297.

Jimenez-Colmenero, F., Sánchez-Muniz, F. J., & Olmedilla-Alonso, B. (2010). Design and development of meat-based functional foods with walnut: Technological, nutritional and health impact. *Food Chemistry*, *123*(4), 959–967.

Jones, P. J., & Jew, S. (2007). Functional food development: Concept to reality. *Trends in Food Science and Technology*, *18*(7), 387–390.

Kailasapathy, K. (2002). Microencapsulation of probiotic bacteria: Technology and potential applications. *Current Issues in Intestinal Microbiology*, *3*(2), 39–48.

Kailasapathy, K. (2009). Encapsulation technologies for functional foods and nutraceutical product development. *CABI Reviews*, 1–19.

Kanugo, A., Goyal, R., Sharma, S., & Gautam, R. K. (2022). Recent advancement of nanotechnology in functional foods: Applications perspectives. T. Bhattacharya, S. Ahmed (Eds.) *Nanotechnology in Functional Foods* (293–328). United Kingdom: Wiley.

Kareb, O., & Aïder, M. (2019). Whey and its derivatives for probiotics, prebiotics, Synbiotics, and functional foods: A critical review. *Probiotics and Antimicrobial Proteins, 11*(2), 348–369.

Karunanandaa, B., Qi, Q., Hao, M., Baszis, S. R., Jensen, P. K., Wong, Y. H. H., … Valentin, H. E. (2005). Metabolically engineered oilseed crops with enhanced seed tocopherol. *Metabolic Engineering, 7*(5–6), 384–400.

Kaur, S., & Das, M. (2011). Functional foods: An overview. *Food Science and Biotechnology, 20*(4), 861–875.

Kim, Y. D., & Morr, C. V. (1996). Microencapsulation properties of gum arabic and several food proteins: Spray-dried orange oil emulsion particles. *Journal of Agricultural and Food Chemistry, 44*(5), 1314–1320.

Krasaekoopt, W., Bhandari, B., & Deeth, H. (2004). The influence of coating materials on some properties of alginate beads and survivability of microencapsulated probiotic bacteria. *International Dairy Journal, 14*(8), 737–743.

Krasaekoopt, W., Pianjareonlap, R., & Kittisuriyanont, K. (2008). Survival of probiotics in fruit juices during refrigerated storage. *Thai Journal of Biotechnology, 8*(1), 129–133.

Li, Y., He, D., Li, B., Lund, M. N., Xing, Y., Wang, Y., … Chen, X. (2021). Engineering polyphenols with biological functions via polyphenol-protein interactions as additives for functional foods. *Trends in Food Science and Technology, 110*, 470–482.

Livney, Y. D. (2010). Milk proteins as vehicles for bioactives. *Current Opinion in Colloid and Interface Science, 15*(1–2), 73–83.

López-Rubio, A., Sanchez, E., Wilkanowicz, S., Sanz, Y., & Lagaron, J. M. (2012). Electrospinning as a useful technique for the encapsulation of living bifidobacteria in food hydrocolloids. *Food Hydrocolloids, 28*(1), 159–167.

Maltesen, M. J., & Van De Weert, M. (2008). Drying methods for protein pharmaceuticals. *Drug Discovery Today: Technologies, 5*(2–3), 81–88.

Marine, D., & Kimball, O. P. (1920). Prevention of simple goiter in man: Fourth paper. *Archives of Internal Medicine, 25*(6), 661–672.

Marsanasco, M., Márquez, A. L., Wagner, J. R., Chiaramoni, N. S., & Alonso, S. D. V. (2015). Bioactive compounds as functional food ingredients: Characterization in model system and sensory evaluation in chocolate milk. *Journal of Food Engineering, 166*, 55–63.

Matos, J., Cardoso, C., Bandarra, N. M., & Afonso, C. (2017). Microalgae as healthy ingredients for functional food: A review. *Food and Function, 8*(8), 2672–2685.

McClements, D. J. (2018). Recent developments in encapsulation and release of functional food ingredients: Delivery by design. *Current Opinion in Food Science, 23*, 80–84.

McClements, D. J., Decker, E. A., & Park, Y. (2008). Controlling lipid bioavailability through physicochemical and structural approaches. *Critical Reviews in Food Science and Nutrition, 49*(1), 48–67.

McClements, D. J., Decker, E. A., Park, Y., & Weiss, J. (2008). Designing food structure to control stability, digestion, release and absorption of lipophilic food components. *Food Biophysics, 3*(2), 219–228.

Misra, S., Pandey, P., & Mishra, H. N. (2021). Novel approaches for co-encapsulation of probiotic bacteria with bioactive compounds, their health benefits and functional food product development: A review. *Trends in Food Science and Technology, 109*, 340–351.

Mollet, B., & Rowland, I. (2002). Functional foods: At the frontier between food and pharma. *Current Opinion in Biotechnology, 13*(5), 483–485.

Morgan, C. A., Herman, N., White, P. A., & Vesey, G. (2006). Preservation of micro-organisms by drying; a review. *Journal of Microbiological Methods*, *66*(2), 183–193.

Murray, M., Dordevic, A. L., Ryan, L., & Bonham, M. P. (2018). An emerging trend in functional foods for the prevention of cardiovascular disease and diabetes: Marine algal polyphenols. *Critical Reviews in Food Science and Nutrition*, *58*(8), 1342–1358.

Napiórkowska, A., & Kurek, M. (2022). Coacervation as a novel method of microencapsulation of essential oils—A review. *Molecules*, *27*(16), 5142.

National Research Council. (1989). *Past experience with genetic modification of plants and their introduction into the environment* (pp. 16–36). Washington, DC: National Academy Press.

Nöthlings, U. T. E., Murphy, S. P., Wilkens, L. R., Henderson, B. E., & Kolonel, L. N. (2007). Flavonols and pancreatic cancer risk: The multiethnic cohort study. *American Journal of Epidemiology*, *166*(8), 924–931.

Yildiz, Ö., & Eyduran, S. P. (2009). Functional components of berry fruits and their usage in food technologies. *African Journal of Agricultural Research*, *4*(5), 422–426.

Perdana, J., Fox, M. B., Siwei, C., Boom, R. M., & Schutyser, M. A. (2014). Interactions between formulation and spray drying conditions related to survival of Lactobacillus plantarum WCFS1. *Food Research International*, *56*, 9–17.

Pereira, H. V. R., Saraiva, K. P., Carvalho, L. M. J., Andrade, L. R., Pedrosa, C., & Pierucci, A. P. T. R. (2009). Legumes seeds protein isolates in the production of ascorbic acid microparticles. *Food Research International*, *42*(1), 115–121.

Plaza, M., Herrero, M., Cifuentes, A., & Ibanez, E. (2009). Innovative natural functional ingredients from microalgae. *Journal of Agricultural and Food Chemistry*, *57*(16), 7159–7170.

Pulatsu, E., & Lin, M. (2021). A review on customizing edible food materials into 3D printable inks: Approaches and strategies. *Trends in Food Science and Technology*, *107*, 68–77.

Ramírez, P., García-Risco, M. R., Santoyo, S., Señoráns, F. J., Ibáñez, E., & Reglero, G. (2006). Isolation of functional ingredients from rosemary by preparative-supercritical fluid chromatography (Prep-SFC). *Journal of Pharmaceutical and Biomedical Analysis*, *41*(5), 1606–1613.

Rathore, S., Desai, P. M., Liew, C. V., Chan, L. W., & Heng, P. W. S. (2013). Microencapsulation of microbial cells. *Journal of Food Engineering*, *116*(2), 369–381.

Rivas, A., Rodrigo, D., Company, B., Sampedro, F., & Rodrigo, M. (2007). Effects of pulsed electric fields on water-soluble vitamins and ACE inhibitory peptides added to a mixed orange juice and milk beverage. *Food Chemistry*, *104*(4), 1550–1559.

Roberfroid, M. B. (2000). Concepts and strategy of functional food science: The European perspective. *The American Journal of Clinical Nutrition*, *71*(6), 1660S–1664S.

Sanguansri, P., & Augustin, M. A. (2006). Nanoscale materials development–a food industry perspective. *Trends in Food Science and Technology*, *17*(10), 547–556.

Shah, N. P. (2007). Functional cultures and health benefits. *International Dairy Journal*, *17*(11), 1262–1277.

Shah, N. P., & Ravula, R. R. (2000). Microencapsulation of probiotic bacteria and their survival in frozen fermented dairy desserts. *Australian Journal of Dairy Technology*, *55*(3), 139–144.

Shahidi, F., & Ambigaipalan, P. (2015). Novel functional food ingredients from marine sources. *Current Opinion in Food Science*, *2*, 123–129.

Shefer,A., & Shefer, S. (2003). A novel encapsulation system providea s controlled release of ingredients. *Food Technology*, *57*, 40–43.

Shiby, V., & Mishra, H. (2013). Fermented milk and milk products as functional foods—A review. *Critical Reviews in Food Science and Nutrition*, *53*(5), 482–496.

Silva, M. A., Albuquerque, T. G., Alves, R. C., Oliveira, M. B. P., & Costa, H. S. (2020). Melon (Cucumis melo L.) by-products: Potential food ingredients for novel functional foods? *Trends in Food Science and Technology*, *98*, 181–189.

Singh, M., Trivedi, N., Enamala, M. K., Kuppam, C., Parikh, P., Nikolova, M. P., & Chavali, M. (2021). A concise review of plant-based meat analogue (PBMA) as a sustainable food. *European Food Research and Technology*, *247*(10), 2499–2526.

Smidsrød, O., & Skja, G. (1990). Alginate as immobilization matrix for cells. *Trends in Biotechnology*, *8*(3), 71–78.

Sobel, R., Versic, R., & Gaonkar, A. G. (2014). Introduction to microencapsulation and controlled delivery in foods. In A.G. Gaonkar, N. Vasisht, A. Ramesh Khare, R. Sobel (Eds.) *Microencapsulation in the food industry* (pp. 3–12). Academic Press, Amsterdam.

Stanton, C., Ross, R. P., Fitzgerald, G. F., & Van Sinderen, D. (2005). Fermented functional foods based on probiotics and their biogenic metabolites. *Current Opinion in Biotechnology*, *16*(2), 198–203.

Storozhenko, S., De Brouwer, V., Volckaert, M., Navarrete, O., Blancquaert, D., Zhang, G. F., … Van Der Straeten, D. (2007). Folate fortification of rice by metabolic engineering. *Nature Biotechnology*, *25*(11), 1277–1279.

Takachi, R., Manami, I., Junko, I., Norie, K., Motoki, I., Shizuka, S., et al. (2008). Fruit and vegetable intake and risk of total cancer and cardiovascular disease: Japan public health center-based prospective study. *American Journal of Epidemiology*, *167*(1), 59–70.

Tanumihardjo, S. A., Bouis, H., Hotz, C., Meenakshi, J. V., & McClafferty, B. (2008). Biofortification of staple crops: An emerging strategy to combat hidden hunger. *Comprehensive Reviews in Food Science and Food Safety*, *7*, 329–334.

Thies, C. (2005). A survey of microencapsulation processes. In S. Benita (Ed.), *Microencapsulation* (p. 1.e20). New York: Marcel Dekker Inc.

Thomopoulos, R., Baudrit, C., Boukhelifa, N., Boutrou, R., Buche, P., Guichard, E., … Tonda, A. (2019). Multi-criteria reverse engineering for food: Genesis and ongoing advances. *Food Engineering Reviews*, *11*(1), 44–60.

Ubbink, J. (2002). Flavor delivery systems: Trends, technologies and applications. *Abstracts of Papers of the American Chemical Society*, *223*, U34–U34.

Ubbink, J., & Kruger, J. (2006). Physical approaches for the delivery of active ingredients in foods. *Trends in Food Science and Technology*, *17*(5), 244–254.

Vaziri, A. S., Alemzadeh, I., Vossoughi, M., & Khorasani, A. C. (2018). Co-microencapsulation of Lactobacillus plantarum and DHA fatty acid in alginate-pectin-gelatin biocomposites. *Carbohydrate Polymers*, *199*, 266–275.

Wilms, P., Daffner, K., Kern, C., Gras, S. L., Schutyser, M. A. I., & Kohlus, R. (2021). Formulation engineering of food systems for 3D-printing applications–A review. *Food Research International*, *148*, 110585.

Woods, V. B., & Fearon, A. M. (2009). Dietary sources of unsaturated fatty acids for animals and their transfer into meat, milk and eggs: A review. *Livestock Science*, *126*(1–3), 1–20.

Wood, J. M. (1999). Osmosensing by bacteria: signals and membrane-based sensors. *Microbiology and molecular biology reviews*, *63*(1), 230–262.

Yadav, H., Jain, S., & Sinha, P. R. (2007). Antidiabetic effect of probiotic dahi containing Lactobacillus acidophilus and Lactobacillus casei in high fructose fed rats. *Nutrition*, *23*(1), 62–68.

Yadav, H., Jain, S., Rastamanesh, R., Bomba, A., Catanzaro, R., & Marotta, F. (2011). Fermentation technology in the development of functional foods for human health: Where we should head. *Fermentation Technology*, *1*(1), 1–2.

Yáñez-Fernández, J., Ramos-Ramírez, E. G., & Salazar-Montoya, J. A. (2008). Rheological characterization of dispersions and emulsions used in the preparation of microcapsules obtained by interfacial polymerization containing Lactobacillus sp. *European Food Research and Technology*, *226*(5), 957–966.

Yuan, Y., Kong, Z. Y., Sun, Y. E., Zeng, Q. Z., & Yang, X. Q. (2017). Complex coacervation of soy protein with chitosan: Constructing antioxidant microcapsule for algal oil delivery. *LWT*, *75*, 171–179.

Zhang, W., Xiao, S., Samaraweera, H., Lee, E. J., & Ahn, D. U. (2010). Improving functional value of meat products. *Meat Science*, *86*(1), 15–31.

Zhao, F. J., & Shewry, P. R. (2011). Recent developments in modifying crops and agronomic practice to improve human health. *Food Policy*, *36*, S94–S101.

Zhu, C., Naqvi, S., Breitenbach, J., Sandmann, G., Christou, P., & Capell, T. (2008). Combinatorial genetic transformation generates a library of metabolic phenotypes for the carotenoid pathway in maize. *Proceedings of the National Academy of Sciences of the United States of America*, *105*(47) 18232–18237

Ziemer, C. J., & Gibson, G. R. (1998). An overview of probiotics, prebiotics and Synbiotics in the functional food concept: Perspectives and future strategies. *International Dairy Journal*, *8*(5–6), 473–479.

CHAPTER 3

Fruits as a Functional Food for Health

Hamid*, **Deepika Kathuria**, **Preethi Ramachandran**, **Sunakshi Gautam**, **Abhimanyu Thakur** and **N S Thakur**

3.1 INTRODUCTION

People are increasingly concerned about their health due to rising rates of metabolic illnesses (obesity, heart diseases, diabetes, arthritis etc.) as well as rising healthcare costs. The influence of nutrition on the prevention of chronic illnesses such as coronary heart disease, cancer and neurological and respiratory disease, as well as ageing, is demonstrated by current scientific research, which establishes a critical relationship between diet and human health. As a result, it is important to consider foods that are recognised as being in the natural plant category as a component of a strategic plan to combat illnesses associated with human nutrition and lifestyle.

The WHO (World Health Organization) advocates for a nutritious diet focused on daily consumption of at least 400 g, or five servings, of a variety of fruits as well as vegetables (2:3), which might help to reduce the chances of evolving noncommunicable diseases and ensure adequate regular consumption of dietary fibre. However, the daily recommended consumption of fruits and vegetables varies around the globe (100–450 g), and many individuals fall short of the recommended amounts (Rodriguez et al. 2020).

Fruits have different characteristics with distinctive nutritional and phytochemical profiles, thus they can play a significant part in maintaining a healthy lifestyle. In the human diet, fruits are an important resource of natural antioxidants, acids, carbohydrates, polyphenolics, flavonoids, minerals, water-soluble vitamins (such as vitamins C and vitamin B complex), amino acids, provitamin A, aroma components, carotenoids, phytosterols, fibre and other phytochemical components. Fruits contain between 70 and 90% water. Fruits pulp and peel don't contain a lot of lipids, but their seeds, which are rarely eaten, are where you'll find the majority of these substances.

* Corresponding author E-mail: hamidfst6789@gmail.com

DOI: 10.1201/9781003315100-3

According to Rodriguez et al. (2020) and Bates, Morris and Crandall (2001), the protein content of fruits varies and is insufficient. Fruits include a significant amount of fibre, which is not digested and goes through the digestive system. Typically, cellulose, hemicellulose and pectic materials make up fibre. Fruits may also include a little amount of starch, although this starch is usually turned into sugars when the fruit ripens.

In reality, since fruits include large amounts of bioactive substances that prevent various harmful physiological processes including metabolic and cardiovascular disorders, fruits and vegetables alone may be termed functional foods. According to Rodriguez et al. (2020) functional foods can be divided into three categories: foods with bioactive substances naturally present (such as dietary fibre), food products with added external bioactive compounds, and extracted food components added to usual meals (such as prebiotics).

Various vitamins (folate, vitamin C, pro-vitamin A), fibres, antioxidant phytochemicals (polyphenolic compounds, flavonoid and carotenoids), and minerals (potassium, calcium and magnesium) are just a few of the many bioactive substances found in fruits that shown to have a numeral of positive impact on health. Its recommended that to experience the best health advantages, one must regularly eat a wide range of plant-based meals since bioactive chemicals differ significantly in content and ratio from fruit to fruit and frequently have processes that are complementary to one another (Liu 2013). In terms of total phenolic content, blackberry and wild blueberry were the highest followed by cranberry, pomegranate, blueberry, raspberry, plum, strawberry, apple and red grape (Wolfe et al. 2008). Various ethnically and geographically unique fruits are utilised in traditional treatments, in addition to commercial fruits to take care of a variety of health issues, such as diabetes and obesity (Babio, Bullo, and Salas-Salvado 2009). Ceylan-Isik et al. (2008) also reported that local fruits with greater antioxidant activity are frequently utilised in traditional Chinese drugs to treat diabetes patients. Additionally, a variety of locally grown fruits with antioxidant activity and possible therapeutic purposes are used in India's Ayurvedic treatments.

Fruit juices high in vitamin C (aonla has 592 mg, orange has 64 mg, lime has 65 mg, zizyphus has 76 mg and papaya has 55 mg/100mL fruit juice) help to virtually quadruple non-heme Fe absorption, which can prevent the body from developing iron-deficiency anaemia (Bhardwaj et al. 2014). The most popular fruit juices include aonla, apple, mango, grapes, bael juice, citrus juices, phalsa, pomegranate and jamun. These juices are popular due to their high vitamin C content, good flavour and colour with the ability to decrease the possibility of some cancers types, including those of the lung, mouth, larynx, pharynx, oesophagus, stomach and colon (Honda, Kai, and Ohi 1999). Numerous fruit juice phytochemicals/bioactives like polyphenolics, limonoids and carotenoids may have an impact on cancer control processes that minimise DNA damage along with aid in DNA repair, hence lowering cancer-causing mutations (Wu et al. 2009; Kyle et al. 2009). These include signal transduction pathway modification, angiogenesis regulation, anti-mutagenic action and anti-inflammatory mechanisms. Apple juice polyphenols (Veeriah et al. 2008), citrus flavonoids (Benavente and Castillo 2008), grape juice anthocyanins

(Iriti and Faoro 2009), limonoids (Poulose, Harris, and Patil 2006) and anthocyanins from various berry juices (Matsunaga et al. 2010) all have the potential to have anti-carcinogenic activities.

In order to prevent cancers of the breast, prostate, colon, skin, oral, lungs, intestine, pancreas, bladder and oesophagus, and leukaemia, pomegranate juice antioxidants and ellagic acid work as anti-angiogenesis, anti-proliferation (cell cycle disruption, growth inhibition and apoptosis) and anti-inflammatory agents (Amin et al. 2009). Significant reductions in total cholesterol, LDL (low-density lipoproteins), LDL/HDL and the total cholesterol to HDL ratio are observed in citrus juice, pomegranate, phalsa and jamun juices (Esmaillzadeh et al. 2006). Fruit juices improve blood circulation by widening blood vessels, which helps diabetics by lowering the risk of clot formation. Additionally, several types of dietary fibre and amino acids included in fruit juices assist in reducing sugar absorption and enhancing insulin sensitivity (Bhardwaj et al. 2014). Natural sources of plant origin are also abundant in a variety of phytochemicals, which have stronger antioxidant activities, boost immunity and protect the body from oxidative stress by assisting in cell growth and repair. Therefore, eating fruits regularly can help to lower the chance of contracting certain ailments.

3.2 BIOACTIVE COMPOUNDS IN FRUITS

Fruits are considered to be a good source of micronutrients such as vitamins, minerals and non-nutrients bioactive components like polyphenolic compounds (Wani et al. 2023) (Table 3.1). It has been reported that more than 5000 distinct phytochemicals have been identified in fruit, and research is still ongoing to discover the remaining ones (Liu 2003). Bioactive compounds are secondary metabolites whose biosynthesis is triggered by light. These compounds are mostly accumulated in the leaves and skin of fruits. The fruit development stage affects the amount of bioactive components (Deepa et al. 2007). Fruits consist of a complex mixture of phenols that belongs to molecular families including benzoic acid derivatives (phenolic acid), flavonoids (flavanones, flavanols, flavones, flavonols, isoflavones, proanthocyanidins, anthocyanins), tannins, lignans and lignins. The two main forms of phenols are phenolic acids and flavonoids (Scalbert et al. 2005). Vasco et al. (2008) proposed the classification of phenols into three groups depending upon concentration i.e., low (<100 mg GAE/100 g), medium (100–500 mg GAE/100 g) and high (>500 mg GAE/100 g). Fruits like apples, grapes, berries and cherries consist of approximately 200–300 mg polyphenols/100 g of fresh fruit (Psndey and Rizvi, 2009; Tabart et al. 2012). The polyphenols derived from these fruits constitute mostly phenolic acid, which is divided into two classes i.e. hydroxybenzoic acid and hydroxycinnamic acid. Vanillic, gallic, syringic, protocatechuic and p-hydroxybenzoic acids are the most prevalent hydroxybenzoic acids whereas p-coumaric, sinapic, ferulic and caffeic acids correspond to hydroxycinnamic acids. Gallic acid is a potent antioxidant found abundantly in red grapes, strawberries, raspberries, grapefruit, cranberries, blackberries, etc., as well as the juices made from these fruits (Selma et al. 2009). These derivatives vary

Table 3.1 Bioactive components in major fruit crops

Main class	Subclass	Chemical compound	Fruit	Quantity	References
Phenolic acid	Hydroxlybenzoic acid	Gallic acid	Strawberry	6.80–24.60	Mahmood et al. (2012)
			Grapes	0.01–4.49	Fu-xiang et al. (2019); Prakash et al. (2020)
			Blueberry	1.8	Može et al. (2011)
			Gooseberry	0.04–2.26	Orsavová et al. (2019)
		Vanillic acid	Strawberry	2.80–16.10	Mahmood et al. (2012)
			Grapes	0.04–0.42	Fu-xiang et al. (2019); Prakash et al. (2020)
			Gooseberry	<0.01	Orsavová et al. (2019)
		Syringic acid	Grapes	0.07–0.79	Fu-xiang et al. (2019); Prakash et al. (2020)
			Gooseberry	0.76–2.09	Orsavová et al. (2019)
	Hydroxycinnamic acids	Chlorogenic acid	Apple	12.17–63.05 mg/100g	Alberti et al. (2014); Francini and Sebastiani (2013)
			Strawberry	11.50–18.20	Mahmood et al. (2012)
			Grapes	0.64–5.34	Prakash et al. (2020)
			Blueberry	25.16–70.00	Gavrilova et al. (2011); Moze et al. (2011)
			Black current	0.79–0.97	Gavrilova et al. (2011)
			Cherry	1.64–4.70	Ceccarelli et al. (2018)
			Gooseberry	0.70–4.93	Orsavová et al. (2019)
		Caffeic acid	Apple	0.74–4.45	Francini and Sebastiani (2013)
			Grapes	0.21–1.56	Fu-xiang et al. (2019); Prakash et al. (2020); Colombo et al. (2019)
			Gooseberry	0.36–0.92	Orsavová et al. (2019)
			Blueberry	0.20	Može et al. (2011)
		Ferulic acid	Strawberry	7.60–24.10	Mahmood et al. (2012)

(*Continued*)

Table 3.1 (Continued)

Main class	Subclass	Chemical compound	Fruit	Quantity	References
			Grapes	0.02–0.61	Prakash et al. (2020)
			Gooseberry	0.04–0.98	Orsavová et al. (2019)
			Blueberry	2.20	Može et al. (2011)
		p-coumaric	Strawberry	1.23–47.50	Mahmood et al. (2012); Milosavljevic et al. (2020)
			Grapes	0.02–0.66	Fu-xiang et al. (2019); Prakash et al. (2020)
			Cherry	17.62–160.68	Ceccarelli et al. (2018)
			Gooseberry	0.10–0.26	Orsavová et al. (2019)
			Blackcurrent	0.89–0.90	Gavrilova et al. (2011)
Flavonoids	Flavonols	Quercetin	Apple	2.14–8.76	Alberti et al. (2014)
			Strawberry	0.59–11.20	Mahmood et al. (2012) Milosavljević et al. (2020)
			Blueberry	0.10–2.17	Gavrilova et al. (2011); Može et al. (2011)
			Cherry	1.09–8.85	Ceccarelli et al. (2018)
		Rutin	Apple	0.2–0.62	Francini and Sebastiani (2013)
			Grapes	0.41–0.85	Fu-xiang et al. (2019); Colombo et al. (2019)
			Blueberry	0.78–3.52	Gavrilova et al. (2011); Moze et al. (2011)
			Blackcurrent	4.24–4.58	Gavrilova et al. (2011)
			Gooseberry	0.59–5.14	Orsavová et al. (2019)
		Kaempferol	Strawberry	18.40–98.10	Mahmood et al. (2012)
			Grapes	0.80–2.18	Fu-xiang et al. (2019)
		Myricetin	Strawberry	12.80–28.50	Mahmood et al. (2012)
	Flavanones	Hesperetin	Grapes	0.03–1.34	Prakash et al. (2020)

(Continued)

Table 3.1 (Continued)

Main class	Subclass	Chemical compound	Fruit	Quantity	References
	Flavan-3-ols	Catechin	Strawberry	2–50	Milosavljevic et al. (2020)
			Cherry	0.53–22	Ceccarelli et al. (2018)
			Apple	9.67–43	Alberti et al. (2014); Francini and Sebastiani (2013)
			Grapes	0.06–13.40	Fu-xiang et al. (2019); Prakash et al. (2020)
			Gooseberry	20.43–26.75	Orsavová et al. (2019)
			Blueberry	1.35–3.01	Gavrilova et al. (2011); Može et al. (2011)
			Blackcurrent	5.07–7.49	Gavrilova et al. (2011)
		Epicatechin	Apple	9.81–26.86	Alberti et al. (2014); Francini and Sebastiani (2013)
			Grapes	0.10–7.23	Fu-xiang et al. (2019); Prakash et al. (2020); Colombo et al. (2019)
			Black current	0.5	Može et al. (2011)
			Cherry	5.07–20.26	Ceccarelli et al. (2018)
			Gooseberry	0.19–220.83	Orsavová et al. (2019)
	Anthocyanins	Pelargonidin	Strawberry	8.11–49.27	Milosavljević et al. (2020)
		Delphinidin	Blueberry	5.16–63.40	Gavrilova et al. (2011); Može et al. (2011)
			Black current	79.19–106.52	Gavrilova et al. (2011)
		Cyanidin	Strawberry	0.16–0.95	Milosavljević et al. (2020)
			Blueberry	0.84–10.30	Gavrilova et al. (2011); Može et al. (2011)
			Black current	70.31–79.88	Gavrilova et al. (2011)
			Cherry	14.48–144.22	Ceccarelli et al. (2018)

(*Continued*)

Table 3.1 (Continued)

Main class	Subclass	Chemical compound	Fruit	Quantity	References
		Malvidin	Blueberry	28.09–100.8	Gavrilova et al. (2011); Može et al. (2011)
		Peonidin	Blueberry	1.26–4.90	Gavrilova et al. (2011); Može et al. (2011)
			Black current	1.27–1.75	Gavrilova et al. (2011)
		Petunidin	Blueberry	5.68–33.40	Gavrilova et al. (2011) Može et al. (2011)
			Black current	2.01–2.34	Gavrilova et al. (2011)
Stilbenes	–	Resveratrol	Grapes	0.01–0.47	Fu-xiang et al. (2019); Prakash et al. (2020)
			Blueberry	0.40	Može et al. (2011)
			Gooseberry	0.15–0.55	Orsavová et al. (2019)
Tannins	–	Ellagic acid	Strawberry	0.79–3.34	Milosavljević et al. (2020)

in methoxylations and hydroxylations patterns of their aromatic rings. Fruits include flavonoids in the form of glycosides or acyl glycosides; however methylated, acylated and sulfate molecules are uncommon and present in trace amounts. Flavonols, a subclass of flavonoids, exhibit a 3-hydroxyflavone backbone featured with unsaturated benzo-c-pyrone (A and C Rings) displaced to a phenyl (B-ring) and as many as seven hydroxyl groups around its skeleton. The most prevalent flavonols found in fruits like berries, grapes, etc. are kaempferol, quercetin and myricetin. Flavanones have a distinct aglycone moiety which interacts with glycosides, which are abundant in citrus fruits. The major flavanones present in grapefruit are narirutin and naringin, hesperidin and narirutin in orange, and eriocitrin in lemon. The fruits with purple, blue and red colours are imparted by a large group of coloured water-soluble pigments called anthocyanins (Olas 2018). They are widely distributed in various fruits such as strawberries, cranberries, bilberries, blackcurrants, blackberries, raspberries, blueberries and chokeberries (You et al. 2011; Olivas-Aguirre et al. 2016). Although there are more than 600 anthocyanins known to exist in nature, only six of them are commonly present in fruits, including pelargonidin, malvidin, delphinidin, peonidin, cyanidin and petunidin (Kong et al. 2003; Smeriglio et al. 2016). The flavonoid class and flavan-3-ols/flavanols subclass are referred to as catechins and are found in several fruits like gooseberries, strawberries, blueberries, cherries, apples and black grapes. In addition, gallocatechin, epicatechin (EC), epicatechin 3-gallate (ECG), epigallocatechin 3-gallate (EGCG) and epicatechin (EC) are the most significant dietary catechins.

Quercetin, a natural polyphenolic compound, is abundantly present in fruits like grapes, apples, cherries, raspberries, blueberries and blackcurrant (Borges et al. 2010; Williamson and Manach 2005). Tannins are another crucial component found in fruits. Tannins are responsible for delineating sensory characteristics (producing an astringent taste) of fruits and their associated products (Puupponen-Pimiä et al.2005). Tannins are available in two forms i.e., hydrolysable tannins including ellagic acid or esters of gallic acid and condensed tannins such as proanthocyanidins (Bautista-Ortín et al. 2012). Tannins stabilise anthocyanins by binding with them and forming a stable copolymer. Hydrolysable tannins are found rarely in fruits such as blueberries, blackberries, raspberries and strawberries (Borges et al. 2010; Bobinaite, Viškelis and Venskutonis 2012). However, ellagitannins, esterified as glucose and ellagic acid, which are found in free and complex form as glucosides, together constitute about 51% of the total phenolic content. Low-molecular-weight substances called stilbenes are found in fruits like berries and grapes. The grapevine uses malonyl-coenzyme A and p-coumaroyl CoA as precursors to the synthesis of resveratrol, a polyphenol belonging to the stilbene family (CoA) (Knutson and Leeuwenburgh 2008). Resveratrol is found in various fruits especially grapes and Vaccinium spp. (cranberry, bilberry and blueberry) (Kasiotis et al. 2013; Sebastia et al. 2012). The first resveratrol molecule was discovered in 1940 in the roots of the white hellebore *Veratrum grandiflorum* O. Loes, and it was later discovered in berry skins, grapevines and *Vitis* spp. (grape) leaves (Kasiotis et al. 2013; De Ligt, Timmers and Schrauwen 2015). Fruits provide a substantial amount of phenolic content in addition to beta-carotene, lycopene and beta-cryptoxanthin, which are precursors to vitamin A.

3.3 NUTRITIONAL PROPERTIES OF FRUITS

3.3.1 General Introduction

The alternative to pharmaceuticals "nutraceutical" is a buzzword these days, and due to their prospective nutritional, safety and therapeutic properties, they have received significant attention (Dutta, Marak, and Meena 2017). The naturally derived compounds with significant bioactivity are referred to as nutraceuticals, which have health-promoting, disease-preventing and medicinal properties and are found in various foods, dietary supplements and herbal products. According to the WHO (World Health Organization), more than 80% of the total world's population relies upon traditional plant-based systems of medicine like phytochemicals, nutritional constituents or as functional foods (Allen 1997; Kasbia 2005). A nutraceutical can be considered the opposite of junk food, and it is a product isolated or purified from foods that is generally sold in medicinal forms not usually associated with food (Mahajan and Rishi 2016). Health Canada defines a nutraceutical as a product isolated or purified from foods and generally sold in medicinal forms not usually associated with food and demonstrated to have a physiological benefit or provide protection against chronic disease (Gupya, Yadav and Patil 2013).

DeFelice (1995) described nutraceutical as any substance which is a food or a part of food that provides medical or health benefits apart from providing basic nutrition. The term "nutraceutical" is derived from two words namely "nutrient" and "pharmaceutical," which means a nourishing food component and medicinal drug respectively. In 1989, the name was coined by Stephen DeFelice, who was founder and chairman of the Foundation for Innovation in Medicine, an American organisation located in Cranford, New Jersey (Radhika, Singh and Shivkumar 2011). Pandey, Verma and Saraf (2010) classified nutraceuticals into two groups, i.e. potential nutraceuticals, which have a promising approach towards a particular health or medicinal benefit, and established nutraceuticals, which are derived from potential nutraceuticals and are based on sufficient clinical tests to demonstrate results.

Dolkar et al. (2017) classified nutraceuticals as natural sources such as carotene (lycopene, lutin, α-carotene, β-carotene, zeoxanthine), polyphenols (anthocyanidine, flavonones, flavones, flavonols, phenolic acids, resveratrol, curcumin), fatty acids like omega-3 fatty acids, conjugated linoleic acid etc., saponins, probiotics/prebiotics/synbiotics (lactobacillus, fructooligosaccharides), phytoestrogen (daidzein, zenistein, lignans), plant stanols/sterols (sterol ester), phenols (caffeic acid, ferulic acid), dietary fibre (insoluble and soluble fibre) and polyols (sugar, alcohols). The different classes of nutraceuticals according to chemical composition include inorganic minerals supplements (minerals like zinc), vitamins supplements (vitamins like ascorbic acid), prebiotics (digestive enzymes), probiotics (*Lactobacillus acidophilus*), digestive enzymes, dietary fibres, cereals and grains (fibres), health drinks like fruit juices, antioxidants such as vitamin C and E, and phytochemicals compounds (carotenoids).

On the other hand, in addition to the fulfilment of essential nutrients (e.g. vitamins and minerals), functional foods provide various health benefits in the form of whole, fortified, enriched or enhanced foods which are consumed on a regular basis and at sufficient levels as part of a varied diet (Hesler 2002). The difference between these two foods, i.e. nutraceuticals and functional foods, is mainly based on the mode of administration and their chemistry or biochemistry, while they both are the same in terms of their biological functions (Tsao and Akhtar 2005). Functional foods help the body to meet the required levels of various nutrients like vitamins, fats, proteins, carbohydrates, etc., which are needed for the healthy survival of a human being. Functional food with properties of prevention and/or treatment of disease(s) and/or disorder(s) can be referred to as nutraceutical food. The list of nutraceutical foods includes various fortified products, dairy products, fruit juices from different fruits etc. (Kalra 2003; Mahajan and Rishi 2016). With the consumption of various nutraceuticals within a variety of functional foods, the efficient functioning of the immune system can be maintained and the supplementation of these functional foods in a regular diet boosts the immune system. A better immune system can help in providing important antiviral prevention against various diseases including COVID-19 (Hamid, Thakur and Thakur 2021).

Nutraceutical/functional foods derived from plants are gaining considerable attention because of their potential nutritional and therapeutic effects along with presumed safety (Pandey et al. 2011; Rajalakshmi et al. 2017). Fruits and vegetables are rich sources of nutraceutical compounds and help in preventing numerous diseases, as they have anti-diabetic, anti-cancerous, antihypertensive, antimicrobial, neuroprotective, anti-inflammatory, antioxidant and antiviral properties, which play important roles in cell detoxification, stimulation of the immune system, cholesterol synthesis and lowering the blood pressure (Sharma, Agarwal and Saxena 2019). Fruits are rich in various nutrients and act as potential sources of many nutraceuticals or functional foods, which can play a tremendous role in this sector. Many of these fruit species possess certain inherent beneficial properties, and in different cultures, they have been used since time immemorial (Dutta, Marak and Meena 2017). The beneficial compounds like polyphenolic components present in various fruits possess biological activities and can be used as pharmaceutical nutraceutical products. These compounds help in promoting antioxidant potential on human health and have additional properties like anti-inflammatory, antifungal, antibacterial and chemopreventive effects, which are associated with the consumption of various fruits in the human diet. In addition to polyphenols, the other compounds commonly occurring in fruits with positive health effects are xanthones, carotenoids and saponins (Dolkar et al. 2017). Fruits are rich in various micronutrients like essential proteins, vitamins (vitamin A, vitamin B_6, vitamin B_{12}, vitamin C, vitamin D, vitamin E, folic acid etc), essential amino acids, fatty acids, minerals (iron, selenium, zinc and copper) and certain phytochemicals have prime importance in a healthy human immune system (Thakur, Joshi and Thakur 2019).

A variety of colourful fruits vary in their composition, and these fruits are accountable for providing a wide range of nutrients/phytochemicals that boost human health (Figure 3.1). Various phytochemicals have potential health benefits; for example lycopene reduces the risk of prostate cancer, anthocyanins delay several diseases associated with ageing, sulfides and polyphenols reduce the risk of cardiovascular diseases, carotenoids, flavonoids, anthocyanins and phenolics reduces the risk of heart disease and cancer, and slow down the ageing process (Dolkar et al. 2017)

Almost all of fruits including berries have nutraceutical properties. Mahima et al. (2013) reviewed the nutraceutical compounds present in different fruits and reported that fruits like berries, pineapple (*Ananas comosus*), plum (*Prunus domestica*), citrus fruits like orange (*Citrus sinensis*), lime/lemon (*Citrus limon*), sour orange/bitter orange (*Citrus × aurantium*), guava (*Psidium guajava*), custard apple/sugar apple (*Annona squamosa*), jambul (*Syzygium cumini*), Indian gooseberry (*Phyllanthus emblica*), mango (*Mangifera indica*), papaya (*Carica papaya*), mangosteen (*Garcinia mangostana*), pomegranate (*Punica granatum*), bananas (*Musa spp.*), bael (*Aegle marmelos*), grapes (*Vitis vinifera*), watermelon (*Citrullus lanatus*) etc., have good nutraceutical potential. Among minor edible fruits that are cultivated or grown in the Indian subcontinent, fruits like amla, bael, carambola, jamun, jack fruit, karonda, pomegranate, passion fruit, phalsa, rambutan, durian, wood apple, Malayan apple, fig, buchanania/charoli, karonda, annona, syzygium and opuntia have potential nutraceutical properties (Ashok, Ravivarman and Kayalvizhi 2020). Among different fruits, berries are predominantly rich in phenolic compounds like phenolic acids (hydroxycinnamic and hydroxybenzoic acid derivates) flavonoids (anthocyanins, cathechins, flavonols etc.), tannins (proanthocyanidins, gallotannins, ellagitannins,), stilbenoids (resveratrol, piceatannol, pterostilbene) and lignans (Rutz et al. 2012; Nile and Park 2014; Rios De Souza et al. 2014). Some citrus fruits and limonoids help in preventing several diseases like heart disease, arterosclerosis and inflammation, as they possess hepatoprotective, neuroprotective, antimicrobial, antioxidative and anti-diabetic properties and further play a crucial role in the prevention

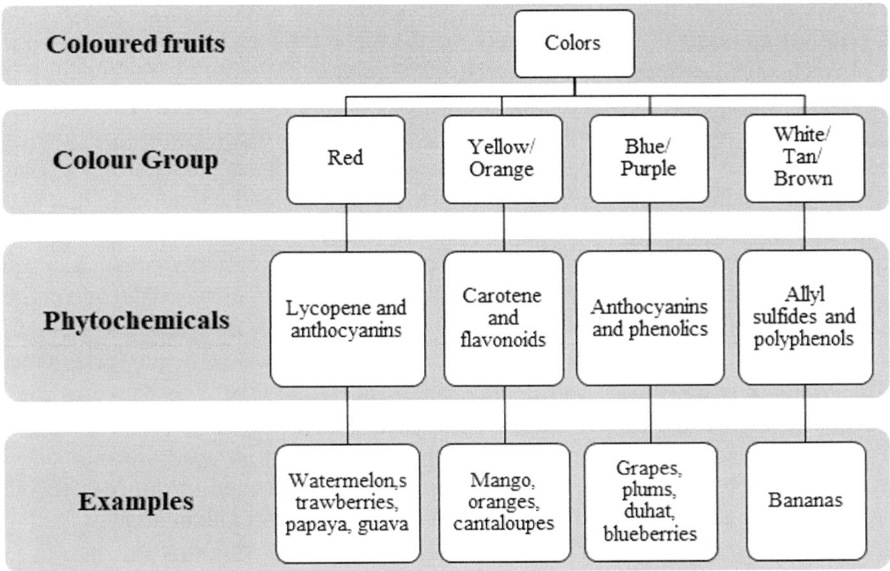

Figure 3.1 Phytochemicals provided by different coloured fruits (adapted from Dolkar et al., 2017)

of different types of cancers (Duarte, Carvalho and Miguel 2016). Pomegranate also contains medicinal properties against cancer, and its extracts are used for the prevention of various types of cancers (Panth, Manandhar and Paudel 2017). Pomegranate and its different parts, such as arils, peel, membrane and seeds, contain different bioactive compounds like ascorbic acid, anthocyanins, gallic acid, ellagic acid, punicalagin, quercetin, etc., which have higher bioactivity due to the potential to reduce free radical generation and the ability to prevent the generation of various diseases (Hamid et al. 2020). With the increasing interest in the search for new natural bioactive compounds, fruits are considered a potential source of nutraceuticals, as they are a rich source of essential oils and their isolated components like monoterpenes (Dell'Agli et al. 2013). As novel therapeutical agents, nutraceuticals as natural products have always played a significant role, and they have immense medicinal properties and profound direct and indirect effects on human health (Sharma et al. 2019).

3.3.2 Antioxidant Potential

A majority of fruits are abundant in antioxidants and phytochemical compounds, and phenolics are the major bioactive compounds that have potent antioxidant activity with health benefits. Sun et al. (2002) studied the antioxidant and antiproliferative activities of common fruits and reported the highest soluble free phenolic content in cranberry (507.0 mg/100 g), followed by apple with 272.1 mg/100 g, red grape with 182.0 mg/100 g, strawberry with 147.8 mg/100 g, lemon with 66.3 mg/100 g, peach with 65.3 mg/100 g, orange with 56.8 mg/100 g, banana with 56.1 mg/100 g, pear with 53.6 mg/100 g and pineapple with 40.4 mg/100 g, while the lowest was reported in grapefruit (30.7 mg/100 g). In terms of antioxidant activity, the highest values have been reported in a phytochemical extract of cranberry i.e. 177.0 μmol/g, followed by apple with 97.6 μmol/g, red grapes with 64.7 μmol/g, strawberry with 64.4 μmol/g, peach with 49.5 μmol/g, lemon with 42.8 μmol/g, pear with 34.2 μmol/g, banana with 32.8 μmol/g, orange with 31.5 μmol/g, grapefruit with 24.7 μmol/g and pineapple with 16.9 μmol/g.

Apple is rich in antioxidant compounds like phenols and carotenoids, and several health benefits can be obtained with its regular consumption. The phenolic compounds possessing strong antioxidant potential that commonly occur in apples are chlorogenic acids, procyanidins, hydroxycinnamic acids, flavonoids, quercetins and anthocyanins (Shehzadi et al. 2020). Bananas have high free radical scavenging activity, which is due to the high content of antioxidants and other phytochemicals like vitamins C and E, as well as flavonoids and β-carotene. The enzymes present in banana are also responsible for increasing its antioxidant capacity (Someya, Yoshiki and Okubo 2002). The bioactive compounds in Citrus spp. have antioxidant action, which has been reported to be higher in the peel than in the fruit tissues (Assefa, Saini and Keum 2017). Citrus fruits have high anti-oxidant activity, as they are a good source of dietary antioxidant compounds like vitamins A, vitamin C, vitamin E, flavonoids, carotenoids, limonoids, coumarins, pectins and mineral elements, which play a vital role in the management of various human degenerative and chronic diseases (Zou et al. 2016). Rocha Ribeiro et al. (2007) carried out research on the pulp composition of four mango cultivars and reported good antioxidant potential in the human diet in terms of total phenolics, carotenoids and ascorbic acid content. The polyphenolic antioxidants mangiferin and glucosyl xanthone present in mango have pharmacological and medicinal properties like immunomodulation, anti-lipid peroxidation, hypotensive, cardiotonic, wound healing, antidiabetic and antidegenerative activities (Shah et al. 2010). Jung et al. (2020) reported that unripe peach fruits are rich in phenolic compounds and can be used as a potentially inexpensive source of antioxidants and can be widely used in various industries including the food industry. Dropped unripe peaches have been reported as excellent sources of bioactive compounds with antioxidant potential which further have health benefits in various food products (Liu, Cao, and Jiang 2015). Pear is rich in bioactive components, which are mostly consumed as fresh fruit and have been used as a traditional remedy with possible medical applications in preventing diseases induced by oxidative stress. Pear fruits are a good source of phenolic compounds, which contributes to their sensory qualities, and possess antioxidant activity owing to various high health benefits (Brahem et al. 2017). Savic et al. (2021) reviewed the chemical composition of pear fruit and mentioned the presence of vitamin C, vitamin E and organic acids (malic, citric, shikimic acids etc.) in addition to other phenolic compounds such as quercitrin, arbutin, isoquercitrin, (–)-epicatechin, astragalin, kaempferol, tannins and triterpenoids including squalene, friedelin, sterols and saponins. Pineapple, which is consumed in many parts of the world, is an important tropical fruit that has good antioxidant activities, as it is a rich source of vitamins like A, B and C as well as several minerals such as phosphorus, calcium and iron with high nutritive value (Hossain and Rahman 2011). The major antioxidant compounds present in major fruits along with antioxidant activity have been listed in Table 3.2.

Table 3.2 Antioxidant compounds and activity of major fruits

Fruit	Antioxidant compounds	Antioxidant activity	References
Apple	TPC: 29.98 and 31.78 mg/100 g (Whole fruit) 520.28/355.22 mg/100 g in seed; 140.69/116.94 mg/100 g in peel; 22.09/37.42 mg/100 g in flesh (Idared/Šampion cv.) 2143, 2056, 2209, 2723, 1600 and 1468 mg GAE/kg FW (Peel of Champion, Generos, Idared, Florina, Golden Delicious and Reinette Simirenko) 290.2 and 219.8 mg/100 g (Red delicious with and without skin) 66.94, 70.90 and 71.88 (Golden Delicious, Fuji rose and Granny Smith)	ABTS radical: 693 and 306 mg Trolox/100 g (Unripe and ripe fruits) 39.6, 47.2, 53.4, 71.7, 29.2 and 31.4 μmol TE/g (Champion, Generos, Idared, Florina, Golden Delicious and Reinette Simirenko) FRAP: 27.5, 26.3, 35.2, 41.6, 25.9 and 24.8 (Peel of Champion, Generos, Idared, Florina, Golden Delicious and Reinette Simirenko) 8.84, 1.93 and 3.95 mg/ml 326–554 and 387–545 (Flesh and peel of Galaval)	Eberhardt, Lee, and Liu (2000); Duda-Chodak Tarko, and Tuszyński (2011); Vasile et al. (2021); Asale et al. (2021); Butkeviciute et al. (2022)
	TFC: 142.7 and 97.6 mg/100 g (Red delicious with and without skin) 15.59, 21.91 and 21.78 (Golden Delicious, Fuji rose and Granny Smith)	DPPH: 122.53, 86.20 and 93.25 EC_{50} μg/ml (Golden Delicious, Fuji rose and Granny Smith)	
	VitaminC: 31.48, 28.83 and 14.97 mg/100 g (Golden Delicious, Fuji rose and Granny Smith)	Antiradical activity: 20–114 and 44–217μM TE/g (Flesh and peel of Galaval)	
Banana	TPC: 119.47, 70.03, 35.88, 34.43 and 7.47, 38.92, 80.14, 9.8 (Peel and pulp of Sagor, Champa, Shail, Bichi) 51 to 54 mg GAE/100 g (Pisang Mas banana) 90.4 mg GAE/100 g (Pisang Awak banana)	DPPH radical: 93.26, 93.54, 94.04, 92.27 and 64.49, 90.83, 91.37, 62.14 (Peel and pulp of Sagor, Champa, Shail, Bichi) 91.37 and 94.04% (Pulp and peel of Shail) 98.19% at 10 mg/ml	Sad et al. (2018); Lim, Lim, and Tee (2007); Alothman, Bhat, and karim. (2009); Choo and Aziz (2010); Baskar et al. (2011); Fatemeh et al. (2012)

(Continued)

Table 3.2 (Continued)

Fruit	Antioxidant compounds	Antioxidant activity	References
Citrus	Vitamin C: 30–60 mg/100 g (Extracts of satsuma mandarin, pomelo, navel orange and tangerine) 110.4 mg/100 g (Orange peel) Vitamin E: 4.50–11.40 mg/kg (Peel and seeds of candied orange, tangerine and lemon) Naringin: 11.90 mg/100 g (Candied shaddock) and Naringenin: 75 mg/l (Sunkist grapefruits) Hesperidin: 4.37–22.0 mg/100 g (Candied orange, lemon and shaddock) Quercetin: 24.09 mg/100 g (Shaddock) Phenolic acids: 180 mg/g DW (flesh of Huyou) and 5060 mg/g DW (peel of Ponkan) 117.26–241.91 mg GA/100 g (13 blood orange cultivars) 77.3, 49.8 and 35.6 mg of GAE/g (Peel of grapefruit lemon and orange) Coumarins: 2.2 mg/100 g (Shaddock) Limonoids: 0–95.46 mg/100 g (Citrus fruits) Carotenoids: 0.021–2.04mg/g dried base (peels of ponkan, wendun and peiyou)	DPPH IC_{50}: 1.1/2.8, 0.6/2.2, 1.7/3.7. 1.4/3.4, 2.1/3.7 and 1.9/3.9 mg/ ml (Peel/tissues of *C. sinensis* var. Washington Navel, *C. reticulata* var. Ponkan, *C. sinensis* var. Sungin, *C. limon, C. paradise* and *C. aurantium*) 76.4, 73.2 and 70.5 (Peel of grapefruit lemon and orange) 21.80 to 357.18, 7.50 to 14.06 and 2.31 to 7.24 μmol TE/mL or ml [Peel, pulp (DW) and juices of 27 Chinese cv.] ABTS: 25.07 to 461.72, 11.62 to 33.17 and 2.01 to 10.04 μmol TE/g or ml [Peel, pulp (DW) and juices of 27 Chinese cv.] H-TAA: 75.22–129.75 mg Trolox/100 g (13 blood orange cultivars) L-TAA: 1.35–1.55 mg Trolox/100 g (13 blood orange cultivars) FRAP: 6.90 to 39.77, 10.46 to 30.72 and juices 1.54 to 4.43 μmol TE/g or ml [Peel, pulp (DW) and juices of 27 Chinese cv.]	Ye (2005); Zhou (2012); Ho et al. (2000); Xu et al. (2008); Wang,Chuang, and Hsu (2008); Ghasemi, Ghasemi, and Ebrahimzadeh (2009); Legua et al. (2021); Elkhatim, Elagib, and Hassan (2018); Chen et al. (2020)

(Continued)

Table 3.2 (Continued)

Fruit	Antioxidant compounds	Antioxidant activity	References
Mango	TPC: 23.92/2.25, 35.14/1.98 and 14.78/ 1.82 mg GAE/g DM (Peel/pulp of Alphonso, Kesar and Totapuri) 116.80 and 122.60mg GAE/g DM (Langra and Chausa) 9mg GAE/g DM (Mango peel extract) 119.30/357.92, 33.19/278.75, 38.61/324.58,40.56/304.58 and 69.44/335.70 (Pulp/peel of Neelum, Bangalora, Banganpalli, Vanraj and Amrapali) TFC: 25.52/13.87, 34.90/9.08 and 14.43/25.56 mg QE/g DW (Peel/pulp of Alphonso, Kesar and Totapuri) 90.89 and 92.55mg CE/g DM (Langra and Chausa) 8.5mg QE/g DM (Mango peel extract) 29.24 and 5.43 mg QE/g DM (Ripe mango peel and pulp) 9.73/310.07, 7.00/76.87, 9.73/118.80, 16.20/199.33 and 37.07/307.60 (Pulp/peel of Neelum, Bangalora, Banganpalli, Vanraj and Amrapali)	ABTS: 1.62/24.77, 1.73/24.81 and 2.54/24.76mg TE/ g DM (Pulp/peel of Alphonso, Kesar and Totapuri) DPPH: 4.35/23.73, 3.30/23.79 and 6.17/23.70 mg BHA/g DM (Pulp/peel of Alphonso, Kesar and Totapuri) FRAP: 2.99/39.16, 2.26/57.24 and 3.09/25.15 mg TE/ g DM (Pulp/peel of Alphonso, Kesar and Totapuri) Total antioxidants: 1.04/4.20, 0.29/1.89, 0.62/5.53, 0.50/4.29 and 0.75/3.84 μmol Trolox/g (Pulp/peel of Neelum, Bangalora, Banganpalli, Vanraj and Amrapali)	Madalageri et al.(2015); Sultana et al. (2012); Gondi et al. (2015); Abdul Aziz et al. (2012);Muralidhara et al. (2019)
Peach	TPC: 1045–1098 GAE μg/mL of sample (Unripe fruit of Changhowon hwangdo and Kanoiwa hakuto) 1209.3/711.7, 1300.9/781.8 and 1354.5/ 881.3 mg GAE/100 g DW (Peel/pulp of Shahpasand, Shireen and Golden); TFC: 599.7/301.3, 677.4/499.7 and 785.5/ 383.7 mg catechin/100 g DW (Peel/pulp of Shahpasand, Shireen and Golden)	DPPH: 93.70–96.65% inhibition (Unripe fruit of Changhowon hwangdo and Kanoiwa hakuto) RP (absorbance values at 700 nm at concentration 0–12.5mg/mL): 0.060–1.99/0.090–2.77, 0.040–1.87/0.050–2.57 and 0.030–1.54/0.070–2.66 (Pulp/peel of Shahpasand, Shireen and Golden) DPPH: 66.8% and 76.5% (peel extract of shahpasand and golden); 43.4% and 49.1% pulp extract of shahpasand and golden)	Jung et al. (2020); Manzoor et al. (2012)

(Continued)

Table 3.2 (Continued)

Fruit	Antioxidant compounds	Antioxidant activity	References
Pear	TPC: 601.50–619.25 and 333.90–355.80 mg GAE/100g (Peel and pulp) 12.98–45.37 and 9.88–22.09 mg/100 g FW (Peel and pulp of San Giovanni Varieties) 13.3/27.3/16.4 mg GAE/g DE (Wild pear's flesh/peel/flesh+peel in Serbia) TFC: 543.50–561.30 and 270.50–290.50 mg CE/100g (Peel and pulp) 1.93/4.82/2.54 mg QE/g DE (Wild pear's flesh/peel/flesh+peel in Serbia) Flavanols: 131.48 to 251.17 and 33.91–129.67 mg/100 g FW (Peel and pulp of San Giovanni Varieties) Flavonols: 30.55–64.33 and 3.93–10.32 mg/100 g FW (Peel and pulp of San Giovanni Varieties)	RP (absorbance values at 12.5 mg/mL conc.): 0.56–0.58 and 0.30–0.32 (peel and pulp extract) DPPH: 49.71–49.94% and 27.89–28.29% (peel and pulp extract) 4.29/0.80/2.85 IC_{50}, mg/mL (Wild pear's flesh/peel/flesh+peel in Serbia) FRAP: 6.46–14.07 and 1.04-2.91 mmol Fe^{2+}/kg (Peel and pulp of San Giovanni Varieties) 618/880/743 μmol Fe(II)/g (Wild pear's flesh/peel/flesh+peel in Serbia) ABTS: 0.76/1.10/0.77 mg AAE/g (Wild pear's flesh/peel/flesh+peel in Serbia) TEAC: 3.38-10.21and 0.57-1.83 mmol Trolox/kg (Peel and pulp of San Giovanni Varieties)	Maleeha et al. (2013); Azzini et al. (2019); Savić et al. (2021)
Pineapple	TPC: 2.6, 13.8 and 51.1 mg/g caffeic acid (Water, ethyl and methanol extracts of ripe pineapple) 126.95, 84.90, 89.01, 71.07, 109.60 and 74.09 mg GA/100g (Imperial, Perola, IAC Fantástico Smooth Cayenne, Gomo-de-Mel and Vitória cultivars) TFC: 39.4, 37.9 and 55.2 mg QE/g (Water, ethyl and methanol extracts of ripe pineapple) Vitamin C: 62.11, 49.79, 43.72, 42.31, 36.74 and 35.88 mg/100g (Imperial, Perola, IAC Fantástico Smooth Cayenne, Gomo-de-Mel and Vitória cultivars)	Antioxidant capacity: 612.1, 1051.8 and 1933.0% as equivalent to ascorbic (Water, ethyl acetate and methanol extracts of ripe pineapple) 21.75, 17.18, 13.29, 12.96 and 13.55 (% inhibition (Imperial, Perola, IAC Fantástico Smooth Cayenne, Gomo-de-Mel and Vitória cultivars) 49,9 and 7211 IC50 (mg/l) in half and full yellow stage	Hossain and Rahman (2011); Ferreira et al. (2016); Hastuti and Lukito (2022)

TPC: Total phenol content; TFC: Total flavonoid content, RP: Reducing power, QE: Quercetin, TE: Trolox equivalent, H-TAA: Hydrophilic total antioxidant activity, L-TAA, Lipophilic total antioxidant activity, DM/DW/DE: Dry matter/dry weight/dry extract, FW: Fresh weight, ABTS-2: 2-azino-bis (3-3thylbenzthiazolin-6sulphonic) acid; DPPH- 2: 2-diphenyl-1-picrylhydrazyl; FRAP: Ferric reducing ability of plasma

3.3.3 Antimicrobial Properties

Plants have wide antimicrobial activity, especially fruits (Table 3.3), which helps in killing microorganisms within the human body (Raza et al. 2019). Most fruits and vegetables contain thousands of phytochemicals that have biological activity, which interact in different ways to maintain well-being and protect against diseases (Pushpaanjali et al. 2020). Antimicrobial activity is very important for food safety and microbiological quality as well as for the prevention of a number of diseases. There is various evidence that proves the beneficial role of fruits in protecting against many diseases. Keeping in mind the harmful effect of chemical-based synthetic medicine on human health, scientists are increasingly focusing on the utilisation of natural foods to treat ailments. This includes the utilisation of fruits as well as vegetables because they are easily available and provide palatability to the food. According to the latest reports, the medical sector recently increased exploration of the phytoconstituent components of fruits due to their antimicrobial properties (Wang et al. 2012). These compounds restrict the activity of microorganisms by penetrating within their cell membrane, causing lysis of cells (Ayala-Zavala et al. 2010). Some reports are summarised here. Baranowska and Bartoszek (2016) proved the antimicrobial property of cranberry. They reported the presence of proanthocyanidins (PAC) type A as a key antimicrobial component. The anti-adherence property of PAC has been utilised to treat/prevent bacterial infection of the urinary tract. According to Wang et al. (2012), the fruits of *Morus alba* possess a 12.5mg/ml minimum inhibitory concentration against *Eschercia coli* bacteria. Goji berry contains phyto-components that showed antimicrobial activity against various gram-negative strains (*Salmonella enterica* subsp. *enterica*, *Klebsiella pneumonia* and *Pseudomonas aeruginosa*) as well as against yeast (*Candida albicans*) only at a concentration of 2mg/mL (Ilic et al. 2020). According to various published reports, antibacterial activity has been observed in raspberry fruits. Fruits having higher concentrations of phenolic compounds, such as apple, aonla, pomegranate and grape, are responsible for protecting against various diseases in humans and also retaining the activity of various microbes to prevent the chance of infection (Raza et al. 2019). Ayala-Zavala et al. (2010) summarised the presence of an optimum concentration of ascorbic acid in apple fruit cv. Gala, which reduced the microbial burden of *Salmonella* ser. *Typhimurium, agona* and *Michigan* up to 0.7 log CFU/g. Liya and Siddique (2018) observed strong antimicrobial activity of apple fruits against *Bacillus cereus* and *E. coli* with 6 to 13 mm zones of inhibition. The presence of eugenol, carvacrol and thymol in grape decreases the growth of yeast, aerobic mesophilic bacteria and mold. Similarly, papaya contains aglycons and carpaine, which worked against *S. aureus*, *E. coli* and *P. aeruginosa*. The authors also reported these properties in strawberry and lemon, which have been used to treat urinary tract infections (Rojas-Grau et al. 2007). Phenolic compounds like flavonoids, flavone glycosides, phenols and organic acids present in orange fruit and peel showed activity against several pathogenic bacteria and fungi (Mitsagga et al. 2021). There are a number of fruits that have antimicrobial activity, and these fruits need to be

Table 3.3 Bioactive compounds of major fruits and their antimicrobial potential

Fruit	Bioactive compounds	Part used	Animal name/model used	Antimicrobial activity	Health effects/nutraceutical properties	References
Fig (*Ficus* sp.)	Flavonoids, terpenes, tannins, alkaloids	fruit	*Pseudomonas sp., Aspergillus niger, Penicillium notatum, Candida* albicans	Antibacterial, antifungal	Anticancer, antidiabetic, antioxidant, antimicrobial, anti-inflammatory	Walia et al. (2022)
Apple (*Malus×domestica)*	Phenolic acid (stilbenes, hydroxycinnamic acids, dihydrochalcones), triterpenes (ursolic, oleanoic, and betulinic acids), flavonoids (flavonols, flavan-3-ols) fiber and vitamin	Fruit	*E. coli, Salmonella, Listeria innocua and Bacillus subtilis, Pseudomonas aeruginosa, Salmonella ebony, Aspergillus niger, Candida albicans*	Antibacterial antifungal	Anti-inflammatory, blood loss, regulate glucose homeostasis, weight loss, lessen the risk of thrombotic stroke and type II diabetes, in addition to the occurrence of lung cancer and cardiovascular disease and coronary fatalities	Farneti et al. (2015) Raphaelli et al. (2019)
Mango	Alkaloids, flavonoids, tannins, saponins, mangiferin, carotenoids, penta-O-galloyl-glucoside, ethyl gallate, vitamin C, dietary fibre,	Fruit, peel, seed	Gram-negative and gram-positive bacteria, *E. coli, Candida albicans, Staphylococcus aureus, Vibrio vulnificus*	Antibacterial, antifungal	Atherosclerosis, diabetes, cancer, cardiovascular diseases, Alzheimer's disease, protect mitochondrial membrane against lipid peroxidation and preserving its integrity, anti-proliferative activities, cataracts and Parkinson's disease	Masibo and He (2009); Kaur et al. (2010); Jahurul et al. (2015); El-Desoukey, Aljor and Alaotibi (2020)
Pomegranate	Phenols, hydrolysable tannins, flavonoids, alkaloids, punicalagin, gallic acid, ellagic acid, esters of core polyol molecules	Fruit, peel	*Staphylococcus epidermidis, Klebsiella pneumonia, Bacillus subtilis, Escherichia coli, Saccharomyces cerevisiae, Proteus vulgaris, Salmonella typhi.*	Food-borne pathogens, spoilage bacteria, antifungal, antiviral (prevent human respiratory syncytial virus, polyhedrosis virus, herpes virus,human immunodeficiency virus poliovirus	Anticancer properties include interference with tumour cell proliferation, invasion, cell cycle, and angiogenesis, which may be attributed to anti-inflammatory properties.	Hamid, Thakur, and Thakur (2021)

(Continued)

Table 3.3 (Continued)

Fruit	Bioactive compounds	Part used	Animal name/model used	Antimicrobial activity	Health effects/nutraceutical properties	References
Aonla	Ascorbic acid, hydrolysable tannins, complex tannins, alkaloids, flavonoids, cardiac glucosides, terepenoids, saponins.	Fruit	*Pencillium notatum, Serratia marcescens, E. coli, Klebsiella pneumoniae, Staphylococcus aureus, Salmonella typhi, Candida albicans, Proteus mirabilis, Pseudomonas aeruginosa, S. paratyphi, Aspergillus niger, Alternaria brassicae*	Antibacterial, antifungal	Diabetes, ulcers, heart diseases, gastrointestinal disorders, improve memory, reduce cholesterol, combat ophthalmic disorders, jaundice, diarrhoea, inflammations	Kumari and Khatkar (2016); Kumar et al. (2020)
Date	Phenolic acid (ferulic, proanthocyanidins, gallic acids, p-coumaric), flavonoids (quercetin, rutin, apigenin), β-carotene, lycopene	Fruit, seed	Gram-negative and gram-positive bacteria, *Pseudomonas phage* (ATCC 14209-B1), *Bacillus cereus, Escherichia coli, Serratia marcescens, Staphylococcus aureus*	Antibacterial, antifungal antiviral	Antioxidant, anticancerous, anti-nephrotoxic, anti-inflammatory.	Maqsood et al. (2019)
Pineapple	Bromelain, favatin, circulin, tannins, catechin	Fruit	*Pseudomonas aeruginosa, Escherichia coli, Klebsiella pneumoniae, Salmonella spp, Staphylococcus aureus,*	Antibiotic, antibacterial	Hay fever, ulcerative colitis, improving the absorption of antibiotics, debridement, relaxing stimulating muscle contractions, cure pulmonary oedema, anti-cancerous, help in weight loss, shortening labour	Ahamed et al. (2016); Hasan et al. (2021)

(Continued)

Table 3.3 (Continued)

Fruit	Bioactive compounds	Part used	Animal name/model used	Antimicrobial activity	Health effects/nutraceutical properties	References
Citrus family	Vitamin C, cardiac glycosides, alkaloids, saponnins, tannins, flavonoids	Fruit, peel	Gram-positive and gram-negative bacteria, *Pseudomonas aeruginosa, Salmonella, Proteus vulgaris, Staphylococcus aureus, S epidermidis, Streptococcus pyogenes, Pseudomonas aeruginosa*, *Shigella flexineri, Bacillus subtilis Escherichia coli*	Antibacterial, antifungal, antiviral	Cancer, immunity booster, regulates metabolism, weight loss, flushes toxins from body	Deshwal and Kaur (2018)
Kiwi	Phenolic acids, lutein, oxycarotenoid, omega-3 fatty acid and alphalinolenic acid, caumarin, umbelliferon, fraxetin, iso-scopoletin, aesculetin, tannins, vitamin C	Fruit, peel	*Porphyromonas gingivalis, Prevotella intermedia, Fusobacterium nucleatum, Aggregatibacter actinomycetemcomitans*	Anti-bacterial, anti-viral	Anti-asthmatic, anti-platelet, anti-nociceptive, anti-tumour, anti-hypertensive	Preeti et al. (2021)
Papaya	Alkaloids, tannins, saponins, glycosides and phenol	Fruit	*Klebsiella pneumoniae, Escherichia coli, Pseudomonas aeruginosa, Proteus mirabilis, Staphylococcus aureus,*	Antibacterial, antibiotic	Respiratory tract ailment and tumour, gastroenteritis, urethritis, otitis media, wound infections	Anibijuwon and Udeze (2009)

further explored provide knowledge to the medical and horticulture sectors. There is immense scope for the potential exploitation of the antimicrobial properties of fruits.

3.3.4 Anticancer Activity

Consuming fruits on a regular basis is strongly connected with reducing the risk of numerous types of commonly occurring cancers. Most of the strong evidence is related to the anticancer properties of compounds present in fruits (Table 3.4) that help in delaying and preventing the occurrence of cancers of the mouth and pharynx, oesophagus, lungs, stomach and colon. However, few studies have been conducted that discuss the effect of fruit consumption in preventing breast, pancreas and bladder cancers.

There are four major mechanisms by which the bioactive components present in fruits help in preventing the risk of cancer. To understand the working of these components, it is necessary to have key knowledge about the chemical carcinogenesis process and various aspects influencing different stages of this progression. The three phases of chemical carcinogenesis are initiation, promotion and progression. Initiation is linked with the initial aspect of cancer induction and refers to abrupt events that create reasons for interaction between carcinogens and DNA, leading to gene mutations. Promotion and progression apply to later phases of carcinogenesis. Promotion is the clonal expansion of genetically altered cells, while progression is the subsequent extrusion and metastatic expansion to other nearby organs. Two different concepts are important in the early phase of carcinogenesis to prevent cancer. One is the interception of elements that react with DNA and the other is the detoxification of potential carcinogens. However, in advanced stages of cancer, treatment is possible if components of food inhibit the propagation of clonal altered cells or interrupt the biology of tumour cells in such a way that their growth and metastasis are hindered.

Phytochemicals having antioxidant properties are effective in interrupting compounds that react with DNA, and this is due to the electrophilic nature of the phytochemicals. Mutagens or carcinogens are deficient in electrons and are attached to electron-rich sources in cells. During cancer development, RNA, DNA and proteins have the extreme nucleophilic tendency to interact with an electrophile, leading to the formation of an adduct, which is formed between the cancer-causing agent and part of the nucleoprotein. Adduction of DNA sometimes causes the carcinogenic process. The phytochemicals of fruits such as phenolics and carotenoids inhibit the development of cancer as they have outstanding antioxidant activity. Phenolics are able to prevent DNA adduction, as they provide alternative targets for the carcinogens to attack. Many phenolic compounds have been tested for anticarcinogenic properties. By employing an in vitro tumour model to compound extract from cranberry, the growth and propagation of breast, colon, prostate and lung cancer could be prevented. Proanthocyanidin oligomers, anthocyanin glycosides, triterpenoids and flavonol are among the phytochemicals that contributed to anticancer properties by inducing tumour cell apoptosis, reducing ornithine decarboxylase activity and decreasing expression of matrix metallo proteinases associated with prostate tumour

Table 3.4 Bioactive compounds of major fruits and their anti-cancerous properties

Fruits	Bioactive compounds	Parts used	Animal name or model used	Health effect/ nutraceutical properties	References
Grapes	Resveratrol (polyphenol)	Berries	Rodent model	Anticancer properties(breast cancer)	Garvin, Ollinger, and Dabrosin (2006)
Strawberry	Ellagic acid, anthocyanins and total phenolics	Methanolic extract of strawberry fruits	Mice model	Chemopreventive potential (leukaemia and breast cancer)	Somasagaraet al. (2012)
Raspberry	Ellagic acid	Fruit	Human pancreatic adenocarcinoma cell lines (assessed by measuring Phosphatidylserine externalisation, caspase activity, mitochondrial membrane potential and DNA fragmentation)	Apoptosis and proliferation in pancreatic cancer cells	Edderkaoui et al. (2008)
Phyllanthus emblica	tannins and flavonoids	Fruit extract	Swiss albino mice	Cancer-preventative and antitumour properties	Banu et al. (2004)
Persimmon	polyphenols	calyx, peel, and flesh from persimmon were extracted in methanol	HT-29 human colon cancer cells, HeLa human cervix adenocarcinoma cells and, PANC-1 human pancreatic cancer cells	Chemotherapeutic property	Kyung-Jin et al. (2011)
Citrus	Flavonoids	Peel	mechanistic model	Chemoprevention	Arroo et al. (2014)
Ficus	Polyphenolic compounds	latex	MDA-MB-231 cells	Anticancer activities	AlGhalban, Khan, and KhattakMohd (2021)
Pomegranate	Anthocyanidins, anthocyanins, flavanols, flavones, flavonones, phenolic acids, and tannins	Peel	Immunocompromised mice bearing human gastric tumours	Anti-cancer effects	Cheshomi et al. (2022)
Kiwifruit	p-coumaric acid & ferulic acid (in the peel), syringic acid & cinnamic acid (in the flesh)	Peel and flesh	Human pancreatic cancer cell line PANC-1 (CRL-1469, isolated from a pancreatic carcinoma of ductal cell origin of a 56-year-old male)	Cytotoxicity against PANC-1 cell line	Azab and Mostafa (2022)
Mango	Polyphenols, terpenoids, carotenoid and phytosterols	Seed kernel extract	Estrogen receptor-positive human breast carcinoma (MCF-7) cells	Anti-tumour properties	Abdullah et al. (2015)

metastasis and anti-inflammatory activities (Neto 2007). Strawberries have an outstanding combination of polyphenolic phytochemicals such as flavonoids, phenolic acids, stilbenes, lignans and tannins. The bioactive constituents have natural anticancer properties as they scavenge free oxygen radicals, decrease ROS concentration and prevent DNA damage (Giampieri et al. 2014). Indian gooseberry has chemoprotective activity and is used in various ayurvedic preparations and traditional herbal practices to treat cancer (Mayachiew and Devahastin 2008).

The modulation of carcinogen metabolism using bioactive components of fruits is the second mechanism by which cancer can be prevented in the early stage. Bioactive components and metabolites of black raspberry have the possibility of modulating T lymphocyte and myeloid cell survival and differentiation. In addition, they also serve as a novel source for inhibiting STAT-mediated signal transduction, thereby suggesting the utility of black raspberry extract in therapeutic intervention and drug development. An *in vitro* study on crude extract from strawberry containing ellagic acid was found to affect human colon (HT-29 and HCT-116), oral (KB and CAL-27) and prostate (LNCaP and DU145) cell lines due to its anti-proliferative effect (Zhang et al. 2008). Anthocyanins found in raspberries are effective in down-regulating cyclooxygenase expression and activity and therefore have antiproliferative action on various cancer cell lines such as MCF-7 (breast), SF-268 (central nervous system, CNS), HCT-116 (colon) and NCI-H460 (lung) (Seeram et al. 2006).

The last two mechanisms by which fruits can impact the cancer process are based on altering the behaviour of cancer cells. Cancer cells develop when abnormal stimulation of certain genes is required by cells, which leads to the production of growth factors causing cells to grow abnormally without cellular restrain. One such initially discovered event is the initiation of oncogenes that distinguishes a cancer cell from a normal cell. Oncogenes are regular cellular genes that turn into cancer cells, allowing the cancer cells to escape factors that regulate cell growth. Citrus fruits are comprised of many monoterpenes, among which limonene is the most intensively studied compound. Limonene was found to inhibit the formation of mammary cancer when fed to rats by repression of the muted *ras* oncogene. This infers that limonene has a tendency to modify the progression of tumours even in the advanced stage of cancer development. Further, the metabolite perillyl alcohol produced from limonene in cherries had chemopreventive properties for pancreatic cancer when tested in hamsters (Wargovich 2000). Phytochemicals with antioxidant properties are able to inhibit cell proliferation and induce cell death. In vitro and in vivo assays of fruit with quercetin have established anticancer effects on different types of cancers, such as prostate, breast, colon and lung cancers, while resveratrol employs action by inhibiting tumour initiation, promotion and progression.

3.3.5 Antidiabetic Activity

Diabetes mellitus (DM) is one of the most common disorders of endocrine and is the third reason for morbidity and mortality, after cardiovascular diseases and cancer. It is the result of a deficiency in the secretion of insulin, resistance to insulin, or both. In DM, following one line of treatment in all patients is difficult because it comprises a diverse group of illnesses of multiple aetiologies that affect the body at multiple organ levels. Generally, DM is classified as type 1 and type 2. In a person suffering from type 1 diabetes (T1DM), the production of the hormone insulin is affected because of the destruction of pancreatic β cells that produce the hormone. However, type 2 diabetes (T2DM) occurs due to the advancing impairment of pancreatic β cells, which affects insulin secretion and diminishes the sensitivity of target tissues to the action of this hormone. Apart from conventional oral and injectable medication, DM can be treated with diet modification, regular exercise, lifestyle changes, weight regulation or herbal or fruit-based therapies (Table 3.5).

Administrating the ethanolic extract of *Terminalia chebula* fruits (200 mg/kg body weight/rat/day) orally for 30 days can significantly reduce the level of glycosylated haemoglobin and glucose in the blood, in streptozotocin (STZ)-induced experimental diabetic rats. By altering the activity of carbohydrates and enzymes that metabolise into glycogen, one can revert back to nearly normal after 30 days of treatment with the extract. The pathological abnormalities in the mitochondria and endoplasmic reticulum of pancreatic β cells were normalised in STZ-induced diabetic rats with treatment with *T. chebula* extract (Kumar et al. 2006). Phenolic components of fruit are important dietary antidiabetic constituents. The antidiabetic activity of different fruit parts (pulp, seed, seed coat, kernel) of indigenous black jamun (*Syzygium cumini* L.) was measured by using an *in vitro* α-amylase inhibition method (Gajera et al. 2017). The α-amylase inhibitors of jamun were effective in lowering the level of postprandial hyperglycemia by restraining starch breakdown, thereby having an efficient antidiabetic strategy. Seeds and kernels of jamun exhibited higher discrete phenolics such as gallic, catechin, ellagic, ferulic acids and quercetin, however, the pulp showed higher concentrations of gallic acid and catechin as α-amylase inhibitors.

Fig fruit has been an essential part of the Mediterranean diet for millennia as a typical protective food in health promotion. Methanolic extract of fig fruits was analysed by liquid chromatography photodiode array quadrupole time-of-flight mass spectrometry (LC-PDA-MS-QT), and the results revealed the presence of 11 phytochemicals that can be classified as anthocyanins, phenolic acids, flavan-3-ols, flavons and flavonols (Wojdyło et al. 2016). Most of the detected phenolic compounds had antidiabetic effects through diverse mechanisms. A significant difference ($p<0.05$) in the inhibitory activity of phenolic compounds against α-amylase and α-glucosidase was recorded. An IC_{50} (mg of dried fruit/mL) inhibition of α-amylase and α-glucosidase ranged from 3.58 to 17.3 and 18.2 to 19.5, respectively. Similarly, in a study by Srivastava et al. (2020) on *Cucumis melo* toluene fraction (CMTF), a significant ($P < 0.05$) decrease in the level of blood glucose (122 mg/dL) was observed on the 28th day in comparison to the positive control group (streptozotocin (STZ)). In diabetic rats induced with alloxan (150 mg/kgb.wt), the oral administration of ethanolic extract of durian (*Duriozibethinus* Murr.) and rambutan (*Nephelium lappaceum* L.) fruit peels for 11 days produced a significant decrease in the level of blood glucose. The highest reduction of glucose levels in blood was in an extract from rambutan fruit peels with a dose of 500 mg/kgb.w, and the percentage reduction was 61.76% (Muhtadi, Primarianti and Sujono 2020).

Table 3.5 Bioactive compounds of major fruits and their antidiabetic potential

Fruits	Bioactive compounds	Parts used	Animal name or model used	Health effect/ nutraceutical properties	References
Passifloraligularis	alkaloids and flavonoids	fruit juice	streptozotocin (STZ, 30 mg/kg body weight) induced diabetic rat model	Controlling diabetes	Palanirajan et al. (2014)
Sweet orange	flavonone and hesperidin	fruit peels extract	rats were induced by alloxan monohydrate (150 mg/kg BW intraperitoneally)	Antidiabetic	Muhtadi et al. (2015)
Red dragon fruit		peel	male white Wistar ratse induced to diabetic conditions by the administration of streptozotocin and nicotinamide dosages at 45 mg/kg body weight and 110 mg/kg body weight, respectively.	Anti-diabetic	Panjaitan and Novitasari (2021)
Jackfruit	Ascorbic acid, β-carotene and lycopene	Fruit	inhibition of hemoglobin glycation method	Antidiabetic and antioxidant activity	Biworo (2015)
Punicagranatum	phenolic compounds	Fruits Aqueous Extract	Alloxan-Diabetic Wistar Rats	Antidiabetic activity	Gharib and Montasser (2019)
Prunus amygdalus	–	Seed coat	Streptozotocin (STZ) induced diabetic rats	Antidiabetic effect	Kumar et al. (2021)
Grapefruit	Naringenin	Juice	male Wistar rats	Antidiabetic Properties	Kannappan and Anuradha (2010)

3.3.6 Traditional Nutraceutical Relevance

Fruits are considered the most traditional approach to honour elders and notable individuals. From ancient times, fruits have been used for curing communicable and non-communicable diseases. They play an essential role in the treatment of several diseases. They are useful in curing diseases caused by nutrient deficiency. The literature available on Ayurveda discusses the role of different fruits in the human diet. Fruits are included in the category of healthy foods. They are food items that have been supplemented with herbal ingredients, vitamins, minerals and nutrients. It is advised that the patients should consume fruits after surgery for fast recovery (Wani et al. 2022).

Traditionally, ripe banana fruit is used to treat gastric problems by neutralising the acid condition and promoting the creation of a thick protective mucus layer (Rajaiah 2010). Many tribes and ethnic groups around the world have discovered the benefits of using wild bananas to treat conditions including high blood pressure, diabetes, anaemia and other respiratory illnesses like fever, cough, tuberculosis and dysentery (Mathew and Negi 2017). The mulberry fruit has been used as a folk treatment in Chinese herbal medicine to treat a variety of illnesses, including diabetes, hypertension, arthritis, anaemia and diseases of the mouth and teeth. Mulberry fruits are also used as laxatives and as emetic, hypoglycemic and worming agents (Calin-Sanchez et al. 2013). Furthermore, desert date fruits have also long been used for therapeutic purposes. In Egyptian folk medicine, the fruit is used as an oral hypoglycemic (Kamel 1998) and an antidiabetic. This fruit aids in the treatment of jaundice, mouth ulcer, whooping cough, skin conditions such as leukoderma, sleeping sickness, etc. (Sarker et al. 2000). In addition, the fruit is used to treat dysentery, constipation, haemorrhoids, stomach aches, yellow fever, syphilis and epilepsy (Ojo et al. 2006). The fruit kernel functions as a vermifuge, a moderate laxative and an antidote to arrow poison (Ojo et al. 2006). Berries are also used in traditional medicine to treat illnesses. Bilberry has been used as a medication to cure diarrhoea and enhance night vision. Urinary infections can be treated with cranberries (Pimia et al. 2005). The red cherry fruit is used to cure toothaches, heart failure, beriberi and dropsy in conventional medicine. Ayurveda, an ancient form of Indian medicine, has acknowledged the therapeutic benefits of gooseberry fruit, including its antiscorbutic (which treats scurvy), diuretic, laxative, antibacterial and anti-dysenteric characteristics. The fruit is frequently used to cure bladder illnesses, gout, piles, hiccups, vomiting, asthma, sore throats and various other ailments (Svoboda 1992). Phyllembin obtained from gooseberry has been found to have a mild depressant action on the central nervous system (Talwar et al. 2008). It has been shown to have aphrodisiac and eyesight-improving benefits. Additionally, it aids in food digestion. It has an astringent flavour and causes a cooling effect on the body. It is highly effective for treating respiratory problems such as asthma, bronchitis and tuberculosis. Diabetes, skin conditions, piles, diarrhoea, discomfort, white discharges, blood disorders, renal issues and cataracts can all benefit greatly from it. It is the primary component of the ayurvedic preparation "Chyavanprasha" and one of the main ingredients of "Thiphalachoorna," Brahma Rasayana and Madumeghachurna, which are digestive and health tonic that delays premature greying and hair loss. Citrus juice has also been used for generations to prevent scurvy in addition to gooseberries. Citrus species' essential oils have potent antibacterial properties that are effective against both bacteria and fungi (Chanthaphon, Chanthachum and Hongpattarakere 2008; Dabbah, Edwards and Moats 1970). Pomegranate is utilised for a variety of purposes in both Ayurveda and traditional medicine. It works well to manage the three doshas: vata, pitta and kapha. The fruit promotes digestion, eases tiredness and quenches thirst. It contains a lot of the antioxidant "ellagitannin-punicalagin" and is a good source of antioxidants. Breast cancer, colon cancer and other types of tumour-genesis can be prevented using pomegranate fruit therapy (Mehta and Lansky 2004; Adams et al. 2006). The bael fruit when matured is used in the treatment of diarrhoea and dysentery, while the ripe fruit is used as a tonic and a laxative, and is considered good for the heart (Kumar et al. 2018). The Chinese consider jackfruit pulp and seeds as a tonic; it is cooling and nourishing as well as helpful in countering the influence of alcohol on the system. The ripe fruit's gum it exudes makes it useful as a laxative. Unripe jackfruit ash is used to treat skin conditions. The ash when combined with coconut oil is used to treat ulcers (Sarala and Krishnamurthy 2014). The starch from the seeds is used to treat biliousness, and the roasted seeds are thought to have aphrodisiac properties. The latex, when combined with vinegar, aids in the healing of glandular swellings, abscesses and snake bites. Wild fruit elephant apple is majorly used in the treatment of fevers and coughs (The Wealth of India 1952; Maniruzzaman 1993). The juice of the chalta fruit helps in the removal of flatulence, increases the quantity of semen and works as a galactagogue, and its external application helps in the suppuration of boils. Jamun fruits are highly valuable in terms of nutrition and medicine. Juice prevents spleen enlargement and is a diuretic. Fresh fruit consumption helps to purify the blood, prevent bad breath, and strengthen the gums and teeth. Fruit juice can relieve throat pain (Shilpa and Krishnakumar 2015). Phlegm, constipation and piles can all be treated with fruits. Sherbet made from fruit juice is excellent for treating diarrhoea and dysentery. It helps with diabetes as well. Fruit should be consumed two hours after a meal (Gayathri et al. 2012). It benefits the liver, which in turn regulates the body's blood sugar. Fruit consumption enhances digestion and staves off fatigue. For centuries, humans have employed fruits for medical purposes in addition to eating them for sustenance due to the presence of secondary metabolites. In addition, wild pomegranate fruit sun-dried by-product (peel) powder, with honey is used as a traditional remedy in some hilly areas of Sirmour district of Himachal Pradesh (India) for curing coughs due to its rich source of phenolics. Hence eating plentiful fruit can help us to stay healthy and keep us away from a variety of health risks.

3.3.7 Other Nutraceutical Properties

Fruits have a large number of bioactive compounds that have biological activities, are neuroprotective and help to control the development of a number of diseases, such as Alzheimer's, where they act in an anti-inflammatory role in addition to exhibiting antimicrobial, antioxidant and antidiabetic potential (Table 3.6). In addition, they exhibit a number of nutraceutical properties that can control the progression of disease. Evidence of these nutraceutical properties comes from various in vitro studies conducted previously on various tropical, subtropical, temperate and minor fruits. Numerous studies show that pomegranate is a rich source of fibre, vitamins A and C, minerals and natural bioactive components such as ellagic acid and anthocyanins. The anti-inflammatory properties of ellagitannin-rich fruits like pomegranate are supported by several studies. Gimenez-Bastida et al. (2021) explained that pomegranate juice and extract seem to encourage modest but positive changes in a number of risk variables, which may help avoid chronic illnesses. In enhancing blood pressure regulation, these foods appear to guard against oxidative stress-related and cardiometabolic disorders. According to earlier research, pomegranate juice or

Table 3.6 Other nutraceutical properties of major fruits

Fruit	Part used	Animal name/ model used	Health effects/ nutraceutical properties	References
Pomegranate	Pomegranate extract	1g/day for 4 weeks: healthy adults	Prebiotic effect: modulation of the gut microbiota, protective factor against obesity, overweight	Li et al (2015); Depommier et al. (2019)
Pomegranate	Juice	500 ml/d for 15 d. healthy mens	Anti-inflammatory effects, lower systolic blood pressure, creatinine and muscle damage parameters	Achraf et al. (2017).
Pomegranate	Fruit extract	2mg for 48 h, CD-1 MICE	Anti-skin-tumour-promoting effects	Afaq et al. (2005)
apple	Apple juice	237 mL of apple juice daily for a month	Preventing neurodegenerative disorders	Remington et al. (2010)
Guava	Red guava	4 g/day, Male BALB/c mice	Protective effects against inflammation and oxidative stress	Li et al (2015)
Persimmon	Persimmon extract	15 mg/kg p.o. per day	Chronic inflammation reduction of collagen-induced arthritis	Gato et al. (2012)
Persimmon	Fruit extracts from the peel and pulp	–	In vitro cerebral Ischemia: encourage cell survival and reduce ROS growth	Forouzanfar et al. (2016)
Watermelon	Watermelon byproducts (seeds)	50 g daily for 40 days	Anti-hypertensive effect	Sajjad et al. (2020)
Grape powder	Whole grapes powder	600mg/ day for 10 weeks	Blood pressure lowering effect- anti-hypertensive effect	Thandapilly et al. (2012)

extract has shown neuroprotective effects linked to the biological activities of ellagic acid as well as its metabolites. In a more recent randomised controlled experiment, 200 healthy individuals drank a glass of juice every day for six to twelve months, which stabilised their ability to memorise visual information in comparison to the control group (Ropacki et al. 2013). Apple contains important phytochemicals such as polyphenols, vitamin C, dietary fibres, and pectin. Flavanols such as catechins, procyanidins and epicatechin; flavonols such as quercetin glycosides; phenolic acids such as gallic, chlorogenic and coumaric acids; and anthocyanins such as cyanidin are the major polyphenols present in apple. Anxiety, agitation and delusions, behaviours associated with the psychological symptoms of dementia, significantly improved, according to Remington et al. (2010), in people with middle-to-late stage Alzheimer's disease who drank 237 (mL) apple juice daily for a month. Due to apple juice's protective effects against oxidative stress, gut bacteria may also be helpful in avoiding neurodegenerative diseases.

Fang, Hsu and Yen (2008) examined jackfruit's anti-inflammatory effects. According to the authors, the flavonoid artocarpesin inhibited the lipopolysaccharide (LPS)-induced synthesis of NO (nitric oxide) and PGE2 (prostaglandin E2) by reducing the expression of the protein-inducible iNOS (nitric oxide synthase) and COX-2 (cyclooxygenase 2). In this work, it was evaluated how LPS-activated RAW 264.7 murine macrophage cells affected the production of proinflammatory mediators. Additionally, jackfruit extract has antitumour properties.

Guava is another fruit that has numerous bioactive compounds that exhibit potential health benefits and plays a role in the management of diabetes, cardiovascular diseases, obesity, oxidative stress and inflammation. Maryanto and Marsono (2019) conducted a study to evaluate the potential of red guava in decreasing the cholesterol quantity in hypercholesterolemic rats connected to the hypocholesterolemic capabilities of guava. They discovered that consuming red guava dramatically decreased the levels of triglycerides (18%), LDL cholesterol (43%) and total cholesterol (32%). However, the consumption of red guava dramatically raised the amount of HDL cholesterol (18%).

Li et al. (2015) demonstrated that red guava can lower the expression of the inflammatory proteins iNOS and NF-B via activation of PPAR, based on the extreme inflammatory responses brought by diabetes. Red guava can also improve hyperinsulinemia and hyperglycemia with kidney metabolic abnormalities in diabetic mice. Additionally, a diet rich in red guava can boost GPx3 expression, reducing diabetes-related oxidation damage, improving lipid metabolism and bringing significant improvements to blood glucose regulation. Insulin resistance, triglycerides, non-esterified fatty acids, cholesterol, and C-reactive protein were observed when blood cholesterol levels were reduced. Kiwi fruit is another fruit that has many health benefits. It helps alleviate constipation and maintains excellent digestive health due to its laxative qualities and the presence of actinidin, a proteolytic enzyme that positively affects the stomach and intestinal digestion of proteins. In addition, it is also a significant source of ascorbic acid, flavonoids and phenolics, which possess antioxidant potential. According to Wu et al. (2019), avocado extract can prevent the dengue virus from replicating by controlling the NF-kB-dependent activation of antiviral interferon response.

Because of the potential bioactive compounds of pineapple, it is a fruit of high interest. Seenak et al. (2021) found that daily pineapple consumption decreased the levels of blood lipids, cardiac risk ratio, atherogenic coefficient and the activity of the liver's enzymes. Daily pineapple consumption also increased cardiac protein carbonyl content and decreased the cardiac malondialdehyde with reduced levels of the pro-inflammatory cytokines interleukin 6 and 1.

Mango is a well-known tropical fruit and contains mangiferin, a xanthone with biological properties including antiviral, anticancerous, antidiabetic, antioxidant, antiaging, hepatoprotective, immunomodulatory and analgesic effects. Maniferin is the main bioactive component of mango fruit (Feng et al. 2019; Imran et al. 2017). It is also a potential therapeutic agent for neurodegenerative disorders. The effects of daily use of passion fruit peel flour in HIV patients were examined by Marques et al. (2016). According to the authors, after 30 days, this (30g) was successful in lowering total cholesterol (by 15.1% compared to the initial value) and triglycerides (by 16.1% compared to the original value) at the end of the trial (90 days). HDL cholesterol levels increased (13.4%) relative to the starting value while LDL cholesterol concentrations decreased (12.7%).

Aonla is another beneficial fruit having high levels of vitamin C, which helps in treating various skin disorders, reduces inflammation and oxidation, and decreases the risk of scurvy. In addition, aonla also contains many polyphenols, such as tannin, which have various health benefits. Amla fruit capsules of 500 mg per capsule daily (once) for 42 days were the focus of a study by Gopa, Bhatt and Hemavathi (2012), which revealed substantial reductions in hyperlipidemia in patients. Further, these authors claim that at the conclusion of the study period, substantial decreases in LDL, total cholesterol and VLDL levels, and a considerable rise in blood HDL levels were also seen. Furthermore, several investigations found that the fruit and extract of *P. emblica* decrease these parameter levels in both hyperlipidemic patients and healthy individuals. The extract increased HDL levels and provided protection against atherosclerosis when given for a period of two to six months. It did this by lowering the level of a crucial inflammatory marker, CRP (C-reactive protein) (Gopa, Bhatt and Hemavathi 2012; Koshy et al. 2012). In order to induce gastrointestinal ulcers in mice, Al-Rehaily et al. (2002) used a variety of techniques, including ligating the pylorus, administering indomethacin and necrotising agent (NaCl 25%, NaOH 0.2 M, 80% ethanol), and inducing hypothermia. These techniques included studying the antisecretory as well as antiulcer activities of aonla extract. Using the pylorus ligated with necrotising agent-intoxicated ulcer techniques, both dosages (500 and 250 mg/kg) decreased gastric output, ulcer index, intraluminal haemorrhage and gastric lesions (hypothermic restraint-induced ulcer method). Thus, the bioactive present in aonla also helps to protect gastrointestinal organs. The mitigation of neurological modifications, notably the biochemical changes seen in carriers of disease (Alzheimer's), is one of the putative protective benefits linked to the amla bioactive molecule. For instance, the neurotoxicity brought on by aluminium chloride was decreased in mice after receiving aonla fruit extract for 60 days (100 mg/kg high in emblicanin A and B) (Bharathi et al. 2018).

Acerola juice contains ascorbic acid, flavonoids, phenolic compounds and various other nutrients. In a study, acerola juice administration was also shown to partially reverse the DNA damage caused by a high-fat diet in the blood, liver, bone marrow and kidney of mice (obese) (Tremonte et al. 2016). Its primary biological active components (rutin and vitamin C), which have a notable impact on lowering oxidative stress resulting in reduced genotoxicity in obesogenic settings, were responsible for these positive effects (Leffa et al. 2014). Various studies reviewed the various bioactive components, such as polyphenols (flavonoids), carotenoids, limonoids and terpenes, in citrus fruit juice as well as in peel and showed that they have possible benefits against obesity, inflammatory illnesses, atherosclerosis, neurological diseases and cancer since they can decrease oxidative stress-related disorders (Saini et al. 2022). The citrus peel flavonoid nobiletin inhibits the inflammation response in LPS-stimulated RAW264.7 cells by increasing autophagy by lowering the levels of cytokine production (inducible COX-2 and iNOS) and activating the Interleukin 6 / signal transducer and activator (Rong et al. 2021).

Gato et al. (2012) also extracted tannin-rich fibre from immature persimmons in order to make bars that would be consumed by 40 participants three times daily, before meals, for 12 weeks. Without modifying the levels of HDL or triglycerides in plasma, it was seen that there was a considerable decline in the total cholesterol with LDL. Young persimmons contain highly polymerised epicatechin, epigallocatechin, epicatechin-3-O-gallate and epigallocatechin-3-O-gallate tannins. As a result, since their daily consumption reduced plasma cholesterol levels in people, tannin-rich fibre could be a valuable product to prevent and cure hypercholesterolemia. The capability of tannins to bind bile acids was associated with much lower plasma levels of HbA1c.

Direito et al. (2019) reported a significant reduction in the oedema volume and radiological abnormalities attributed to bone CIA in rats with collagen-induced arthritis (CIA), which served as evidence of persimmon extract's anti-inflammatory effect. Rats with CIA experience less tissue damage and chronic inflammation when given persimmon extract (15 mg/kg p.o. per day), an effect that is most likely due to the extract's strong antioxidant properties. It was determined that fresh persimmon fruit extract has a suppressive impact on human neutrophil oxidative burst. In both instances, the extract demonstrated the capacity to operate as an intracellular antioxidant and to obstruct neutrophil activity, preventing the formation of harmful reactive oxygen species that would have amplified the inflammatory signals already in motion.

Forouzanfar et al. (2016) conducted a study on in vitro cerebral ischemia and reported that GOSD (glucose-oxygen-serum deprivation)-induced PC12 cell injury was caused by the deprivation of glucose, oxygen and serum. Persimmon extracts protected PC12 cells from this injury by an antioxidant mechanism and by reducing intracellular ROS generation. An H2DCF-DA fluorescent probe was used to measure the intracellular ROS levels and the effects of PuHE and PeHE on the production of ROS after GOSD in PC12 cells and showed that pretreatment for two hours with *Diospyros kaki* fruit extracts from the peel (Pe) and pulp (Pu) were able to support cell survival with lower ROS growth during GOSD stress.

Sajjad et al. (2020) administered 50 g of watermelon byproduct (seeds) daily to male and female rats to assess their anti-hypertensive effects for 40 days. The findings show a significant decline (13–16 mmHg; 8–10 mmHg) in systolic and diastolic blood pressure as a result of the presence of phytochemicals and cations (K^+, Ca^+, and Mg^+) that provide potential for vasodilation and antioxidants. According to Finbarrs, Chinedu and Ojo (2018), ethanolic watermelon seed extract dramatically increases immunoreactivity to neurofilament protein antibodies, which enhances Wistar rats' cognitive function. According to one study by Thandapilly et al. (2012), administering whole grape powder for ten weeks helped a 20-week-old spontaneously hypertensive rat's blood pressure drop significantly (SHR).

3.4 RECENT ADVANCEMENTS IN THE DEVELOPMENT OF FUNCTIONAL FOODS

Recent advances and progress in the field of development of functional foods include modern technologies such as genomics, genetic expressions, the use of liposome technology and many more. In addition, nanotechnology is currently playing a significant role in improving functional food. This technology has proved its potential as an effective drug delivery system. A recent nanotechnology approach utilised in the food industry by scientists is reverse engineering, which is explained elsewhere in this book.

The encapsulation approach is widely utilised to infuse active compounds into the food matrix in order to improve its functional qualities. Among the various techniques, encapsulation with a particle size below 100 nm, known as nano-encapsulation, plays a vital part in effectively delivering the bioactive compounds in the food. This nano-encapsulation drug delivery system includes micro and nanoemulsions, solid lipid nanoparticles and liposomes. Liposome technology is the recent means of encapsulation of functional ingredients (Dias-Souza, Soares, and Dos-Santos 2017). Liposomes are good carriers of bioactive compounds, act as drug delivery vehicles and help in better delivery of these compounds through encapsulation. By using this technology, the encapsulation of lipid-soluble and water-soluble components and their incorporation within a food product become very easy (Dias et al. 2017). This technology is extensively used due to its capacity to bind with a wide range of compounds, and it offers stability to encapsulated constituents and is employed widely in the production of functional food and nutraceutical products (Ajeeshkumar et al. 2021). Proliposome technology along with micronised sucrose has been extensively used in the functional food sector (Silva et al. 2017). This technology is already being used in food sectors such as fruit juices and baked goods. Liposomes loaded with alpha-tocopherol, ascorbic acid and chlorogenic acid have been used to enhance protection from microbes, physical properties and, most importantly, nutritional value (Dima et al. 2019). In addition, microencapsulation has a prominent role in reducing the growth of pathogenic microorganisms, which flourish on the food surface by supplying inhibitory ingredients (Bigliardi and Galati 2013). The other recent technique involved in encapsulation is co-encapsulation, which consists of two phases: the continuous phase (oil/organic solution) and the dispersed phase (cell polymer suspension). By using the co-encapsulation technique, both bioactive compounds and probiotic bacteria can be encapsulated in a single product, thus having a synergistic effect on health (Misra, Pandey and Mishra 2021). Han et al. (2020) utilised the co-encapsulation technique for *Lactobacillus bulgaricus* and *L. Paracasei* with sodium alginate and whey protein isolate.

Reverse engineering/multi-criteria reverse engineering technology is developing in the food sector at a very slow pace but is showing effective results in the development of personalised food (Thomopoulos et al. 2019). "Reverse" means that the first step of this process is ascertaining consumer preference, which is usually the last step of the development of functional food products (Dima et al. 2019). This means that the consumer is the main link for the design and preparation of functional food products. Reverse engineering is helpful in choosing the right condition for the development process of food. It allows the right options for the design to be chosen as well as any redesign of the food, food processes, packaging etc. (Thomopoulos et al. 2019).

Foodomics employs the most recent omics approaches to enhance consumers' health and well-being, and utilises chemometrics, biostatics and other tools of bioinformatics for the investigation of food and nutritional information to establish a connection between food components, diet, health, food safety and the prevention of diseases (Li et al. 2020). Nutrigenomics has been exploited extensively in the food sector. The interaction between the human genome and food has been recognised as an innovative method to change disease conditions. A personalised diet has the potential to offer a wide window of opportunities in the area of functional food (Bigliardi and Galat 2013). The involvement of three different aspects – human health, food, and nutrition at different molecular levels – helps in providing the information for the production of food products with functional characteristics and nutraceuticals as part of a personalised diet (Li et al. 2020).

3D printing in the food sector, the utilisation of 3D-printing techniques, revolves around the development of food products with any shape, dimensions, nutritional composition or flavours. This technique is a promising tool for the incorporation of sensitive bioactive compounds to develop healthy functional food products (Tomasevic et al. 2021). This technique can be utilised for the development of functional food, personalised diets and food supplements, which would target a large population by altering the taste and texture of food to make it more palatable (Portanguen et al. 2019). Recent advances in the extension of 3D printing evolved a newer technique, known as 4D printing, for the customisation of new food products. Instead of storing the information, 4D printing allows the colour, nutrition and flavour of a printed product to be released when the user consumes it (Teng et al. 2021).

There is a wide scope for the development of functional food by using these techniques. There is a lack of research and publications in this sector, but these techniques are important and need to be explored more in the future.

Hilachuk et al. (2021) summarise several functional food patents (Table 3.7) and their development. The database of the National Institute of Intellectual Property (NPI), which records patent applications, shows patents in areas such as prebiotic, probiotic, gluten-free, lactose-free/low lactose and fortified food. Of these, the most research has been carried out on probiotics followed by fortified, enriched food products and gluten-free products. In July 2000, DJ Stewart and his coworkers took a patent (US20006087353) for the method of esterification and hydrogenation of plant sterol to increase dispensability and solubility. They increased the stability as well as the potency of plant sterol for its effective utilisation in nutraceuticals, foods and beverages. In 2007, SJ Furcich patented a technology (US20070042102) on the methods

Table 3.7 Patents on functional foods and their claims proved in food industry

Patent	Inventors	Patent number	Place and year)	Findings/Claims
Method for obtaining a means of antimicrobial activity (combination of dried leaves of wild apple with wild apple fruit pulp)	Nesterova et al. (2017)	RU2639119C1	Russia (2017)	• Extract was effective against *E coli*, *Bacillus anthracoides*, *B subtilis*, *Salmonella typhimurium*, *Shigella sonnei*, *Staphylococcus aureus*, *S epidermis*
Grape extract, dietary supplement thereof, and processes therefor	Anil et al. (2009)	US 2007/0071871 A1	United state (2009)	• Product comprises polyphenolic compounds • Claimed as a nutraceutical product
A process for the preparation of antimicrobial food preservative by using silver nitrate	Nirmala (2011)	IN186123	India (2011)	• Silver nitrate act as antimicrobial compounds at optimum concentration and can be used in food preservation
A process for the preparation of soup powder formulation based on Indian gooseberry (*Phyllunthus emblical*)"	Sukumar et al. (2009)	IN195797	India (2009)	• Dehydrated material may be effected in a conventional mill to obtain the desired soup powder of particle size ranging between 170 and 200 microns • Soup powder formulation retains 66–68-weight % of vitamin C
Compositions for treating and preventing hyperlipidemia	Liu and Shengmin (2007)	WO07137449	China (2007)	• Claims the positive effects for treatment of hyperlipidemia by the use of functional components (flavones, sterols, beta-glucan and procyanidins) in appropriate combination.

(*Continued*)

Table 3.7 (Continued)

Patent	Inventors	Patent number	Place and year)	Findings/Claims
A process for the preparation of ripe banana powder	Pasupuleti et al. (2017)	IN222015	India (2017)	• Process for the preparation of banana • The banana pulp is pasteurised at a specific temperature (65–95°C) and cooled immediately to inactivate the enzymes
Natural colourant and methods thereof	Syed and Amin (2017)	EP 2 393 881 B1	Malaysia (2017)	• A method for manufacturing natural colourant, wherein the natural colourant is obtained from red dragon fruit • The natural colourant is a bioactive ingredient, wherein the natural colourant includes betalains
Functional food composition and methods	Doron Friedman Gideon Strassmann Ascher Shmulewitz	EP2605756A2	Europe (2014)	• The functional food is produced by mixing a lipid solvent including a bioactive agent with a semi-solid or liquid food base carrier until the formation of an emulsion that includes solid lipid particles loaded with the bioactive agent.

of forming phytosterol-fortified cocoa powder, and the developed cocoa powder was enriched with plant sterol. Patent number US20087329429, filed by MJ Chimel in September 2008, was with regard to the process of development of healthy dark and white chocolate bars with cocoa proanthocyanidins and sterol/stanol (Kamal-Eldin and Moazzami 2009). Friedman and his coworkers in 2014 patented (EP2605756A2) a technology for the development of functional food and also patented the functional food. Yadav et al. (2006) filed a patent (patent application no. 201711042760) for the utilisation of tomato and tomato processing by-products for the development of fibre, mineral and antioxidant-rich novel bakery products.

3.5 CONCLUSION

Fruits contain many polyphenols including flavonoids. Other widely occurring fruit constituents with beneficial health effects include xanthones, carotenoids and saponins. Fruits are also good sources of a number of micronutrients, including fatty acids, iron, selenium, zinc, copper, necessary proteins with all essential amino acids, vitamins A, B_6, B_{12}, C, D and E, and specific phytochemicals that are crucial for a strong immune system. The composition of a wide range of colourful fruits varies, and these fruits are responsible for supplying a mixture of nutrients with phytochemicals that improve human health. According to many in vivo and in vitro studies, fruits are rich in nutraceutical compounds that help in the defence of a range of diseases. These compounds may include those that are anti-diabetic, anti-cancer, antihypertensive, neuroprotective, anti-inflammatory, antioxidant, antimicrobial and antiviral. They also help with cell detoxification, cholesterol synthesis and blood pressure regulation. However, due to a lack of awareness of the regular consumption of fruits, consumers do not understand this concept. Most of the population consumes fruits only occasionally, even though India produces a large number of regional types and varieties of fruits. The alternate trend of consumption needs to shift toward functional food preparation and consumption. Because of their suggested safety and possible medicinal and nutritional benefits, plant-derived nutraceuticals and functional foods are receiving a lot of attention these days. Overall, many people choose fresh fruits for health benefits, but fruits also provide a wider scope for the development of functional and nutraceutical products for inclusion as alternative foods in the diet for consumers.

Functional food has grown significantly in popularity recently and holds great promise for the future. Many sectors have moved toward the creation of functional and nutraceutical products during the past five years, reflecting the growing consumer demand for variety in plant-based foods. Fruit and functional products made from fruit have huge potential. A significant opportunity for the industry to produce novel functional and nutraceutical products has been created by the growing demand, usage patterns, non-perishability and growth potential of plant-based functional products. Customers seek out functional goods that are full of nutrients and bioactive substances. However, the general advice is to consume fresh fruits in the prescribed amounts in order to improve one's health status. A campaign should be launched to raise awareness of this issue among young people and other age groups. The scientific community is expected to develop high-quality functional and nutraceutical products with a continuously changing behaviour toward novel food products in order to attain market acceptability, among other things.

REFERENCES

Abdul Aziz, N. A., L. M. Wong, R. Bhat, and L. H. Cheng. 2012. "Evaluation of processed green and ripe mango peel and pulp flours (*Mangifera indica* var. Chokanan) in terms of chemical composition, antioxidant compounds and functional properties." *Journal of the Science of Food and Agriculture* 92(3): 557–563. https://doi.org/10.1002/jsfa.4606.

Abdullah, A. S., A. S. Mohammed, A. Rasedee, and M. E. Mirghani. 2015. "Oxidative stress-mediated apoptosis induced by ethanolic mango seed extract in cultured estrogen receptor positive breast cancer MCF-7 cells." *International Journal of Mocelular Science* 16(2): 3528–3536. https://doi.org/10.3390/ijms16023528.

Achraf, A., C. Hamdi, M. Turki, O. Abdelkarim, F. Ayadi, A. Hoekelmann, S. Yaich, and N. Souissi. 2017. "Natural pomegranate juice reduces inflammation, muscle damage and increase platelets blood levels in active healthy Tunisian aged men." *Alexandria Journal of Medicine* 54(1): 45–48. http://doi.org/10.1016/j.ajme.2017.03.005.

Adams, L. S., N. P. Sreeram, B. B. Agarwal, Y. Takada, and D. Heber. 2006. "Pomegranate ellagitannins and punicalagin suppress inflammatory cell signaling in colon cancer cells." *Journal of Agricultural and Food Chemistry* 54(3): 980–985. https://doi.org/10.1021/jf052005r.

Afaq, F., M. Saleem, C. G. Krueger, J. D. Reed, and H. Mukhtar. 2005. "Anthocyanin and hydrolyzable tannin-rich pomegranate fruit extract modulates MAPK and NF-kappa B pathways and inhibits skin tumorigenesis in CD-1 mice." *International Journal of Cancer* 113(3): 423–433. https://doi.org/10.1002/ijc.20587.

Ahamed, T. S., V. V. Priya, R. G. R. V. Gayathri, and R. V. Geetha. 2016. "Evaluation of anti microbial activity of pineapple extract against selected oral pathogen." *Journal of Pharmaceutical Sciences and Research*8(6): 491–504.

Ajeeshkumar, K. K., P. A. Aneesh, N. Raju, M. Suseela, C. N. Ravishankar, and S. Benjakul. 2021. "Advancements in liposome technology: Preparation techniques and applications in food, functional foods, and bioactive delivery: A review." *Comprehensive Reviews in Food Science and Food Safety* 20(2): 1280–1306. https://doi.org/10.1111/1541-4337.12725.

Alberti, A., A. A. F. Zielinski, D. M. Zardo, I. M. Demiate, A. Nogueira, and L. I. Mafra. 2014. "Optimisation of the extraction of phenolic compounds from apples using response surface methodology." *Food Chemistry* 149: 151–158. https://doi.org/10.1016/j.foodchem.2013.10.086.

AlGhalban, F. M., A. A. Khan, and N. K. Khattak Mohd. 2021. "Comparative anticancer activities of Ficuscarica and Ficus salicifolia latex in MDA-MB-231 cells." *Saudi Journal of Biological Sciences* 28(6): 3225–3234. https://doi.org/10.1016/j.sjbs.2021.02.061.

Allen, L. V. 1997. "Nutritional products." In: Covington, T. R., Berardi, R. R., Young, L. L., Kendall, S. C., and Hickey, M. J., editors. *Handbook of Nonprescription Drugs*. Washington, DC: American Pharmaceutical Association.

Alothman, M., R. Bhat, and A. A. Karim. 2009. "Antioxidant capacity and phenolic content of selected tropical fruits from Malaysia, extracted with different solvents." *Food Chemistry* 115(3): 785–788. https://doi.org/10.1016/j.foodchem.2008.12.005.

Al-Rehaily, A. J., T. S. Al-Howirin, M. O. Al-Sohaibani, and S. Rafatullah. 2002. "Gastroprotective effects of 'Amla' Emblica officinalis on in vivo test models in rats." *Phytomedicine* 9(6): 515–522. https://doi.org/10.1078/09447110260573146.

Amin, A. R. M. R., O. Kucuk, F. R. Khuri, and D. M. Shin. 2009. "Perspectives for cancer prevention with natural compounds." *Journal of Clinical Oncology* 27(16): 2712–2719. https://doi.org/10.1200/JCO.2008.20.6235.

Anibijuwon, I. I., and A. O. Udeze. 2009. "Antimicrobial activity of *Carica papaya* (pawpaw leaf) on some pathogenic organisms of clinical origin from South-Western Nigeria." *Ethnobotanical Leaflets* 7: 4–10.

Anil, J. S., W. Haibo, and A. Steve. 2009. "Grape extract, dietary supplement thereof, and processes therefor." US Patent 2007/0071871. A1.

Arroo, R. R. J., K. Beresford, A. S. Bhambra, M. Boarder, R. Budriesi, Z. Cheng, M. Micucci, K. C. Ruparelia, S. S. Surichan, and S. P. Androutsopoulos. 2014. "Phytoestrogens as natural prodrugs in cancer prevention: Towards a mechanistic model." *Phytochemistry Reviews* 13(4): 853–866. https://doi.org/10.1007/s11101-014-9355-3.

Asale, Y., E. Dessalegn, D. Assefa, and M. Abdisa. 2021. "Phytochemicals and antioxidant activity of different apple cultivars grown in South Ethiopia: Case of the wolayta zone." *International Journal of Food Properties* 24(1): 354–363. https://doi.org/10.1080/10942912.2021.1885440.

Ashok, A. D., J. Ravivarman, and K. Kayalvizhi. 2020. "Nutraceutical values of minor fruits on immunity development to combat diseases." *International Journal of Current Microbiology and Applied Sciences* 9(6): 1303–1311. https://doi.org/10.20546/ijcmas.2020.906.162.

Assefa, A. D., R. K. Saini, and Y. S. Keum. 2017. "Extraction of antioxidants and flavonoids from yuzu (Citrus junos Sieb ex Tanaka) peels: A response surface methodology study." *Journal of Food Measurement and Characterization* 11(2): 364–379. https://doi.org/10.1007/s11694-016-9405-1.

Ayala-Zavala, J. F., C. Rosas-Domínguez, C. Vega-Vega, and G. A. González-Aguilar. 2010. "Antioxidant enrichment and antimicrobial protection of fresh-cut fruits using their own byproducts: Looking for integral exploitation." *Journal of Food Science* 75(8): 175–181. https://doi.org/10.1111/j.1750-3841.2010.01792.x.

Azab, E. F., and H. S. Mostafan. 2022. "Phytochemical analysis and antioxidant defense of kiwifruit (Actinidia deliciosa) against pancreatic cancer and AAPH-induced RBCs hemolysis." *Food Science and Technology* 42: 1–10.

Azzini, E., G. Maiani, A. Durazzo, M. S. Foddai, F. Intorre, and E. Venneria. 2019. "*S. Giovanni* varieties (*Pyrus communis* L.): Antioxidant properties and phytochemical characteristics." *Oxidative Medicine and Cellular Longevity*: 6714103. https://doi.org/10.1155/2019/6714103.

Babio, N., M. Bulló, and J. Salas-Salvadó. 2009. "Mediterranean diet and metabolic syndrome: The evidence." *Public Health Nutrition* 12(9A): 1607–1617. https://doi.org/10.107.S136890009990449.

Banu, S. M., K. Selvendiran, J. P. V. Singh, and D. Sakthisekaran. 2004. "Protective effect of *Emblica officinalis* ethanolic extract against 7,12-dimethylbenz(a)anthracene (DMBA) induced genotoxicity in Swiss albino mice." *Human and Experimental Toxicology* 23(11): 527–531. https://doi.org/10.1191/0960327104ht484oa.

Baranowska, M., and A. Bartoszek. 2016. "Antioxidant and antimicrobial properties of bioactive phytochemicals from cranberry." *Advances in Hygiene and Experimental Medicine* 70: 1460–1468. https://doi.org/10.5604/17322693.1227896.

Baskar, R., S. Shrisakthi, B. Sathyapriya, R. Shyampriya, R. Nithya, and P. Poongodi. 2011. "Antioxidant potential of peel extracts of banana varieties (*Musa sapientum*)." *Food and Nutrition Sciences* 2(10): 1128–1133. https://doi.org/10.4236/fns.2011.210151.

Bates, R. P., J. R. Morris, and P. G. Crandall. 2001. *Principles and practices of small-and mediumscale fruit juice processing*. Rome: Food and Agriculture Organization (FAO). http://wwfao.org/docrep/005y251ey251e00htm.

Bautista-Ortín, A. B., P. Rodríguez-Rodríguez, R. Gil-Muñoz, E. Jiménez-Pascual, N. Busse-Valverde, A. Martínez-Cutillas, J. M. Lopez-Roca, and E. Gómez-Plaza. 2012. "Influence of berry ripeness on concentration, qualitative composition and extractability of grape seed tannins." *Australian Journal of Grape and Wine Research* 18(2): 123–130. https://doi.org/10.1111/j.1755-0238.2012.00178.x.

Benavente-Garcia, O., and J. Castillo. 2008. "Update on uses and properties of citrus flavonoids: New findings in anticancer, cardiovascular, and anti-inflammatory activity." *Journal of Agricultural and Food Chemistry* 56(15): 6185–6205. https://doi.org/10.1021/jf8006568.

Bharathi, M. D., and A. J. Thenmozhi. 2018. "Attenuation of aluminum-induced neurotoxicity by tannoid principles of Emblica officinalis in Wistar rats." *International Journal of Nutrition, Pharmacology, Neurological Disease* 8: 35. https:doi.org/10.4103/ijnpnd.ijnpnd_23_18.

Bhardwaj, R. L., U. Nandal, A. Pal, and S. Jain. 2014. "Bioactive compounds and medicinal properties of fruit juices." *Fruits* 69(5): 391–412. https:doi.org/10.1051/fruits/2014027.

Bigliardi, B., and F. Galati. 2013. "Innovation trends in the food industry: The case of functional foods." *Trends in Food Science and Technology* 31(2): 118–129. https:doi.org/10.1016/j.tifs.2013.03.006.

Biworo, A. 2015. "Antidiabetic and antioxidant activity of jackfruit (*ArtocarpusHeterophyllus*) extract." *Journal of Medical and Biological Engineering* 4(4): 318–323. https:doi.org/10.12720/jomb.4.4.318-323.

Bobinaite, R., P. Viškelis, and P. R. Venskutonis. 2012. "Variation of total phenolics, anthocyanins, ellagic acid and radical scavenging capacity in various raspberry (*Rubus* spp.) cultivars." *Food Chemistry* 132(3): 1495–1501. https://doi.org/10.1016/j.foodchem.2011.11.137.

Borges, G., A. Degeneve, W. Mullen, and A. Crozier. 2010. "Identification of flavonoid and phenolic antioxidants in black currants, blueberries, raspberries, red currants, and cranberries." *Journal of Agricultural and Food Chemistry* 58(7): 3901–3909. https:doi.org/10.1021/jf902263n.

Brahem, M., C. M. G. C. Renard, S. Eder, M. Loonis, R. Ouni, M. Messaoud, and C. Le Bourvellec. 2017. "Characterization and quantification of fruit phenolic compounds of European and Tunisian pear cultivars." *Food Research International* 95: 125–133. https://doi.org/10.1016/j.foodres.2017.03.002.

Butkeviciute, A., V. Abukauskas, V. Janulis, and D. Kviklys. 2022. "Phenolic content and antioxidant activity in apples of the 'galaval' cultivar grown on 17 different rootstocks." *Antioxidants* 11(2): 266. https:doi.org/10.3390/antiox11020266.

Calin-Sanchez, A., J. Martinez-Nicolas, S. Munera-Picazo, A. Carbonell-Barrachina, P. Legua, and F. Hernandez. 2013. "Bioactive compounds and sensory quality of black and white mulberries grown in Spain." *Plant Foods for Human Nutrition* 68(4): 370–377. https:doi.org/10.1007/s11130-013-0382-9.

Ceccarelli, D., C. Talento, S. Favale, E. Caboni, and F. Cecchini. 2018. "Phenolic compound profile characterization by Q-TOF LC/MS in 12 Italian ancient sweet cherry cultivars." *Plant Biosystems - An International Journal Dealing with all Aspects of Plant Biology* 152(6): 1346–1353. https://doi.org/10.1080/11263504.2018.1478899.

Ceylan-Isik, A. F., R. M. Fliethman, L. E. Wold, and J. Ren. 2008. "Herbal and traditional Chinese medicine for the treatment of cardiovascular complications in diabetes mellitus." *Current Diabetes Reviews* 4(4): 320–328. https:doi.org/10.2174/157339908786241142.

Chanthaphon, S., S. Chanthachum, and T. Hongpattarakere. 2008. "Antimicrobial activities of essential oils and crude extracts from tropical *Citrus* spp. against food-related microorganisms." *Songklanakarin Journal of Science and Technology* 30(1): 125–131.

Chen, Q., D. Wang, C. Tan, Y. Hu, B. Sundararajan, and Z. Zhou. 2020. "Profiling of flavonoid and antioxidant activity of fruit tissues from 27 Chinese local citrus cultivars." *Plants* 9(2): 196. https://doi.org/10.3390/plants9020196.

Cheshomi, H., A. R. Bahrami, H. Rafatpanah, and M. M. Matin. 2022. "The effects of ellagic acid and other pomegranate (*Punicagranatum* L.) derivatives on human gastric cancer AGS cells." *Human and Experimental Toxicology* 41: 1–16. https://doi.org/10.1177/09603271211064534.

Choo, C. L., and N. A. A. Aziz. 2010. "Effects of banana flour and beta-glucan on the nutritional and sensory evaluation of noodles." *Food Chemistry* 119(1): 34–40. https://doi.org/10.1016/j.foodchem.2009.05.004.

Colombo, F., C. D. Lorenzoa, L. Regazzoni, M. Fumagallia, E. Sangiovannia, L. Peres de Sousac, L. Bavarescod, D. Tomasie, A. Bossof, G. Aldinib, P. Restania, and M. Dell'Aglia. 2019. "Phenolic profile and anti-inflammatory activity of sixteen Table Grape (*Vitis vinifera* L.) varieties." *Food and Function* 10(4): 1797–1807. https://doi.org/10.1039/C8FO02175A.

Dabbah, R., M. V. Edwards, and A. W. Moats. 1970. "Antimicrobial action of some citrus fruit oils on selected food borne bacteria." *Journal of Applied Microbiology* 19(1): 27–31. https://doi.org/10.1128/am.19.1.27-31.1970.

De Ligt, M., S. Timmers, and P. Schrauwen. 2015. "Resveratrol and obesity: Can resveratrol relieve metabolic disturbances?." *Biochimica et Biophysica Acta (BBA) – Molecular Basis of Disease*1852 (6): 1137–1144. https://doi.org/10.1016/j.bbadis.2014.11.012.

Deepa, N., C.Kaur, B. George, B. Singh, and H.Kapoor. 2007. "Antioxidant constituents in some sweet pepper (*Capsicum annuum* L.) genotypes during maturity." *LWT – Food Science and Technology* 40(1): 121–129. https://doi.org/10.1016/j.lwt.2005.09.016.

DeFelice, S. L. 1995. "The Nutraceutical Revolution: Its impact on food industry R&D." *Trends in Food Science and Technology* 6(2): 59–61. https://doi.org/10.1016/S0924-2244(00)88944-X.

Dell Agli, M., C. Di Lorenzo, M. Badea, E. Sangiovanni, L. Dima, E. Bosisio, and P. Restani. 2013. "Plant food supplements with anti-inflammatory properties: A systematic review (I)." *Critical Reviews in Food Science and Nutrition* 53(4): 403–413. https://doi.org/10.1080/10408398.2012.691916.

Depommier, C., A. Everard, C. Druart, H. Plovier, M. Van Hul, S. Vieira-Silva, G. Falony, J. Raes, D. Maiter, N. M. Delzenne, M. de Barsy, A. Loumaye, M. P. Hermans, J. P. Thissen, W. M. de Vos, and P. D. Cani. 2019. "Supplementation with Akkermansia muciniphila in overweight and obese human volunteers: A proof-of-concept exploratory study." *Nature Medicine* 25(7): 1096–1103. https://doi.org/ 10.1038/s41591-019-0495-2.

Deshwal, K. V., and B. Kaur. 2018. "Antimicrobial activity of citrus fruits on certain pathogenic microorganism." *Journal of Plant Development Sciences* 10(8): 481–483.

Dias-Souza, M. V., D. L. Soares, and V. L. Dos-Santos. 2017. "Comparative study of free and liposome-entrapped chloramphenicol against biofilms of potentially pathogenic bacteria isolated from cooling towers." *Saudi Pharmaceutical Journal* 25(7): 999–1004. https://doi.org/10.1016/j.jsps.2017.03.003.

Dima, C., E. Assadpour, S. Dima, and S. M. Jafari. 2019. "Bioactive-loaded nanocarriers for functional foods: From designing to bioavailability." *Current Opinion in Food Science* 33: 21–29. https://doi.org/10.1016/j.cofs.2019.11.006.

Direito, R., J. Rocha, A. T. Serra, A. Fernandes, M. Freitas, E. Fernandes, R. Pinto, R. Bronze, B. Sepodes, and M. E. Figueira. 2019. "Anti-inflammatory effects of persimmon (Diospyros kaki L.) in experimental rodent rheumatoid arthritis." *Journal of Dietary Supplements* 17(6): 663–683. https://doi.org/10.1080/19390211.2019.1645256.

Dolkar, B. P., V. K. Wali, V. Sharma, and R. A. Shah. 2017. "Fruits as nutraceuticals." *Ecology, Environment and Conservation* 23(2): S113–S118.

Duarte, A., C. Carvalho, and M. G. Miguel. 2016. "Bioactive compounds of citrus as health promoters." In: Silva, L., and Silva, B., editors. *Natural Bioactive Compounds from Fruits and Vegetables as Health Promoters*. Bentham Science Publishers. 1: 29–97. DOI: 10.2174/97816810823941160101

Duda-Chodak, A., T. Tarko, and T. Tuszyński. 2011. "Antioxidant activity of apples – An impact of maturity stage and fruit part." *Acta Scientiarum Polonorum, Technologia Alimentaria* 10(4): 443–454.

Dutta, F., M. C. Marak, and A. K. Meena. 2017. "Nutraceuticals from major fruit crops." *International Journal of Agriculture Innovations and Research* 5(6): 898–902.

Eberhardt, M. V., C. Y. Lee, and R. H. Liu. 2000. "Antioxidant activity of fresh apples." *Nature* 405(6789): 903–904. http://doi.org/10.1038/35016151.

Edderkaoui, M., I. Odinokova, I. Ohno, I. Gukovsky, V. L. Go, S. J. Pandol, and A. S. Gukovskaya. 2008. "Ellagic acid induces apoptosis through inhibition of nuclear factor kappa B in pancreatic cancer cells." *World Journal of Gastroenterology* 14(23): 3672–3680. https://doi.org/10.3748/wjg.14.3672.

El-Desoukey, R. M. A., N. M. Aljor, and A. D. Alaotibi. 2020. "The phytochemical and antimicrobial effect of mango (*Mangifera indica* L.) peel extracts on some animal pathogens as eco-friendly." *Acta Scientific Microbiology* 3(4): 34–39.

Elkhatim, K. A. S., R. A. A. Elagib, and A. B. Hassan. 2018. "Content of phenolic compounds and vitamin C and antioxidant activity in wasted parts of Sudanese citrus fruits." *Food Science and Nutrition* 6(5): 1214–1219. https://doi.org//10.1002/fsn3.660.

Esmaillzadeh, A., F. Tahbaz, I. Gaieni, H. AlaviMajd, and L. Azadbakht. 2006. "Cholesterol-lowering effect of concentrated pomegranate juice consumption in type II diabetic patients with hyperlipidemia." *International Journal of for Vitamin and Nutrition Research* 76(3): 147–150. https://doi.org/10.1024/0300-9831.76.3.147.

Fang, S. C., C. L. Hsu, and G. C. Yen. 2008. "Anti-inflammatory effects of phenolic compounds isolated from the fruits of Artocarpus heterophyllus." *Journal of Agricultural and Food Chemistry* 56(12): 4463–4468. https:doi.org/10.1021/jf800444g.

Farneti, B., D. Masuero, F. Costa, P. Magnago, M. Malnoy, G. Costa, U. Vrhovsek, and F. Mattivi. 2015. "Is there room for improving the nutraceutical composition of apple?" *Journal of Agricultural and Food Chemistry* 63(10): 2750–2759. https://doi.org/10.1021/acs.jafc.5b00291.

Fatemeh, S. R., R. Saifullah, F. M. A. Abbas, and M. E. Azhar. 2012. "Total phenolics, flavonoids and antioxidant activity of banana pulp and peel flours: Influence of variety and stage of ripeness." *International Food Research Journal* 19(3): 1041.

Feng, S. T., Z. Z. Wang, Y. H. Yuan, H. M. Sun, N. H. Chen, and Y. Zhang. 2019. "Mangiferin: A multipotent natural product preventing neurodegeneration in Alzheimer's and Parkinson's disease models." *Pharmacological Research* 146: 104336. https://doi.org/10.1016/j.phrs.2019.104336.

Ferreira, E. A., H. E. Siqueira, E. V. V. Boas, V. S. Hermes, and A. D. O. Rios. 2016. "Bioactive compounds and antioxidant activity of pineapple fruit of different cultivars." *Revista Brasileria de Fruticultura* 38(3): e-146. https://doi.org/10.1590/0100-29452016146.

Finbarrs, E., F. Chinedu, and O. P. Ojo. 2018. "Effect of ethanolic seed extract of *Citrus lanatus* (watermelon) on immunoreactivity of hippocampal neurofilament of adult Wistar rats." *Era's Journal of Medical Research* 10: 5. https://doi.org/10.24041/ejmr2018.54.

Forouzanfar, F., S. Torabi, V. R. Askari, E. Asadpour, and H. R. Sadeghnia. 2016. "Protective effect of Diospyros kaki against glucose- oxygen-serum deprivation-induced PC12 cells injury." *Advance in Pharmacology of Science*: 3073078. https://doi.org/10.1155/2016/3073078.

Francini, A., and L. Sebastiani. 2013. "Phenolic compounds in apple (*Malus* x *domestica* Borkh.): Compounds characterization and stability during postharvest and after processing." Antioxidants 2 (3): 181–193. https://doi.org/10.3390/antiox2030181.

Fu-Xiang, L., L. Fu-hua, Y. Ya-Xuan, Y. Ran, and M. Jian. 2019. "Comparison of phenolic profiles and antioxidant activities in skins and pulps of eleven grape cultivars (*Vitis vinifera* L.)." *Journal of Integrative Agriculture* 18(5): 1148–1158. https://doi.org/10.1016/S2095-3119(18)62138-0.

Gajera, H. P., S. N. Gevariya, D. G. Hirpara, S. V. Patel, and B. A. Golakiya. 2017. "Antidiabetic and antioxidant functionality associated with phenolic constituents from fruit parts of indigenous black jamun (*Syzygium cumini* L.) landraces." *Journal of Food Science and Technology* 54(10): 3180–3191. https://doi.org/10.1007/s13197-017-2756-8.

Garvin, S., K. Ollinger, and C. Dabrosin. 2006. "Resveratrol inducesapoptosis and inhibits angiogenesis in human breast cancerxenografts in vivo." *Cancer Letters* 231(1): 113–122. https://doi.org/10.1016/j.canlet.2005.01.031.

Gato, N., A. Kadowaki, N. Hashimoto, S. I. Yokoyama, and K. Matsumoto. 2012. "Persimmon fruit tannin-rich fiber reduces cholesterol levels in humans." *Annals of Nutrition and Metabolism* 62(1): 1–6. https://doi.org/10.1159/000343787.

Gavrilova, V., M. Kajdzanoska, V. Gjamovski, and M. Stefova. 2011. "Separation, characterization and quantification of phenolic compounds in blueberries and red and black currants by HPLC-DAD-ESI-MSn." *Journal of Agriculture and Food Chemistry* 59(8): 4009–4018. https://doi.org/10.1021/jf104565y.

Gayathri, A., P. M. Benish Rose, J. Saranya, P. Eganathan, P. Sujanapal, and K. P. Ajay. 2012. "Antimicrobial, antioxidant, anticancer activities of *Syzygium caryophyllatum* (L.) Alston." *International Journal of Green Pharmacy* 6(4): 285–288. https://doi.org/10.4103.0973-8258.108210.

Gharib, E., and K. S. Montasser. 2019. "Study of the antidiabetic activity of *Punicagranatum L.* fruits aqueous extract on the alloxan-diabetic Wistar rats." *Iranian Journal of Pharmaceutical Research* 18(1): 358–368.

Ghasemi, K., Y. Ghasemi, and M. A. Ebrahimzadeh. 2009. "Antioxidant activity, phenol and flavonoid contents of 13 citrus species peels and tissues." *Pakistan Journal of Pharmaceutical Sciences* 22(3): 277–281.

Giampieri, F., J. M. Alvarez-Suarez, L. Mazzoni, T. Y. Forbes-Hernandez, M. Gasparrini, A. M. Gonzalez-Paramas, C. Santos-Buelga, J. L. Quiles, S. Bompadre, B. Mezzetti, and M. Battino. 2014. "Polyphenol-richstrawberry extract protects human dermal fibroblasts against hydrogenperoxide oxidative damage and improves mitochondrial functionality." *Molecules* 19(6): 7798–7816. https://doi.org/10.3390/molecules19067798.

Giménez-Bastida, J. A., M. A. Ávila-Gálvez, J. C. Espín, and A. González-Sarrías. 2021. "Evidence for health properties of pomegranate juices and extracts beyond nutrition: A critical systematic review of human studies." *Trends in Food Science and Technology* 114: 410–423. https://doi.org/10.1016/j.tifs.2021.06.014.

Gondi, M., S. A. Basha, J. J. Bhaskar, P. V. Salimath, and U. J. S. Prasada Rao. 2015. "Anti diabetic effect of dietary mango (*Mangifera indica* L.) peel in streptozotocin induced diabetic rats." *Journal of the Science of Food and Agriculture* 95(5): 991–999. https://doi.org/10.1002/jsfa.6778.

Gopa, B., J. Bhatt, and K. G. Hemavathi. 2012. "A comparative clinical study of hypolipidemic efficacy of Amla (*Emblica officinalis*) with 3-hydroxy-3-methylglutaryl-coenzyme-A reductase inhibitor simvastatin." *Indian Journal of Pharmacology* 44(2): 238–242. https://doi.org/10.4103/0253-7613.93857.

Gupya, S. K., G. K. Yadav, and S. M. M. Patil. 2013. "Nutraceutical – A bright scope and opportunity of Indian healthcare market." *The Pharmaceutical Innovation* 2(4): 478–481.

Hamid, H., A. Thakur, and N. S. Thakur. 2021. Role of functional food components in COVID-19 pandemic: A review. *Annals of Phytomedicine* 10(1): S240–S250. https://doi.org/ 10.21276/ap.covid19.2021.10.1.22.

Hamid, T., N. S. Thakur, C. Sharma, K. Bhatt, and A. A. Khan. 2020. "Pomegranate and its wild genotypes: Nutraceutical opportunities and challenges." *Annals of Phytomedicine* 9(1): 32–43. https://doi.org/10.21276/ap.2020.9.1.5.

Han, C., Y. Xiao, E. Liu, Z. Su, X. Meng, and B. Liu. 2020. "Preparation of Ca-alginate whey protein isolate microcapsules for protection and delivery of *L. bulgaricus* and *L.paracasei*." *International Journal of Biological Macromolecules* 163: 1361–1368. https://doi.org/ 10.1016/j.ijbiomac.2020.07.247.

Hasan, A. N., T. Saha, and T. Ahmed. 2021. "Antibaterial activity of the extracts of pineapple and pomelo against five different pathogenic bacterial isolates." *Stamford Journal of Microbiology* 11(1): 1–6. https://doi.org/10.3329/sjm.v11i1.57143.

Hastuti, R. L., and P. I. Lukito. 2022. "Vitamin C levels and antioxidant activity of pineapple wet candied based on the level of pineapples ripeness (*Ananas comosus* var. queen) as a functional food product." *International Journal of Applied Pharmaceutics* 14(1): 27–31.

Hesler, C. M. 2002. "Functional foods: Benefits, concerns and challenges-A position paper from American council on science and health." *The Journal of Nutrition* 132(12): 3772–3781. https://doi.org/10.1093/jn/132.12.3772.

Hilachuk, D., C. Crisostimo, A. J. A. Silva, and D. Paula. 2021. "Patenting activity on functional foods: A Brazilian scenario." *Journal of Technology Management and Innovation* 16(2): 70–81. https://doi.org/10.4067/S0718-27242021000200070.

Ho, P. C., D. J. Saville, P. F. Coville, and S. Wanwimolruk. 2000. "Content of CYP3A4 inhibitors, naringin, naringenin and bergapten in grapefruit and grapefruit juice products." *Pharmaceutica Acta Helvetiae* 74(4): 379–385. https://doi.org/10.1016/s0031-6865(99)00062-x.

Honda, T., I. Kai, and G. Ohi. 1999. "Fat and dietary fiber intake and colon cancer mortality: A chronological comparison between Japan and United States." *Nutrition and Cancer* 33(1): 95–99. https://doi.org/10.1080/01635589909514754.

Hossain, M. A., and S. M. M. Rahman. 2011. "Total phenolics, flavonoids and antioxidant activity of tropical fruit pineapple." *Food Research International* 44(3): 672–676. https://doi.org/10.1016/j.foodres.2010.11.036.

Ilic, T., M. Dodevska, M. Marceti, D. Bozic, I. Kodranov, and B. Vidovic. 2020. "Chemical characterization, antioxidant and antimicrobial properties of goji berries cultivated in Serbia." *Foods* 9(11): 1614. https://doi.org10.3390/foods9111614.

Imran, M., M. S. Arshad, M. S. Butt, J. H. Kwon, M. U. Arshad, and M. T. Sultan. 2017. "Mangiferin: A natural miracle bioactive compound against lifestyle related disorders." *Lipids in Health and Disease* 16(1): 84. https://doi.org/10.1186/s12944-017-0449-y.

Iriti, M., and F. Faoro. 2009. "Bioactivity of grape chemicals for human health." *Natural Product Communications* 4(5): 611–634.

Jahurul, M. H. A., I. S. M. Zaidu, K. Ghafoor, F. Y.Al-Juhaimi, K.Nyam, N. A. N.Norulaini, F.Sahena, and O. A. K. Mohd. 2015. "Mango (*Mangifera indicaL.*) by-products and their valuable components: A review." *Food Chemistry* 183: 173–180. https://doi.org/10.1016/j.foodchem.2015.03.046.

Jung, K. N., S. Y. Kim, G. W. Lee, D. Kim, Y. S. Park, S. K. Dhungana, J. H. Kim, and D. H. Shin. 2020. "Quality characteristics and antioxidant activity of unripe peach

(*Prunus persica* L. Batsch) extracts with distilled water coupled with ultrasonication and prethanol-A. *International Journal of Fruit Science* 20(2): S111–S122. https://doi.org/10.1080/15538362.2019.1709111.

Kalra, E. K. 2003. "Nutraceutical-definition and introduction." *AAPS PharmSci* 5(3): 2–3. https://doi.org/10.1208/ps050325.

Kamal-Eldin, A., and A. Moazzami. 2009. "Plant sterols and stanols as cholesterol-lowering ingredients in functional foods." *Recent Patents on Food, Nutrition and Agriculture* 1(1): 1–14. https://doi.org/10.2174/2212798410901010001.

Kamel, M. S. 1998. "A furostanol saponin from fruits of *Balanitesaegyptiaca*." *Phytochemistry* 48(4): 755–757. https://doi.org/10.1016/s0031-9422(97)01015-7.

Kannappan, S., and C. V. Anuradha. 2010. "Naringenin enhances insulin-stimulated tyrosine phosphorylation and improves the cellular actions of insulin in a dietary model of metabolic syndrome." *European Journal of Nutrition* 49(2): 101–109. https://doi.org/10.1007/s00394-009-0054-6.

Kasbia, G. S. 2005. "Functional foods and Nutraceuticals in the management of obesity." *Nutrition and Food Science* 35(5): 344–351. https://doi.org/10.1108/00346650510625557.

Kasiotis, K. M., H. Pratsinis, D. Kletsas, and S. A. Haroutounian. 2013. "Resveratrol and related stilbenes: Their anti-aging and anti-angiogenic properties." *Food and Chemical Toxicology* 61: 112–120. https://doi.org/10.1016/j.fct.2013.03.038.

Kaur, J., X. Rathinam, M. Kasi, K. M. Leng, R. Ayyalu, S. Kathiresan, and S. Subramaniam. 2010. "Preliminary investigation on the antibacterial activity of mango (*Mangifera indicaL*: Anacardiaceae) seed kernel." *Asian Pacific Journal of Tropical Medicine* 3(9): 707–710. https://doi.org/10.1016/s1995-7645(10)60170-8.

Knutson, M. D., and C. Leeuwenburgh. 2008. "Resveratrol and novel potent activators of SIRT1: Effects on aging and age-related diseases." *Nutrition Reviews* 66(10): 591–596. https://doi.org/10.1111/j.1753-4887.2008.00109.x.

Kong, J. M., L. S. Chia, N. K. Goh, T. F. Chia, and R. Brouillard. 2003. "Analysis and biological activities of anthocyanins." *Phytochemistry* 64(5): 923–933. https://doi.org/10.1016/s0031-9422(03)00438-2.

Koshy, S. M., Z. Bobby, A. P. Hariharan, and S. M. Gopalakrishna. 2012. "Amla (*Emblica officinalis*) extract is effective in preventing high fructose diet-induced insulin resistance and atherogenic dyslipidemic profile in ovariectomized female albino rats." *Menopause* 19(10): 1146–1155. https://doi.org/10.1097/gme.0b013e31824e5bf7.

Kumar, A., P. Singh, R. P. Singh, and R. N. Kewat. 2020. "Assessment of antimicrobial activity of underutilized fruits (aonla, bael, ber, jackfrui and kaitha)." *International Journal of Current Microbiology and Applied Sciences* 9(10): 1026–1032. https://doi.org/10.20546/ijcmas.2020.910.123.

Kumar, G. P. S., P. Arulselvan, D. S. Kumar, and S. P. Subramanian. 2006. "Antidiabetic activity of fruits of Terminalia chebulaon streptozotocin induced diabetic rats." *Journal of Health Science* 52(2): 283–291. https://doi.org/10.1248/jhs.52.283.

Kumar, H. S. Y., K. Hippargi, S. K. Nataraj, B. S. Shivakumar, S. Hullur, M. Ganapathi, and A. K. Kamble. 2018. "Nutraceutical and medicinal values of minor fruits in Western Ghats of South India." *Journal of Pharmacognosy and Phytochemistry* SP3: 404–408.

Kumar, V., R. Sachan, M. Rahman, K. Sharma, F. A. Al-Abbasi, and F. Anwar. 2021. "Prunus Amygdalus extract exert antidiabetic effect via inhibition of DPP-IV: In-silico and in-vivoapproaches." *Journal of Biomolecular Structure and Dynamics* 39(11): 1–15. https://doi.org/10.1080/07391102.2020.1775124.

Kumari, P., and B. S. Khatkar. 2016. "Assessment of total polyphenols, antioxidants and antimicrobial properties of aonla varieties." *Journal of Food Science and Technology* 53(7): 3093–3103. https://doi.org/10.1007/s13197-016-2282-0.

Kyle, J. A., L. Sharp, J. Little, G. G. Duthie, and G. McNeill. 2009. "Dietary flavonoid intake and colorectal cancer: A case-control study." *British Journal of Nutrition* 7: 1–8. https://doi.org/10.1017/S0007114509991784.

Kyung-Jin, J., L. Jeung-Min, L. Seung-Cheol, and P. Hae-Ryong. 2011. "Anticancer activity of persimmon (*Diospyros kaki* L.) calyx extracts on human cancer cells." *Journal of Medicinal Plants Research* 5(12): 2546–2550. https://doi.org/10.5897/JMPR.9000136.

Leffa, D. D., J. da Silva, F. Daumann, A. L. F. Dajori, L. M. Longaretti, A. P. Damiani, and V. M. de Andrade. 2014. "Corrective effects of acerola (*Malpighia emarginata* DC.) juice intake on biochemical and genotoxical parameters in mice fed on a high-fat diet." *Mutation Research/Fundamental and Molecular Mechanisms of Mutagenesis* 770: 144–152. https://doi.org/10.1016/j.mrfmmm.2013.11.005.

Legua, P., H. Modica, I. Porras, A. Conesaa, and A. Continella. 2021. "Bioactive compounds, antioxidant activity and fruit quality evaluation of eleven blood orange cultivars." *Journal of the Science of Food and Agriculture* 102: 2960–2971. https://doi.org/10.1002/jsfa.11636.

Li, P. Y., C. C. Hsu, M. C. Yin, Y. H. Kuo, F. Y. Tang, and C. Y. Chao. 2015. "Protective effects of red guava on inflammation and oxidative stress in streptozotocin-induced diabetic mice." *Molecules* 20(12): 22341–22350. https://doi.org/10.3390/molecules201219831.

Li, S., Y. Tian, P. Jiang, Y. Lin, X. Liu, and H. Yang. 2020. "Recent advances in the application of metabolomics for food safety control and food quality analyses." *Critical Reviews in Food Science and Nutrition* 61(9): 1448–1469. https://doi.org/10.1080/10408398.2020.1761287.

Li, Z., S. M. Henning, R. P. Lee, Q. Y. Lu, P. H. Summanen, G. Thames, K. Corbett, J. Downes, C. H. Tseng, S. M. Finegold, and D. Heber. 2015. "Pomegranate extract induces ellagitannin metabolite formation and changes stool microbiota in healthy volunteers." *Food and Function* 6(8): 2487–2495. https://doi.org/10.1039/C5FO00669D.

Lim, Y. Y., T. T. Lim, and J. J. Tee. 2007. "Antioxidant properties of several tropical fruits: A comparative study." *Food Chemistry* 103(3): 1003–1008. https://doi.org/10.1016/j.foodchem.2006.08.038.

Liu, H. J., Cao, and W. Jiang. 2015. Evaluation of physiochemical and antioxidant activity changes during fruit on-tree ripening for the potential values of unripe peaches." *Scientia Horticulture* 193: 32–39. https://doi.org/10.1016/j.scienta.2015.06.045.

Liu, J., and W. U. Shengmin. 2007. "Compositions for treating and preventing hyperlipidemia." China Patent WO07137449.

Liu, R. H. 2013. "Dietary bioactive compounds and their health implications." *Journal of Food Science* 78(1): 18–25. https://doi.org/10.1111/1750-3841.12101.

Liu, R. H. 2013. "Health-promoting components of fruits and vegetables in the diet." *Advances in Nutrition* 4(3): 384–392. https://doi.org/10.3945/an.112.003517.

Liya, S. J., and R. Siddique. 2018. "Determination of antimicrobial activity of some commercial fruit (apple, papaya, lemon and strawberry) against bacteria causing urinary tract infection." *European Journal of Microbiology and Immunology* 8(3): 95–99. https://doi.org/10.1556/1886.2018.00014.

Madalageri, D., P. C. Bharati, V. Orsat, V. Raghavan, and U. Kage. 2015. "Antioxidant activity in pulp and peel of three mango varieties."*Journal of Horticultural Sciences* 10(2): 199–209.

Mahajan, M., and M. Rishi. 2016. "Functional foods and nutraceuticals: An overview." *Research and Reviews in Biotechnology and Biosciences* 3(1): 1–11.

Mahima, A. K. Verma, R. Tiwari, K. Karthik, S. Chakraborty, R. Deb, and K. Dhama. 2013. "Nutraceuticals from fruits and vegetables at a glance: A review." *Journal of Biological Sciences* 13(2): 38–47.

Mahmood, T., F. Anwar, M. Abbas, and N. Saari. 2012. "Effect of maturity on phenolics (phenolic acids and flavonoids) profile of strawberry cultivars and mulberry species from Pakistan." *International Journal of Molecular Sciences* 13(4): 4591–4607. https://doi.org/10.3390/ijms13044591.

Maleeha, M., F. Anwar, I. A. Bhatti, and A. Jamil. 2013. "Variation of phenolics and antioxidant activity between peel and pulp parts of pear (*Pyrus communis* L.) fruit." *Pakistan Journal of Botany* 45(5): 1521–1525.

Maniruzzaman, F. M., and S. Udvid. 1993. *A Compendium of Plants in Bangladesh*, 1st ed. Dhaka: Bangla Academy: 270.

Manzoor, M., F. Anwar, Z. Mahmood, U. Rashid, and M. Ashraf. 2012. "Variation in minerals, phenolics and antioxidant activity of peel and pulp of different varieties of peach (*Prunus persica* L.) fruit from Pakistan." *Molecules* 17(6): 6491–6506. https://doi.org/10.3390/molecules17066491.

Maqsood, S., O. Adiamo, M. Ahmad, and P. Mudgil. 2019. "Bioactive compounds from date fruit and seed as potential nutraceutical and functional food ingredients." *Food Chemistry*: 125522. https://doi.org/10.1016/j.foodchem.2019.125522.

Marques, S. S. F., L. R. M. Feio, A. U. O. Sabaa-Srur, R. Luo, P. Shejwalkar, K. Hara, T. Dobbs, and R. E. Smith. 2016. "Evaluation of the effects of passion fruit peel flour (*Passiflora edulis* fo. *flavicarpa*) on metabolic changes in HIV patients with lipodystrophy syndrome secondary to antiretroviral therapy." *Revista Brasileria de Farmacognosia* 26: 420–426. https://doi.org/10.1016/j.bjp.2016.03.002.

Maryanto, S., and Y. Marsono. 2019. "The effect of guava on the improvement of lipid profile in hypercholesterolemic rats." *IOP Conference Series: Earth and Environmental Science* 276: 012054. https://doi.org/10.1088/1755-1315/276/1/012054.

Masibo, M., and Q. He. 2009. "Mango bioactive compounds and related nutraceutical properties—A review." *Food Reviews International* 25(4): 346–370. https://doi.org/10.1080/87559120903153524.

Mathew, S. H., and P. S. Negi. 2017. "Traditional uses, phytochemistry and pharmacology of wild banana (*Musa acuminate* Colla): A review." *Journal of Ethnopharmacology* 196: 124–140. https://doi.org/10.1016/j.jep.2016.12.009.

Matsunaga, N., K. Tsuruma, M. Shimazawa, S. Yokota, and H. Hara. 2010. "Inhibitory actions of bilberry anthocyanidins on angiogenesis." *Phytotherapy Research* 24(1): 42–47. https://doi.org/10.1002/ptr.2895.

Mayachiew, P., and S. Devahastin. 2008. "Antimicrobial and antioxidantactivities of Indian gooseberry and galangal extracts." *LWT – Food Science and Technology* 41(7): 1153–1159. https://doi.org/10.1016/j.lwt.2007.07.019.

Mehta, R., and E. P. Lansky. 2004. "Breast cancer chemopreventive properties of pomegranate (*Punica granatum*) fruit extracts in a mouse mammary organ culture." *European Journal of Cancer Prevention* 13(4): 345–348.

Milosavljevic, D. M., D. R. Mutavdzic, K. Radotic, J. M. Milivojevic, V. M. Maksimovic, and J. J. D. Maksimovic. 2020. "Phenolic profiling of 12 strawberry cultivars using different spectroscopic methods." *Journal of Agriculture and Food Chemistry* 68(15): 4346–4354. https://doi.org/10.1021/acs.jafc.9b07331.

Misra, S., P. Pandey, and H. N. Mishra. 2021. "Novel approaches for co-encapsulation of probiotic bacteria with bioactive compounds, their health benefits and functional food product development: A review." *Trends in Food Science and Technology*. https://doi.org/10.1016/j.tifs.2021.01.039.

Mitsagga, C., K. Petrotos, and I. Giavasis. 2021. "Antimicrobial properties of lyophilized extracts of olive fruit, pomegranate and orange peel extracts against foodborne pathogenic and spoilage bacteria and fungi in vitro and in food matrices." *Molecules* 26(22): 7038. https://doi.org/10.3390/molecules26227038.

Moze, S., T. Polak, L. Gasperlin, D. Koron, A. Vanzo, N. P. Ulrih, and V. Abram. 2011. "Phenolics in Slovenian bilberries (*Vaccinium myrtillus* L.) and blueberries (*Vaccinium corymbosum* L.)." *Journal of Agriculture and Food Chemistry* 59(13): 6998–7004. https://doi.org/10.1021/jf200765n.

Muhtadi, H., T. Azizah, A. Suhendi, and K. H. Yen. 2015. "Antidiabetic and antihypercholesterolemic activities of Citrus sinensis peel: In vivo study." *National Journal of Physiology, Pharmacy and Pharmacology* 5(5): 382–385.

Muhtadi, A., U. Primarianti, and T. A. Sujono. 2020. "Antidiabetic activity of durian (*Durio zibethinus* Murr.) and Rambutan (*Nephelium Lappaceum* L.) fruit peels in alloxan diabetic rats." *Procedia Food Science* 3: 255–261. https://doi.org/10.1016/j.profoo.2015.01.028.

Muralidhara, B. M., G. L. Veena, A. K. Bhattacherjee, and S. Rajan. 2019. "Antioxidants in ripe peel and pulp of twelve mango (*Mangifera indica*) cultivars." *Indian Journal of Agricultural Sciences* 89(10): 1580–1584.

Nesterova, N. V., Samylina, and A. Irina. 2017. "Method for obtaining of a means with antimicrobial activity (combination of dried leaves of wild apple with wild apple fruit pulp)." India Patent RU2639119C1.

Neto, C. C. 2007. "Cranberry and its phytochemicals: A review of in vitro anticancer studies." *Journal of Nutrition* 137: 186–193.

Nile, S. H., and S. W. Park. 2014. "Edible berries: Bioactive components and their effect on human health." *Nutrition* 30(2): 134–144.

Nirmala, A. S. 2011. "A process for the preparation of antimicrobial food preservative." India Patent IN186123.

Ojo, O. O., M. S. Nadro, and I. O. Tella. 2006. "Protection of rats by extracts of some common Nigerian trees against acetaminophen-induced hepatotoxicity." *African Journal of Biotechnology* 5: 755–760.

Olas, B. 2018. "Berry phenolic antioxidants–implications for human health?" *Frontiers in Pharmacology* 9: 1–14. https://doi.org/10.3389/fphar.2018.00078.

Olivas-Aguirre, F. J., J. Rodrigo-García, N. D. R. Martínez-Ruiz, A. I. Cardenas-Robles, S. O. Mendoza-Díaz, E. Alvarez-Parrilla, G. A. Gonzalez-Aguilar, L. A. De la Rosa, A. Ramos-Jimenez, and A. Wall-Medrano. 2016. "Cyanidin-3-O-glucoside: Physical-chemistry, foodomics and health effects." *Molecules* 21(9): 1264. https://doi.org/10.3390/molecules21091264.

Orsavova, J., I. Hlavacova, J. Mlcek, L. Snopek, and L. Misurcova. 2019. "Contribution of phenolic compounds, ascorbic acid and vitamin E to antioxidant activity of currant (*Ribes* L.) and gooseberry (*Ribes uvacrispa* L.) fruits." *Food Chemistry* 284(30):323–333. https://doi.org/10.1016/j.foodchem.2019.01.072.

Palanirajan, A., M. Deivasigamani, K. G. Velliyur, and D. Kanakasabapathi. 2014. Antioxidant and antidiabetic effect of aqueous fruit extract of *Passifloraligularis* Juss. on streptozotocin induced diabetic rats." *International Scholarly Research Notices*. Article ID 130342. http://doi.org/10.1155/2014/130342.

Pandey, K. B., and S. I. Rizvi. 2009. “Plant polyphenols as dietary antioxidants in human health and disease.” *Oxidative Medicine and Cellular Longevity* 2(5): 270–278. https://doi.org/10.4161/oxim.2.5.9498.

Pandey, M., R. K. Verma, and S. A. Saraf. 2010. “Nutraceuticals: New era of medicine and health.” *Asian Journal of Pharmaceutical and Clinical Research* 3: 11–15.

Pandey, N., R. M. Meena, S. K. Rai, and S. Pandey. 2011. “Meicinal plants derived nutraceuticals: A re-emerging health. *International Journal of Pharmacy and Biological Sciences* 2(4): 419–441.

Panjaitan, R. G. P., and Novitasari. 2021. “Anti-diabetic activity of the red dragon fruit peel (hylocereuspolyrhizus) in ethanol extract against diabetic rats.” *Pharmacognosy Journal* 13(5): 1079–1085. http://doi.org/10.5530/pj.2021.13.140.

Panth, N., B. Manandhar, and K. R. Paudel. 2017. “Anticancer activity of *Punica granatum* (pomegranate): A review.” *Phytotherapy Research* 31(4): 568–578. https://doi.org/10.1002/ptr.5784.

Pasupuleti, V., G. K. Shyamrao, M. Evelin, and J. Jacob. 2017. “A process for the preparation of ripe banana powder.” India Patent IN222015.

Pimia, R. P., L. Nohynek, H. Alakomi, and K. Oksman-Caldentey. 2005. “Bioactive berry compounds-novel tools against human pathogens.” *Applied Microbiology and Biotechnology* 67(1): 8–18.

Portanguen, S., P. Tournayre, S. Jason, A. Thierry, and M. Pierre-Sylvain. 2019. “Toward the design of functional foods and biobased products by 3D printing: A review.” *Trends in Food Science and Technology.* S092422441830308X.. https://doi.org/10.1016/j.tifs.2019.02.023.

Poulose, S. M., E. D. Harris, and B. S. Patil. 2006. “Antiproliferative effects of citrus limonoids against human neuroblastoma and colonic adenocarcinoma cells” 56(1): 103–112. https://doi.org/10.1207/s15327914nc5601_14.

Prakash, O., A. Supriya, and V. B. Kudachikar. 2020. “Physicochemical changes, phenolic profile and antioxidant capacities of colored and white grape (*Vitis vinifera* L.) varieties during berry development and maturity.” *International Journal of Fruit Science* 20(3): 1773–1783. https://doi.org/10.1080/15538362.2020.1833809.

Preeti, K. B., N. K. Sowmya, G. V. Gayathri, and D. S. Mehta. 2021. “Antibacterial activity of *Actinidia deliciosa* (kiwi fruit) extract against period onto pathogenic bacteria: An in-vitro study.” *World Journal of Pharmacy and Pharmaceutical Sciences* 10(1): 1212–2020.

Pushpaanjali, G., R. V. Geetha, R. S. Leslie, and T. Lakshmi. 2020. “Recent advances in nutraceutical and functional food.” *Indian Journal of Forensic Medicine and Toxicology* 14(4): 5238–5243.

Puupponen-Pimia, R., L. Nohynek, H. L. Alakomi, and K. M. Oksman-Caldentey. 2005. “Bioactive berry compounds—Novel tools against human pathogens.” *Applied Microbiology and Biotechnology* 67(1): 8–18.

Radhika, P. R., R. B. M. Singh, and T. Shivkumar. 2011. “Nutraceuticals: An area of tremendous scope.” *IJRAP*2: 410–415.

Rajaiah, R.2010. *How the Banana Goes to Heaven*. Chennai: Westland.

Rajalakshmi, P., S. V. Kumar, G. Subhashini, V. Vadivel, and M. Pugalenthi. 2017. “Underutilized fruits derived nutraceuticals: A rejunuvator.” *Indian Journal of Scientific Research* 13(1): 46–53.

Raphaelli, C. O., G. Dannenberg, G. O. Dalmazo, E. S. Pereira, M. Radunz, M. Vizzotto, A. M. Fiorentini, E. A. Gandra, and L. Nora. 2019. “Antibacterial and antioxidant properties of phenolic-rich extracts from apple (*Malus domestica* cv. Gala).” *International Food Research Journal* 26(4): 1133–1142.

Raza, H., M. A. Shehzad, A. Baloach, and R. Muhmmad. 2019. "Antioxidant and antimicrobial activity of fruit juices." *International Journal of Agriculture and Sustainable Development* 1: 94–102.

Remington, R., A. Chan, A. Lepore, E. Kotlya, and T. B. Shea. 2010. "Apple juice improved behavioral but not cognitive symptoms in moderate-to-late stage Alzheimer's disease in an open-label pilot study." *American Journal of Alzheimer's Disease and Other Dementias* 25(4): 367–371. https://doi.org/10.1177/1533317510363470.

Rios De Souza, V., P. A. Pereira, T. L. da Silva, L. C. de Oliveira Lima, R. Pio, and F. Queiroz. 2014. "Determination of bioactive compounds, antioxidant activity and chemical composition of Brazilian blackberry, red raspberry, strawberry, blueberry and sweet cherry fruits." *Food Chemistry* 156: 362–368. https://doi.org/10.1016/j.foodchem.2014.01.125.

Rocha-Ribeiro, S. M., J. H. Queiroz, M. E. Lopes-Ribeiro, F. M. Campos, and H. M. Pinheiro-Santana. 2007. "Antioxidant in mango (*Mangifera indica* L.) pulp." *Plant Foods for Human Nutrition* 62(1): 13–17.

Rodríguez, L. G. R., V. M. Z. Gasga, M. Pescuma, C. V. Nieuwenhove, F. Mozzi, and J. A. S. Burgos. 2020. "Fruits and fruit by-products as sources of bioactive compounds. Benefits and trends of lactic acid fermentation in the development of novel fruit-based functional beverages." *Food Research International*. https://doi.org/10.1016/j.foodres.2020.109854.

Rojas-Grau, M. A., R. M. Raybaudi-Massilia, R. C. Soliva-Fortuny, R. J. Avena-Bustillos, T. H. McHugh, and O. Martın-Belloso. 2007. "Apple puree-alginate edible coating as carrier of antimicrobial agents to prolong shelf-life of fresh-cut apples." *Postharvest Biology and Technology* 45(2): 254–264. https://doi.org/10.1016/j.postharvbio.2007.01.017.

Rong, X., J. Xu, Y. Jiang, F. Li, Y. L. Chen, Q. P. Dou, and D. P. Li. 2021. "Citrus peel flavonoid nobiletin alleviates lipopolysaccharide-induced inflammation by activating IL-6/STAT3/FOXO3a-mediated autophagy." *Food and Function* 12(3): 1305–1317. https://doi.org/10.1039/D0FO02141E.

Ropacki, S. A., S. M. Patel, and R. E. Hartman. 2013. "Pomegranate supplementation protects against memory dysfunction after heart surgery: A pilot study." *Evidence-Based Complementary and Alternative Medicine: eCAM*: 932401. https://doi.org/10.1155/2013/932401.

Rutz, J. K., G. B. Voss, and R. C. Zambiazi. 2012. "Influence of the degree of maturation on the bioactive compounds in blackberry (*Rubus* spp.) cv. Tupy." *Food and Nutrition Sciences*3: 1453–1460. http://doi.org/10.4236/fns.2012.310189.

Sad, A. A., M. M. Hoquea, and W. Zzamana. 2018. "Physicochemical and antioxidant properties of banana varieties and sensorial evaluation of jelly prepared from those varieties available in Sylhet region." *International Journal of Food Studies* 7(2): 89–97. https://doi.org/10.7455/ijfs/7.2.2018.a8.

Saini, R. K., A. Ranjit, K. Sharma, P. Prasad, X. Shang, K. G. M. Gowda, and Y. S. Keum. 2022. "Bioactive compounds of citrus fruits: A review of composition and health benefits of carotenoids, flavonoids, limonoids, and terpenes." *Antioxidants* 11(2): 239. https://doi.org/10.3390/antiox11020239.

Sajjad, S., B. Israr, F. Ali, and I. Pasha. 2020. "Investigating the effect of phytochemicals rich watermelon seeds against hypertension." *Pakistan Journal of Agricultural Sciences* 57: 1157–1164. https://doi.org/10.21162/PAKJAS/20.9231.

Sarala, P., and S. Krishnamurthy. 2014. "Monkey jack: Underutilized edible medicinal plant, nutritional attributes and traditional foods of Western Ghats, Karnataka, India." *Indian Journal of Traditional Knowledge* 13: 508–518.

Sarker, S. D., B. Bartholomew, and R. J. Nash. 2000. Alkaloids from *Balanites aegyptiaca*." *Fitoterapia* 71(3): 328–330. https://doi.org/10.1016/S0367-326X(99)00149-5.

Savic, A., A. Alimpic, J. Zivkovic, K. Savikin, S. Jaric, P. D. Marin, and S. Duletic-Lausevic. 2021. "Phenolic composition, and antioxidant and antineurodegenerative potential of methanolic extracts of fruit peel and flesh of pear varieties from Serbia." *Polish Journal of Food and Nutrition Sciences* 71(2): 225–236. http://doi.org/10.31883/pjfns/138320.

Scalbert, A., C. Manach, C. Morand, C. Rémésy, and L. Jiménez. 2005. "Dietary polyphenols and the prevention of diseases." *Critical Reviews in Food Science and Nutrition* 45(4): 287–306. https://doi.org/10.1080/1040869059096.

Sebastia, N., A. Montoro, J. Mañes, and J. M. Soriano. 2012. "A preliminary study of presence of resveratrol in skins and pulps of European and Japanese plum cultivars." *Journal of the Science of Food and Agriculture* 92(15): 3091–3094. https://doi.org/10.1002/jsfa.5759.

Seenak, P., S. Kumphune, W. Malakul, R. Chotima, and N. Nernpermpisooth. 2021. "Pineapple consumption reduced cardiac oxidative stress and inflammation in high cholesterol diet-fed rats." *Nutrition and Metabolism* 18(1): 36.

Seeram, N. P., L. Adams, Y. Zhang, L.Rupo, D. Sand, H. Scheuller, and D. Heber. 2006. "Blackberry, black raspberry, blueberry, cranberry, red raspberry and strawberry extracts inhibit growth stimulate apoptosis of human cancer in vitro." *Journal of Agriculture and Food Chemistry* 54: 9329–9339. https://doi.org/10.1021/jf061750g.

Selma, M. V., J. C. Espin, and F. A. Tomas-Barberan. 2009. "Interaction between phenolics and gut microbiota: Role in human health." *Journal of Agricultural and Food Chemistry* 57(15): 6485–6501. https://doi.org/10.1021/jf902107d.

Shah, K. A., M. B. Patel, R. J. Patel, and P. K. Parmar. 2010. "*Mangifera indica* (mango)." *Pharmacognosy Reviews* 4(7): 42–48.

Sharma, S., R. Agarwal, and R. Saxena. 2019. "Natural sources of nutraceutical compounds such as fresh fruits and vegertables and from their waste products: A review." *Science and, Technology and for Development* 8(10): 276–293.

Shehzadi, K., Q. Rubab, L. Asad, M. Ishfaq, B. Shafique, M. M. AliNawaz Ranjha, S. Mahmood, G. Mueen-Ud-Din, T. Javaid, and B. Sabtain. 2020. "A critical review on presence of polyphenols in commercial varieties of apple peel, their extraction and health benefits." *Open Access Journal Biogeneric Science and Research* 6(2): 18. https://doi.org/10.46718/JBGSR.2020.06.000141.

Shilpa, K. J., and G. Krishnakumar. 2015. "Nutritional, fermentation and pharmacological studies of *Syzygium caryophyllatum* (L.) Alston and *Syzygium zeylanicum* (L.) DC fruits." *Cogent Food and Agriculture* 1(1): 1018694. https://doi.org/10.1080/23311932.2015.1018694.

Silva, G. S., C. G. Jange, J. S.Rocha, M. A. Chaves, and S. C. Pinho. 2017. "Characterisation of curcumin-loaded proliposomes produced by coating of micronised sucrose and hydration of phospholipid powders to obtain multilamellar liposomes." *International Journal of Food Science and Technology* 52(3): 772–780. https://doi.org/10.1111/ijfs.13334.

Smeriglio, A., D. Barreca, E. Bellocco, and D. Trombetta. 2016. "Chemistry, pharmacology and health benefits of anthocyanins." *Phytotherapy Research* 30(8): 1265–1286. https://doi.org/10.1002/ptr.5642.

Somasagara, R. R., M. Hegde, K. K. Chiruvella, A. Musini, B. Choudhary, and C. R. Sathees. 2012. "Extracts of strawberry fruits induce intrinsic pathway of apoptosis in breast cancer cells and inhibits tumor progression in mice." *PLOS ONE* 7(10): 1–12. https://doi.org/10.1371/journal.pone.0047021.

Someya, S., Y. Yoshiki, and K. Okubo. 2002. "Antioxidant compounds from bananas (*Musa cavendish*)." *Food Chemistry* 79(3): 351–354. https://doi.org/10.1016/S0308-8146(02)00186-3.

Srivastava, A. K., A. Mukerjee, and A. Tripathi. 2020. "Antidiabetic and antihyperlipidemic activities of *Cucumis melo* var. *momordica* fruit extract on experimental animals." *Futur Journal of Pharmaceutical Science* 6(1): 92.

Sukumar, D., H. Jayapal, and K. B. Kodangala. 2009. "A process for the preparation of soup powder formulation based on Indian gooseberry (*Phyllunthus emblical*)." India Patent IN195797.

Sultana, B., Z. Hussain, M. Asif, and A. Munir. 2012. "Investigation on the antioxidant activity of leaves, peels, stems bark, and kernel of mango (*Mangifera indica* L.)." *Journal of Food Science* 77(8): 849–852. https://doi.org/10.1111/j.1750-3841.2012.02807.x.

Sun, J., Y. F. Chu, X. Wu, and R. H. Liu. 2002. "Antioxidant and antiproliferative activities of common fruits." *Journal of Agricultural and Food Chemistry* 50(25): 7449–7454. https://doi.org/10.1021/jf0207530.

Svoboda, R. E. 1992. "Ayurvedic life health and longevity." *Penguin Books India Ltd*: 161.

Syed, M., and B. Amin. 2017. "Natural colorant and methods thereof." European Patent EP 2 393 881. B1.

Tabart, J., T. Franck, C. Kevers, J. Pincemail, D. Serteyn, J. O. Defraigne, and J. Dommes. 2012. "Antioxidant and anti-inflammatory activities of *Ribes nigrum* extracts." *Food Chemistry* 131(4): 1116–1122. https://doi.org/10.1016/j.foodchem.2011.09.076.

Talwar, G. P., S. A. Dar, and M. K. Rai. 2008. "A novel polyherbal microbicide with inhibitory effect on bacterial, fungal and viral genital pathogens." *International Journal of Antimicrobial Agents* 32(2): 180–185. https://doi.org/10.1016/j.ijantimicag.2008.03.004.

Teng, X., M. Zhang, and A. S. Mujumdar. 2021. "4D printing: Recent advances and proposals in the food sector." *Trends in Food Science and Technology* 110: 349–363. https://doi.org/10.1016/j.tifs.2021.01.076.

Thakur, A., V. K. Joshi, and N. S. Thakur. 2019. "Immunology and its relation with food components: An overview." *International Journal of Food and Fermentation Technology* 9(1): 1–16. https://doi.org/10.30954/2277-9396.01.2019.3.

Thandapilly-Sijo, J., L. LeMaistre-Jillian, L. Louis-Xavier, M. Anderson-Christopher, N. Thomas, and D. Anderson-Hope. 2012. "Vascular and cardiac effects of grape powder in the spontaneously hypertensive rat." *American Journal of Hypertension* 25(10): 1070–1076. https://doi.org/10.1038/ajh.2012.98.

The Wealth of India. 1952. CSIR, New Delhi, India. 3: 64–67.

Thomopoulos, R., C. Baudrit, N. Boukhelifa, R. Boutrou, P. Buche, E. Guichard, and V. Guillard. 2019. "Multi-criteria reverse engineering for food: Genesis and ongoing advances." *Food Engineering Reviews* 11(1): 44–60.

Tomasevic, I., P. Putnik, F. Valjak, B. Pavlic, B. Sojic, A. B. Markovinovic, and D. B. Kovacevic. 2021. "3D printing as novel tool for fruit-based functional food production." *Current Opinion in Food Science* 41: 138–145. https://doi.org/10.1016/j.cofs.2021.03.015.

Tremonte, P., E. Sorrentino, M. Succi, L. Tipaldi, G. Pannella, E. Ibañez, and R. Coppola. 2016. "Antimicrobial effect of *Malpighia Punicifolia* and extension of water buffalo steak shelf-life." *Journal of Food Science* 81(1): 97–105. https://doi.org/10.1111/1750-3841.13141.

Tsao, R., and M. H. Akhtar. 2005. "Nutraceuticals and functional foods: Current trend in phytochemicals, antioxidant research." *Journal of Food, Agriculture and Environment* 3(1): 10–17.

Vasco, C., J. Ruales, and A. Kamal-Eldin. 2008. "Total phenolic compounds and antioxidant capacities of major fruits from Ecuador." *Food Chemistry* 111(4): 816–823. https://doi.org/10.1016/j.foodchem.2008.04.054.

Vasile, M., A. Bunea, C. R. Ioan, B. C. Ioan, S. Socaci, and M. Viorel. 2021. "Phytochemical content and antioxidant activity of *Malus domestica* Borkh Peel extracts." *Molecules (Basel, Switzerland)* 26(24): 7636. https://doi.org/10.3390/molecules26247636.

Veeriah, S., C. Miene, N. Habermann, T. Hofmann, S. Klenow, J. Sauer, F. Böhmer, S. Wölfl, and B. L. Pool-Zobel. 2008. "Apple polyphenols modulate expression of selected genes related to toxicological defence and stress response in human colon adenoma cells." *International Journal of Cancer* 122: 2647–2655. https://doi.org/10.1002/ijc.23440.

Walia, A., N. Kumar, R. Singh, H. Kumar, V. Kumar, R. Kaushik, and A. P. Kumar. 2022. "Bioactive compounds in ficus fruits, their bioactivities, and associated health benefits: A review." *Journal of Food Quality*. https://doi.org/10.1155/2022/6597092.

Wani, S. A., J. Kour, R.-U. Haq, and M. S. Elshikh (Eds.). 2023. *Fruits and Their Roles in Nutraceuticals and Functional Foods*, 1st ed. CRC Press. ISBN 9781032194462. https://doi.org/10.1201/9781003259213.

Wani, S. A., H. R. Naik, J. Wagay, N. A.Ganie, M. Z. Mulla, and B. N. Dar. 2022. Mentha: A review on its bioactive compounds and potential health benefits. *Quality Assurance and Safety of Crops and Foods* 14(4): 154–168.

Wang, W., Z. Yuangang, F. Yujie, and E. Thomas. 2012. "In vitro antioxidant and anti-microbial activity of extracts from *Morus alba* L. Leaves, stems and fruits." *The American Journal of Chinese Medicine* 40(2): 349–356. https://doi.org/10.1142/S0192415X12500279.

Wang, Y. C., Y. C. Chuang, and H. W. Hsu. 2008. "The flavonoid, carotenoid and pectin content in peels of citrus cultivated in Taiwan." *Food Chemistry* 106(1): 277–284. https://doi.org/10.1016/j.foodchem.2007.05.086.

Wargovich, M. J. 2000. "Anticancer properties of fruits and vegetables." *HortiScience* 35(4): 573–574.

Williamson, G., and C. Manach. 2005. "Bioavailability and bioe_cacy of polyphenols in humans." II. Review of 93 intervention studies. *American Journal of Clinical Nutrition* 81: 243–255. https://doi.org/10.1093/ajcn/81.1.230S.

Wojdyło, A., P. Nowicka, A. A. Carbonell-Barrachina, and F. Hernández. 2016. "Phenolic compounds, antioxidant and antidiabetic activity of different cultivars of *Ficuscarica* L. fruits." *Journal of Functional Foods* 25: 421–432. https://doi.org/10.1016/j.jff.2016.06.015.

Wolfe, K. L., X. M. Kang, X. J. He, M. Dong, Q. Y. Zhang, and R. H. Liu. 2008. "Cellular antioxidant activity of common fruits." *Journal of Agricultural and Food Chemistry* 56(18): 8418–8426. https://doi.org/10.1021/jf801381y.

Wu, C. H., C. F. Wu, H. W. Huang, Y. C. Jao, and G. C. Yen. 2009. "Naturally occurring flavonoids attenuate high glucose-induced expression of proinflammatory cytokines in human monocytic THP-1 cells." *Molecular Nutrition and Food Research* 53(8): 984–995. https://doi.org/10.1002/mnfr.200800495.

Wu, Y. H., C. K. Tseng, H. C. Wu, C. K. Wei, C. K. Lin, I. S. Chen, H. S. Chang, and J. C. Lee. 2019. "Avocado (*Persea americana*) fruit extract (2*R*,4*R*)-1,2,4-trihydroxyheptadec-16-yne inhibits dengue virus replication via upregulation of NF-κB–dependent induction of interferon responses." *Scientific Reports* 9: 1–10.

Xu, G., X.Ye, D. Liu, Y. Ma, and J. Chen. 2008. "Composition and distribution of phenolic acids in Ponkan (*Citrus poonensis* Hort. ex Tanaka) and Huyou (*Citrus paradisi* Macf. Changshanhuyou) during maturity." *Journal of Food Composition and Analysis* 21(5): 382–389. https://doi.org/10.1016/j.jfca.2008.03.003.

Yadav, S. K., R. S. Sangwan, D. Mehta, and P. Dwivedi. 2006. "Utilization of tomato and tomato processing by-products for the development of fiber, minerals and antioxidant rich novel products." India Patent IN201711042760.

Ye, X. 2005. "Citrus processing and comprehensive utilization." *China Light Industry Press* (in Chinese).

You, Q., B. Wang, F. Chen, Z. Huang, X. Wang, and P. G. Luo. 2011. "Comparison of anthocyanins and phenolics in organically and conventionally grown blueberries in selected cultivars." *Food Chemistry* 125(1): 201–208. https://doi.org/10.1016/j.foodchem.2010.08.063.

Zhang, Y., N. P. Seeram, R. Lee, L. Feng, and D. Heber. 2008. "Isolation and identification of strawberry phenolics with antioxidant and human cancer cell antiproliferative properties." *Journal of Agriculture and Food Chemistry* 56(3): 670–675. https://doi.org/10.1021/jf071989c.

Zhou, Z. Q. 2012. *Citrus Fruits Nutrition*. Beijing, China: Science Press.

Zou, Z., W. Xi, Y. Hua, C. Nie, and Z. Zhou. 2016. "Antioxidant activity of citrus fruits." *Food Chemistry* 196: 885–896. https://doi.org/10.1016/j.foodchem.2015.09.072.

CHAPTER 4

Vegetables as a Functional Food for Health

Samuel Ayofemi Olalekan Adeyeye* and **Ayenampudi Surendra Babu**

4.1 INTRODUCTION

Vegetables are edible plant matter and are consumed by humans and other animals as food; they include the flowers, fruits, stems, leaves, roots, and seeds of the plant (Sinha et al. 2010; Vainio & Bianchini 2003). Different vegetables are grown all over the world depending on climatic conditions while some are grown in greenhouses and protected environments in less suitable locations (Sinha et al. 2010; Vainio & Bianchini 2003). China remains the largest producer of vegetables globally. Vegetables are cultivated at both subsistence and commercial levels. Vegetables may be eaten both raw or in cooked form and play a significant role in human nutrition because they are low in fat and carbohydrates. They are also good sources of phytochemicals such as vitamins, minerals and dietary fibre (Sinha, et al. 2010; Vainio & Bianchini 2003; Wani et al. 2022).

Vegetables are very rich in vitamins, especially the antioxidant vitamins A, C and E. Their inclusion in the diet could improve human health and reduce the incidence of cancer, stroke, cardiovascular disease and other chronic ailments (Terry 2011; Büchner 2010).

However, some vegetables contain toxins and antinutrients, which may affect physiological functions and inhibit the absorption of nutrients in them by binding with proteins and mineral elements. The common anti-nutritional factors in vegetables are α-solanine, α-chaconine (Finotti et al. 2006), enzyme inhibitors of cholinesterase, protease, amylase, etc., cyanide and cyanide precursors, oxalic acid, tannins and others (Terry 2011; Büchner 2010). Toxins in vegetables serve as natural defences to ward off insects, predators and fungi that might attack them. Legumes like beans contain phytohaemagglutinin, and cassava roots contain cyanogenic glycoside as do bamboo shoots. However, some toxins and most antinutrients in vegetables can be

* Corresponding author E-mail: adeyeyes@hindustanuniv.ac.in

 DOI: 10.1201/9781003315100-4

removed by adequate cooking and fermentation. However, green potatoes contain glycoalkaloids and should be avoided to prevent food poisoning (Centre for Food Safety 2015).

Hygiene is critical for vegetables to be eaten raw, and there is a need for proper cleaning, handling and storage of such vegetable products to limit contamination (Centers for Disease Control and Prevention 2013). In the United States, nearly half of gastrointestinal infections caused by norovirus are attributed to leafy vegetables. These vegetables are consumed without heat processing, and contamination occurs during handling by an infected food handler. This chapter discusses vegetables as a functional food for health. It also explains the bioactive compounds in vegetables, the nutraceutical properties of vegetables and recent advancements in the development of functional vegetables.

4.2 BIOACTIVE COMPOUNDS IN VEGETABLES

Recently, as a result of job schedules and busy lifestyles, unhealthy eating habits are creating more lifestyle-related chronic health diseases. However, there has been increased awareness about healthy food choices of fresh produce with important phytochemicals for healthy life. Phytochemicals, also known as phytonutrients, are present in abundance in vegetables with immense health benefits (Sharma et al. 2021). They are helpful in reducing certain diseases such as high blood pressure, diabetes, cancer etc. Phytochemicals are present in abundance in vegetables such as fruits, beans, grains, herbs nuts and seeds (Wani et al. 2022; Sharma et al. 2021; Huang et al. 2016).

Phytochemicals help plants to resist fungi, bacteria and plant virus infections, and insects and pest infestation. Common phytochemicals are carotenoids, polyphenols, terpenoids, phytosterols, limonoids, glucosinolates and flavonoids like anthocyanidins, flavones and flavonols (Sharma et al. 2021; Huang et al. 2016). The phytochemical protective role could be due to their antioxidant activity since the pathogenesis of many chronic diseases is a result of the overproduction of oxidants (reactive oxygen species and reactive nitrogen species) in the human body. Phytochemicals stimulate the immune system by modulating the function of various other metabolic reactions in the body, but they are not essential nutrients and some particularly anti-nutritional factors among them may interfere with the absorption of other nutrients in the body if not regulated (Sharma et al. 2021; Chukwuebuka and Chinenye 2015). They also act as anti-inflammatory, anti-allergic, anti-bacterial, anti-fungal, anti-spasmodic, hypolipidemic, hepatoprotective, neuroprotective, anti-ageing, diuretic, immunomodulator and carminative agents (Sharma et al. 2021; Prakash et al. 2012).

Vegetables supply important dietary ingredients such as carbohydrates, proteins, minerals, vitamins and various bioactive substances (Sharma et al. 2021; Kader et al. 2000). Apart from being sources of micro and macronutrients, vegetables are rich in many bioactive compounds that help them to act as protective foods (Sharma et al. 2021; Dias 2012; Boeing et al. 2012). Vegetables are sources of essential diet ingredients particularly in underdeveloped and developing countries because of their importance in managing malnutrition. Diets in developing countries are more centred on carbohydrate-based diets, which do not have sufficient essential nutrients to meet the body's requirements. The foods of animal origin that solve the problem of malnutrition are too costly in those countries, and vegetables play a key role that could assist in reducing malnutrition and solving various health challenges by offering balanced diets containing an abundance of bioactive compounds. Diets containing a variety of vegetables will help in providing the required nutrients to achieve sustainable health goals and protect the body against many diseases (Sharma et al. 2021; Baldermann et al. 2016).

Vegetables contain pigments which constitute the sensorial appeal as well as sources of bioactive compounds which are important to human health. The green colour of leafy vegetables contains chlorophyll, which is a potent antioxidant. Chlorophyll helps the immune system, purifies the blood and prevents hypertension, detoxifies the body, prevents diabetes and cancer, and could help prevent bad odours and breath due to the presence of magnesium salts. Chlorophylls have been found to be potent antimutagens, antioxidants, anticarcinogens and photosensitisers (Sharma et al. 2021; Inanc 2020; Suparmi et al. 2016). Chlorophyll has a porphyrin structure with magnesium moiety similar to haemoglobin with one difference that the central atom is iron. Chlorophylls can improve the formation of red blood cells, help in weight loss, heal damaged skin and remove toxins, and it reduces cut inflammation and the formation of free radicals (Sharma et al. 2021; Alsuhaibani et al. 2017).

Orange- and yellow-coloured vegetables are very rich in carotenoids (Sharma et al. 2021; Mezzomo and Ferreira 2016). Carotenoids contain beta-carotene (vitamin A precursor), lutein, zeaxanthin and lycopene. Vitamin A helps in the development of eye pigment, and vegetables are major sources of vitamin A particularly coloured fruits and vegetables such as carrots. Vitamin A is very important in the healthy development of the foetus/child during pregnancy and during lactation, particularly to boost the immune system (Sharma et al. 2021; Grune et al. 2010). Vitamin A helps in the formation of the macula, which is the most sensitive part of the retina that possesses the highest concentration of photoreceptors for sharp vision and perceiving colour. Lutein, zeaxanthin and meso-zeaxanthin are components of carotenoids that have control over the formation of macular pigment with their antioxidant and anti-inflammatory properties, and it thereby preserves the photoreceptor from oxidation from high-energy blue light which can lead to macular degeneration (Sharma et al. 2021; Scripsema et al. 2015).

Vegetables that possess blue, purple and red pigments are rich in anthocyanins, which are water-soluble (Sharma et al. 2021; Ramos et al. 2014; Clifford et al. 2015). Anthocyanins perform several functions such as protection against solar exposure and ultraviolet radiation as a result of their free radical scavenging activity, thereby preventing oxidative damage to DNA, proteins, lipids and other macromolecules due to reactive oxygen species (ROS). Other vegetables such as cauliflower, cucumbers, onion and garlic are good sources of dietary fibre and aid bowel movement and prevent high cholesterol. Onion and garlic are also good sources of antioxidant compounds such as polyphenols and flavonoids like quercetin. Polyphenols are produced by plants as secondary metabolites, and they protect plants against ultraviolet radiation or aggression by pathogens (Sharma et al. 2021; Lee 2017). Polyphenols are also antioxidants and protect the human body from oxidative stress and other related debilitating sicknesses like cancers, coronary heart disease and inflammation (Sharma et al. 2021).

4.3 NUTRITIONAL PROPERTIES OF VEGETABLES

4.3.1 General

Vegetables are considered the portion of a plant that can be consumed, such as the bulbs, tubers, roots, leaves, stem and the sweet and fleshy product of a tree or other plant that contains seeds (Division of Agriculture and Natural Resources, University of California). Consumption of vegetables is essential, as they promote health and play a pivotal role in the protection and functioning of various organs of the body, as they are an excellent source of phytochemicals. Phytochemicals are the bioactive compounds in fruits, vegetables, grains and other plant foods that may provide several health benefits as well as regular nutrition (Liu 2004). These components include beneficial antioxidants, natural colourants, minerals and vitamins. Coloured vegetables (yellow/green/purple/blue/red/orange) imply their antioxidant properties and their potential ability to minimise the risk of various complaints and illnesses such as constipation, haemorrhoids, cancer, stroke and heart diseases (Abobatta 2021). Beetroot and eggplant are such examples. Red-coloured vegetables such as tomato, radish and red cabbage decrease the risk of cancer and improve heart health. Yellow- and orange-coloured vegetables including carrot, pumpkin and sweet potato possess carotenoids that enhance the functioning of the eyes. Further, green-coloured vegetables (broccoli, spinach, lettuce, cucumber etc.) exhibit anticancer properties (Hoejskov 2014). These vegetable crops are regarded as functional and nutraceutical foods, as they contain ample amounts of many phytonutrients (Zhao 2007). Nutraceuticals are substances seen as a natural component of foods that have health benefits in preventing or treating one or more diseases or enhancing physiological performance and reduction of health risks (Ramya and Patel 2019). The term nutraceutical was coined from the two words "nutrition" and "pharmaceutical" by Dr. Stephen de Felice (Biesalski 2001).

4.3.2 Antioxidant Potentials

Antioxidants are chemical compounds that protect the body from harmful oxidative reactions. They neutralise or counteract the free radicals and reactive oxygen species generated during redox reactions by scavenging them, thereby preventing damage to cellular components and the chemical stability of the cells (Singh et al. 2012). Phytochemicals such as carotenoids, ascorbates, lipoic acids, tocopherols, lycopene and polyphenols are potential natural antioxidants with free radical scavenging activity (Kondratyuk and Pezzuto 2004). Various vegetables offer a wide range of antioxidants to fight against free radicals (Shetty et al. 2013). Tomatoes are one of the most produced and consumed vegetables around the globe and are a rich source of lycopene, β-carotene, folate, chlorogenic acid, potassium, vitamin C, rutin, phenolics, tocopherol and xanthophylls, which possess high antioxidative capacities (George et al. 2004; Collins et al. 2022).

Rivero et al. (2022) reported the antioxidant potential of various tomato cultivars from Cuba (Roma VF-p73 (Ro), Rilia (Ri), Campbell-28 (Ca), Vyta (Vy), CIMA (Ci) and Germany (Suso F1 Hybride RZ (Su), Marmande (Ma), goldene konigin (GK) and yellow pearshaped (YP). The FRAP analysis (ferric reducing antioxidant potential) of fresh tomato and its puree ranged between 21.9 and 47.9 mmol Fe+2/Kg, and 39.9 to 74.2 mmol Fe+2/Kg, respectively. The antioxidant potential of tomato samples was attributed to the presence of various phytochemicals such as lycopene, ascorbic acid, total flavonoids and hydrophilic phenolics (Rivero et al. 2022) (Table 4.1).

Other vegetables such as iceberg and butterhead lettuce are the most used vegetables in salad preparations. Lettuce of three different colours, green, dark red,

Table 4.1 Summary of antioxidant potentials/properties of vegetables

Vegetable	Scientific name	Bioactive compounds	Mechanism of action (MOA)	Assay type	Reference
Red-lettuce cultivars	*Lactuca sativa* L.	Polyphenols particularly anthocyanin, flavone and phenolic acid	-	*In vitro* (DPPH and ABTS scavenging activity)	Qin et al. (2018)
Tomato cultivars	*Lycopersicon esculentum* Mill.	Lycopene, Ascorbic acid, Total flavonoids and hydrophilic phenolics	-	*In vitro* (FRAP assay)	Rivero et al. (2022)
Sweet potato varieties (orange-flesh, purple-flesh and yellow-flesh)	*Ipomoea batatas*	Carotenoid and total phenolic content	-	*In vitro* (DPPH assay)	Ooi et al. (2021)
Eggplant fruit cultivars (black, purple and white)	*Solanum melongena* L	Total phenolic compounds, flavonoids and anthocyanins	-	*In vitro* (DPPH and ORAC assay)	Colak et al. (2022)

DPPH: 1,1-diphenyl-2-picrylhydrazyl; ABTS: 2,2′-azino-bis(3-ethylbenzothiazoline-6-sulfonic acid); FRAP: ferric reducing antioxidant potential; ORAC: oxygen radical absorbance capacity and red, are reported to have higher antioxidant activity owing to its higher anthocyanin content (Kim et al. 2016). According to Qin et al. (2018), dark-red lettuce (B-2 variety) exhibited potential ABTS (2,2′-azino-bis(3-ethylbenzothiazoline-6 -sulfonic acid)) and DPPH (1,1-diphenyl-2-picrylhydrazyl) scavenging activity and antioxidant activity, which is ascribed to the higher anthocyanin, flavone and phenolic acid content. Other vegetables such as sweet potatoes are also reported to be a rich source of antioxidants such as carotenoids, particularly β-carotene (Alam et al. 2020). Broccoli presents excellent antioxidant properties due to its higher levels of flavonoids and vitamins C and K (dos Reis et al. 2015).

4.3.3 Antimicrobial Properties

Numerous studies have demonstrated the antimicrobial potential of vegetables due to the excess amounts of antimicrobial constituents. Vegetables are similar to medicinal plants, representing no toxic effect, or a lower toxic effect, and possess the capacity to produce numerous secondary metabolites of relatively complex structures that exhibit antimicrobial properties (Kubo et al. 2004; Matasyoh et al. 2009). The antimicrobial properties of plant foods are because of the secondary metabolic compounds such as tannins, terpenoids, alkaloids and flavonoids (Mabou et al. 2016). Ekom et al. (2021) reported that the presence of antimicrobial compounds in capsicum, such as alkaloids, polyphenols, flavonoids, anthocyanins, tannins, triterpenes, anthraquinones and saponins, which exerted antimicrobial activity against three strains and four isolates of *Staphylococcus aureus* and *Escherichia coli* species (Table 4.2). According to Khoo et al. (2017), purple and red vegetables exhibit mild antimicrobial activity. Purple and red vegetables are reported to have anthocyanins, which have been shown to have antimicrobial properties. Earlier research showed that fresh leafy green vegetables contain abundant levels of phytochemicals, particularly polyphenols (e.g., phenolic acids, flavonoids and other aromatic compounds), which are known to exhibit demonstrate antimicrobial activity (Gutierrez et al. 2008; Kim et al. 2013). Salamatullah et al. (2021) stated that the beetroot pulp extract of 50% exhibited potent antimicrobial activity against a wide range of Gram -Ve than Gram +Ve bacterial species.

4.3.4 Anticancer Activity

Cancer is associated with a group of diseases that represents abnormalities in the dynamics of the normal cell cycle leading to uncontrolled cell proliferation. As the cell turns carcinogenic, the developed tumour becomes capable of targeting other noncancerous cells creating a "tumour microenvironment," further triggering the generation of new blood vessels for nutritional requirements and thereby propagating across the organs, generating metastasis. These are common characteristics seen in various types of cancers and are expressed as the so-called "hallmarks of cancer" (Hanahan et al. 2022). Regular consumption of vegetables in the diet has been significantly associated with the reduction of several common cancers (Steinmetz and Potter 1991). Research evidence has confirmed the positive impact of vegetable

Table 4.2 Summary of antimicrobial properties of vegetables

Vegetable	Scientific name	Bioactive compounds	Mechanism of action (MOA)	Assay type	Reference
Capsicum	*Capsicum annuum* L	Alkaloids, polyphenols, alkaloids, flavonoids, anthocyanins, anthraquinones tannins, triterpenes and saponins	**Tested Microorganisms:** Antimicrobial activity of *Capsicum annuum* was tested against three strains and four isolates of *Staphylococcus aureus* and *Escherichia coli* **MOA:** Antimicrobial activity of *Capsicum annuum* was attributed to the alteration of the microbial cell membrane and leakage of cellular components leading to the microbial cell death	*In vitro*	Ekom et al. (2021)
Cabbage, acclimatised broccoli, cauliflower, and black radish	*Brassica oleracea* L and *Rhaphanus sativus* L	Glucosinolates and hydrolysed substances including isothiocyanates, allylthiocyanates and allylcyanides	**Tested Microorganisms:** A total of 14 bacterial and fungal strains including *Streptococcus*, *Staphylococcus*, *Escherichia coli*, *Pseudomonas*, *Enterococcus*, *Candida* and *Bacillus* **MOA:** Affecting cytoplasmic membrane, reducing the surface charge and altering the hydrophobicity thus damaging the cellular integrity	*In vitro*	Muntean et al. (2021)
Beetroot	*Beta vulgaris*	Not specified	**Tested Microorganisms** *Listeria monocytogenes* **MOA:** Reduction of intracellular ATP level and dissipation of proton motive force components	*In vitro*	Gong et al. (2022)

consumption against various cancers such as mouth, oesophagus, pharynx, lung, stomach and colon. Other cancers, including pancreas, breast and bladder, were also controlled by the consumption of various vegetables (Wargovich 2000). Veiga et al. (2022) studied the cytotoxicity of quercetin-rich extracts of yellow onion on adrenocortical carcinoma cell lines (H295R). Quercetin was reported to affect the cell cycle by arresting the cell in the G1 phase and the prevalence of the cells in the G0/Sub-G1 phase (Table 4.3). Brandi et al. (2005) demonstrated the anticancer activity of cabbage juice against breast cancer cells compared with other mammalian cell lines. The extracts of curly kale exhibited an antiproliferative effect over human colon cancer cell lines (Olsen et al. 2012). Similar results were noticed for horseradish,

Table 4.3 Summary of anticancer activity/properties of vegetables

Vegetable	Scientific name	Bioactive compounds	Mechanism of action (MOA)	Assay type	Reference
Yellow onion	*Allium cepa*	Quercetin	Quercetin was reported to affect the cell cycle by arresting the cell in the G1 phase and the prevalence of the cells in the G0/Sub-G1 phase	*In vivo*	Veiga et al. (2022)
Moringa leaf varieties (Conventional PKM-1 (Periyakulam-1), PKM-2 (Periyakulam-2), ODC (Oddanchatram), and Jaffna)	*Moringa Oleifera L*	Kaempferol, niazimicin, Beta-sitosterol -3-O-beta-D-glu copyranoside, 4(α-l-rhamnosyloxy)-benzyl isothiocyanate	Moringa leaf extract was hypothesised to induce apoptosis. Further, the bioactive compounds of moringa were believed to target the cell cycle by accumulating the cells at the sub-G1 phase	*In vivo*	Farooq et al. (2021)
Purple carrot	*Daucus carota*	Ascorbic acid, citric acid, catechin, peonidin and its derivates, cyanidin and its derivatives, caffeoylquinic acid, caffeoyl N-tryptophan, procyanidin dimer type B	The anticancer activity of purple carrot extract is ascribed to reducing Ki-67 immuno-expression in rats	*In vivo*	Soares et al. (2018)

inhibiting colon cancer and breast cancer cells (Weil et al. 2005; Bo et al. 2016). Desai et al. (2020) noticed that a diet rich in garlic and onions may reduce the risk of breast cancer.

4.3.5 Antidiabetic Activity

Diabetes mellitus (DM) is a chronic non-communicable disease categorised by hyperglycemia, indicating an excessive blood sugar level. It is a metabolic syndrome where the body produces inadequate insulin levels or becomes insensitive to the hormonal secretion from β-cells of pancreatic islets, which control blood glucose levels. Common symptoms of DM include polyuria (frequent urination), polydipsia (greater thirst), and polyphagia (excess hunger) (Laulloo et al. 2021). Despite several advancements in the antidiabetic synthetic drug systems, there has been growing interest in plant food-based approaches as antidiabetic remedies due to the presence of a complex range of natural phytochemicals that are non-toxic and of a safe nature (Venkataraman et al. 2018). The leaf extract of moringa was found to reduce the blood glucose level, insulin and cytokine in type-2 diabetic groups. This antidiabetic phenomenon of moringa leaf extract was attributed to modulating hyperinsulinemia, PPARy, and inflammatory cytokines (Anwer et al. 2021) (Table 4.4). In a study, Perez et al. (2019) demonstrated the α-amylase inhibitory activity of two commercial bitter gourd cultivars (MCC and MCM). The α-amylase and α-glucosidase inhibitory of the better gourd extract ranged between 66% to 69%, and IC50 values ranged from 0.26 and 0.29 mg/mL. The antidiabetic potential of bitter gourd extract was because of the higher levels of triterpenes and 27 saponins, 1 flavonoid and 1 sterol present (Sun et al. 2020). The *in vivo* study confirmed the antidiabetic potential of bottle gourd extract when administered to alloxan-induced diabetic mice. The results showed a 52% reduction in fasting blood glucose level compared to the standard drug, metformin (49%) (Zeng et al. 2016).

4.3.6 Other Nutraceutical Properties

4.3.6.1 Bone Health

Osteoporosis is a bone disease that is a socioeconomic threat to the entire world (Rachner et al. 2011). Nutritious food plays a vital role in maintaining bone health and the prevention of osteoporosis (Ahmadieh and Arabi 2011). Dietary factors such as protein, calcium, vitamin D, dairy and soya products were proven to have a positive correlation with bone health. However, recent evidence confirmed that the several components in vegetables positively affect bone health and decrease the risk of osteoporosis (Ahmadieh and Arabi 2011; Lambert et al. 2015). Amaranth leaves (*Amaranthus mangostanus* L) were reported to have a positive effect on bone health and can be used for the treatment of postmenopausal osteoporosis. An *in vivo* animal study demonstrated the anti-osteoporosis activity of amaranth by inducing expression of osteoclast-specific genes, TRAP, cathepsin K, NF-activated T-cells (NFATc1) and Dc-stamp in RAW 264.7 cells (Jeong et al. 2020).

Table 4.4 Summary of antidiabetic activity/properties of vegetables

Vegetable	Scientific name	Bioactive compounds	Mechanism of action (MOA)	Assay type	Reference
Green bitter gourd	*Momordica charantia*	Gallic acid, chlorogenic acid, catechin, caffeic acid, p-coumaric acid and ferulic acid	The antidiabetic effect of bitter gourd extract was attributed to its ability to stimulate the proliferation and the spontaneous recovery of pancreatic β-cells, increasing the cell count and function	*In vivo*	Eweda et al. (2021)
Moringa leaf	*Moringa Oleifera L*	Not specified	Moringa leaf was reported to exert antidiabetic activity by increasing insulin secretion from pancreatic β-cells and down-regulation of insulin receptors or dysfunction in post-receptors signalling	*In vivo*	Anwer et al. (2021)
Aerial yam	*Dioscorea bulbifera*	Alkaloids, cardiac glycosides, carbohydrates, flavonoids, proteins, saponins, tannins and terpenoids	The antidiabetic effect of yam extract was assumed to be the effect of flavonoids on the pancreatic β-cells leading to their regeneration and secretion of excess insulin	*In vivo*	Akinyele et al. (2021)

4.4 ANTI-AGEING PROPERTIES

Aeing is a constant process of the human life cycle associated with natural alteration in mechanical-biological processes. This metabolic process is related to many factors including the time and energy of biological process. Further, the ageing process can be ascribed to stem cell senescence, oxidative stress and cell longevity. Ageing is linked to several diseases including cancer, neurodegenerative disease and other diseases caused due to oxidative stress. Furthermore, certain phytochemicals derived from vegetable extracts are shown to delay the ageing process (Mechchate et al. 2022). Svendsen et al. (1994) investigated the anti-ageing properties of garlic on morphology, long-term growth characteristics and macromolecular synthesis of human skin fibroblasts. The results of the study showed that the inclusion of garlic extract in a standard cell culture medium exhibited anti-ageing, youth-preserving and positive effects on human fibroblast cells. Amelia et al. (2021) observed that there is an increase in collagen, hydration, elasticity, sebum and pigmentation in male rats applied with red spinach extract. Application of a 10% concentration of spinach extract showed improved skin moisture content and skin collagen levels in animal experimentation.

4.5 RECENT ADVANCEMENTS IN THE DEVELOPMENT IN THE DEVELOPMENT OF FUNCTIONAL VEGETABLES

The major challenge for agriculture in the 21st century is to feed nutritious and healthy food to a growing population. The human diet requires certain vitamins, minerals and nutraceutical compounds to maintain a healthy lifestyle. Not everyone chooses a diet that is dense in essential nutrients required for the human body (Chekol et al. 2021). Genetic engineering can be a technology for developing nutritionally superior plants and thereby plant-based food products. For instance, the idea of blending blueberries with tomatoes was to enhance the levels of anthocyanins that people can receive in their diet. Anthocyanin is an antioxidant that fights against cancer and other severe health disorders (Chekol et al. 2021). Romer et al. (2000) produced β-carotenoid enriched tomato fruits using genetic engineering, which resulted in a threefold increase (up to 45%) in total carotenoid content. Similarly, Lu et al. (2006) developed GM cauliflower with enhanced levels of β-carotene. In another study, resveratrol content (56 μg/g fresh leaf weight) and γ -tocopherol were increased in lettuce plant by genetic modification (Cho et al. 2005; Liu et al. 2006). In a recent study, Kim et al. (2019) developed transgenic sweet potato lines with enhanced γ -tocopherol levels. Li et al. (2021) have demonstrated that the gene expression of DcBCH1 can increase the antioxidant capacity and synthesis of carotenoid content in carrot.

4.6 CONCLUDING REMARKS AND FUTURE DIRECTIONS

Globally, vegetables are important food commodities that are crucial for food and nutrition security programmes because they are very important in supplying vital protective nutrients such as minerals, vitamins and antioxidants required for the overall well-being of people as well as providing energy-rich food. Vegetables play a key role in providing taste, palatability and fibre, increasing appetite, and neutralising stomach acids. Vegetables have a vital role to play in addressing micronutrient malnutrition referred to as "hidden hunger" when taking into account their high consumption. Several factors such as increases in cases of diabetes, coronary heart diseases, cancer and increased stress levels are affecting consumer's choices and preferences as they move from processed food products to fresh food products like vegetables, resulting in increasingly healthy and high-quality diets. Vegetables contain phytochemicals with potent antioxidant activities, and they are sources of edible colour. Vegetables have low consumption levels in developing countries because of poverty and a lack of awareness of their nutritional and medicinal importance. Utilisation of vegetables particularly among the rural poor and vulnerable social groups in developing countries would bring better nutrition and improved health. Nutrition should be an essential driver of plant breeding programmes so that efforts are geared toward breeding crops that are adapted to meet the nutritional needs of the people. Low-cost techniques should be developed for vegetable production and processing, particularly deeply pigmented vegetables, to improve the nutritional status of people. Furthermore, research and post-harvest studies are needed on the effect of processing on the stability and bioavailability of the phytochemicals and natural pigments in vegetables to improve their consumption in traditional diets and cuisines.

REFERENCES

Abobatta, W. F. 2021. Nutritional and healthy benefits of fruits. *Diabetes*, 40(2): 31979–31983.

Ahmadieh, H., & Arabi, A. 2011. Vitamins and bone health: Beyond calcium and vitamin D. *Nutrition Reviews*, 69(10): 584–598.

Akinyele, N. K., Beatrice, O. E. O., Adeniyi, O. F., Adetoun, E. M., & Oluboade, O. O. 2021. Studies of the anti-hyperglycemic and antioxidant activities of the extract of aerial yam (*Dioscorea bulbifera*). *Journal of Medicinal Plants Research*, 5(10): 503–514.

Alam, M. K., Sams, S., Rana, Z. H., Akhtaruzzaman, M., & Islam, S. N. 2020. Minerals, vitamin C, and effect of thermal processing on carotenoids composition in nine varieties orange-fleshed sweet potato (*Ipomoea batatas* L.). *Journal of Food Composition and Analysis*, 92: 103582.

Alsuhaibani, A. M., ALkehayez, N. M., Alshawi, A. H., & Al-Faris, N. A. 2017. Effects of chlorophyll on body functioning and blood glucose levels. *Asian Journal of Clinical Nutrition*, 9(2): 64–70. 10.3923/ajcn.2017.64.70.

Amelia, Girsang, E., Nasution, A. N., & Ginting, C. N. 2021. Anti-aging effectiveness of red spinach extract ointment (*Amaranthus tricolor* L.) against collagen, elasticity, hydration, sebum, and pigment levels in Wistar rats. *Proceedings of the 2021 IEEE International Conference on Health, Instrumentation & Measurement, and Natural Sciences (InHeNce)*, Medan, Indonesia: IEEE; 2021: 1–6.

Anwer, T., Safhi, M. M., Makeen, H. A., Alshahrani, S., Siddiqui, R., Sivakumar, S. M., & Alam, M. F. 2021. Antidiabetic potential of Moringa oleifera Lam. leaf extract in type 2 diabetic rats, and its mechanism of action. *Tropical Journal of Pharmaceutical Research*, 20(1): 95–103.

Baldermann, S., Blagojevic, L., Frede, K., Klopsch, R., Neugart, S., & Neumann, A. 2016. Are neglected plants the food for the future? *Critical Reviews in Plant Sciences*, 35(2): 106–119. 10.1080/07352689.2016.1201399.

Biesalski, H. K. 2001. Nutraceuticals: The link between nutrition and medicine. In *Nutraceuticals in Health and Disease Prevention*, 17–42. CRC Press.

Bo, P., Lien, J. C., Chen, Y. Y., Yu, F. S., Lu, H. F., Yu, C. S., & Chung, J. G. 2016. Allyl isothiocyanate induces cell toxicity by multiple pathways in human breast cancer cells. *The American Journal of Chinese Medicine*, 44(2): 415–437.

Boeing, H., Bechthold, A., Bub, A., Ellinger, S., Haller, D., & Kroke, A. 2012. Critical review: Vegetables and fruit in the prevention of chronic diseases. *European Journal of Nutrition*, 51(6): 637–663. 10.1007/s00394-012-0380-y.

Brandi, G., Schiavano, G. F., Zaffaroni, N., De Marco, C., Paiardini, M., Cervasi, B., & Magnani, M. 2005. Mechanisms of action and antiproliferative properties of Brassica oleracea juice in human breast cancer cell lines. *The Journal of Nutrition*, 135(6): 1503–1509.

Büchner, F. L., Bueno-de-Mesquita, H., Bas, R., Martine, M., Overvad, K., Dahm, C. C., Hansen, L., Tjønneland, A., Clavel-Chapelon, F., Boutron-Ruault, M.-C. 2010. Variety in fruit and vegetable consumption and the risk of lung cancer in the European prospective investigation into cancer and nutrition. *Cancer Epidemiology, Biomarkers and Prevention*, 19(9): 2278–2286. 10.1158/1055-9965.EPI-10-0489.

Centre for Food Safety. 2015. Naturally occurring toxins in vegetables and fruits. *Risk Assessment in Food Safety*. Retrieved 2015-03-30.

Chekol, C. 2021. The health effects of genetically modified foods: A brief review. *International Journal of Nutritional Sciences*, 6(1): 1047.

Cho, E. A., Lee, C. A., Kim, Y. S., Baek, S. H., Reyes, B. G., & Yun, S. J. 2005. Expression of γ-tocopherol methyltransferase transgene improves tocopherol composition in lettuce (Latuca sativa L.). *Molecules and Cells (Springer Science & Business Media BV)*, 19(1): 16–22.

Chukwuebuka, E., & Chinenye, I. J. 2015. Biological functions and anti-nutritional effects of phytochemicals in living system. *IOSR Journal of Pharmacy and Biological Sciences*, 10: 10–19. 10.9790/3008-10231019.

Clifford, T., Howatson, G., West, D. J., & Stevenson, E. J. 2015. The potential benefits of red beetroot supplementation in health and disease. *Nutrients*, 7(4): 2801–2822. 10.3390/nu7042801.

Colak, N., Kurt-Celebi, A., Gruz, J., Strnad, M., Hayirlioglu-Ayaz, S., Choung, M. G., & Ayaz, F. A. 2022. The phenolics and antioxidant properties of black and purple versus white eggplant cultivars. *Molecules*, 27(8): 2410.

Collins, E. J., Bowyer, C., Tsouza, A., & Chopra, M. 2022. Tomatoes: An extensive review of the associated health impacts of tomatoes and factors that can affect their cultivation. *Biology*, 11(2): 239.

Desai, G., Schelske-Santos, M., Nazario, C. M., Rosario-Rosado, R. V., Mansilla-Rivera, I., Ramírez-Marrero, F., & Mu, L. 2020. Onion and garlic intake and breast cancer, a case-control study in Puerto Rico. *Nutrition and Cancer*, 72(5): 791–800.

dos Reis, L. C. R., de Oliveira, V. R., Hagen, M. E. K., Jablonski, A., Flôres, S. H., & de Oliveira Rios, A. 2015. Carotenoids, flavonoids, chlorophylls, phenolic compounds and antioxidant activity in fresh and cooked broccoli (Brassica oleracea var. Avenger) and cauliflower (Brassica oleracea var. Alphina F1). *LWT – Food Science and Technology*, 63(1): 177–183.

Dias, J. S. 2012. Nutritional quality and health benefits of vegetables: A review. *Food and Nutrition Sciences*, 3(10): 1354–1374. 10.4236/fns.2012.310179.

Ekom, S. E., Tamokou, J. D. D., & Kuete, V. 2021. Antibacterial and therapeutic potentials of the Capsicum annuum extract against infected wound in a rat model with its mechanisms of antibacterial action. *BioMed Research International*. Article ID 4303902. 10.1155/2021/4303902.

Eweda, S. M., Ali, M. A., Abd El-Bary, H. M., El-Sokkary, N. H., Helmy, M. H., & Kamel, M. A. N. 2021. Bitter gourd extract improves glucose homeostasis and lipid profile via enhancing insulin signaling in the liver and skeletal muscles of diabetic rats. *Asian Pacific Journal of Tropical Biomedicine*, 11(8): 344.

Finotti, E., Bertone, A., & Vivanti, V. 2006. Balance between nutrients and anti-nutrients in nine Italian potato cultivars. *Food Chemistry*, 99(4): 698. 10.1016/j.foodchem.2005.08.046.

Gong, S., Jiao, C., & Guo, L. 2022. Antibacterial mechanism of beetroot (Beta vulgaris) extract against Listeria monocytogenes through apoptosis-like death and its application in cooked pork. *LWT*, 165: 113711.

Gutierrez, J., Bourke, P., Lonchamp, J., & Barry-Ryan, C. 2009. Impact of plant essential oils on microbiological, organoleptic and quality markers of minimally processed vegetables. *Innovative Food Science and Emerging Technologies*, 10(2): 195–202.

Grune, T., Lietz, G., Palou, A., Ross, A. C., Stahl, W., & Tang, G. 2010. Betacarotene is an important vitamin A source. *Journal of Nutrition*: 2268–2285, 10.3945/jn.109.119024 .ants.

Hanahan, D. 2022. Hallmarks of cancer: New dimensions. *Cancer Discovery*, 12(1): 31–46.

Hoejskov, P. S. 2014. Importance of fruit and vegetables for public health and food safety. In *Proceedings of the Pacific Regional Workshop on Promotion of Fruit and Vegetables for Health*, Nadi, Fiji, 20–23 October 2014.

Huang, Y., Xiao, D., Burton-Freeman, B. M., & Edirisinghe, I. 2016. Chemical changes of bioactive phytochemicals during thermal processing. *Reference Module in Food Science*, 1–9. 10.1016/b978-0-08-100596-5.03055-9.

Inanc, A. L. 2011. Chlorophyll: Structural properties, health benefits and its occurrence in virgin olive oils. *Akademik Gıdatr*, 9: 90–344.

Jeong, Y. H., Hur, H. J., Lee, A. S., Lee, S. H., & Sung, M. J. 2020. Amaranthus mangostanus inhibits the differentiation of osteoclasts and prevents ovariectomy-induced bone loss. *Evidence-Based Complementary and Alternative Medicine: eCAM*. doi: 10.1155/2020/1927017

Kader, A. A., Perkins-veazie, P., & Lester, G. E. 2000. Nutritional quality of fruits, nuts, and vegetables and their importance in human health. *USDA, Agriculture Research Service*, 1: 1–7.

Khoo, H. E., Azlan, A., Tang, S. T., & Lim, S. M. 2017. Anthocyanidins and anthocyanins: Colored pigments as food, pharmaceutical ingredients, and the potential health benefits. *Food and Nutrition Research*, 61(1): 1361779.

Kim, M. J., Moon, Y., Tou, J. C., Mou, B., & Waterland, N. L. 2016. Nutritional value, bioactive compounds and health benefits of lettuce (Lactuca sativa L.). *Journal of Food Composition and Analysis*, 49: 19–34.

Kim, S. E., Lee, C. J., Ji, C. Y., Kim, H. S., Park, S. U., Lim, Y. H., & Kwak, S. S. 2019. Transgenic sweet potato plants overexpressing tocopherol cyclase display enhanced α-tocopherol content and abiotic stress tolerance. *Plant Physiology and Biochemistry*, 144: 436–444.

Kim, S. J., Cho, A. R., & Han, J. 2013. Antioxidant and antimicrobial activities of leafy green vegetable extracts and their applications to meat product preservation. *Food Control*, 29(1): 112–120.

Kondratyuk, T. P., & Pezzuto, J. M. 2004. Natural product polyphenols of relevance to human health. *Pharmaceutical Biology*, 42(supl): 46–63.

Kubo, I., Fujita, K. I., Nihei, K. I., & Nihei, A. 2004. Antibacterial activity of akyl gallates against Bacillus subtilis. *Journal of Agricultural and Food Chemistry*, 52(5): 1072–1076.

Lambert, H., Frassetto, L., Moore, J. B., Torgerson, D., Gannon, R., Burckhardt, P., & Lanham-New, S. 2015. The effect of supplementation with alkaline potassium salts on bone metabolism: A meta-analysis. *Osteoporosis International*, 26(4): 1311–1318.

Laulloo, S. B. J., Bhowon, M. G., & Jalloo, Y. 2021. Literature review: Anti-diabetic potential of some selected edible vegetables in tropical region. *Longhua Chinese Medicine*, 4: 33.

Lee, B. J. 2017. Do white vegetables have health benefits ? What's your favorite vegetable? *CR Consumer Reports*, 2–4.

Li, T., Liu, J. X., Deng, Y. J., Xu, Z. S., & Xiong, A. S. 2021. Overexpression of a carrot BCH gene, DcBCH1, improves tolerance to drought in Arabidopsis thaliana. *BMC Plant Biology*, 21(1): 1–13.

Lin, D., & Zhao, Y. 2007. Innovations in the development and application of edible coatings for fresh and minimally processed fruits and vegetables. *Comprehensive Reviews in Food Science and Food Safety*, 6(3): 60–75.

Liu, R. H. 2004. Potential synergy of phytochemicals in cancer prevention: Mechanism of action. *The Journal of Nutrition*, 134(12): 3479S–3485S.

Liu, S., Hu, Y., Wang, X., Zhong, J., & Lin, Z. 2006. High content of resveratrol in lettuce transformed with a stilbene synthase gene of Parthenocissus Henryana. *Journal of Agricultural and Food Chemistry*, 54(21): 8082–8085.

Lu, S., Van Eck, J., Zhou, X., Lopez, A. B., O'Halloran, D. M., Cosman, K. M., & Li, L. 2006. The cauliflower or gene encodes a DnaJ cysteine-rich domain-containing protein that mediates high levels of β-carotene accumulation. *The Plant Cell*, 18(12): 3594–3605.

Mabou, F. D., Jean, D. D. T., Ngnokam, D., Voutquenne-Nazabadioko, L., Kuiate, J. R., & Bag, P. K. 2016. Complex secondary metabolites from Ludwigia leptocarpa with potent antibacterial and antioxidant activities. *Drug Discoveries and Therapeutics*, 10(3): 141–149.

Matasyoh, J. C., Maiyo, Z. C., Ngure, R. M., & Chepkorir, R. 2009. Chemical composition and antimicrobial activity of the essential oil of Coriandrum sativum. *Food Chemistry*, 113(2): 526–529.

Mechchate, H., El Allam, A., El Omari, N., El Hachlafi, N., Shariati, M. A., Wilairatana, P., & Bouyahya, A. 2022. Vegetables and their bioactive compounds as anti-aging drugs. *Molecules*, 27(7): 2316.

Mezzomo, N., & Ferreira, S. R. S. 2016. Carotenoids functionality, sources, and processing by supercritical technology: A review. *Journal of Chemistry*, 16, 10.1155/2016/3164312.

Muntean, D., Ştefănuţ, M. N., Căta, A., Buda, V., Danciu, C., Bănică, R., & Ienaşcu, I. M. 2021. Symmetrical antioxidant and antibacterial properties of four Romanian cruciferous extracts. *Symmetry*, 13(5): 893.

Olsen, H., Grimmer, S., Aaby, K., Saha, S., & Borge, G. I. A. 2012. Antiproliferative effects of fresh and thermal processed green and red cultivars of curly kale (Brassica oleracea L. convar. acephala var. sabellica). *Journal of Agricultural and Food Chemistry*, 60(30): 7375–7383.

Ooi, S. F., Sukri, S. A. M., Zakaria, N. N. A., & Harith, Z. T. 2021. Carotenoids, phenolics and antioxidant properties of different sweet potatoes (*Ipomoea batatas*) varieties. *IOP Conference Series: Earth and Environmental Science*, 756(1): 012077.

Perez, J. L., Jayaprakasha, G. K., Crosby, K., & Patil, B. S. 2019. Evaluation of bitter melon (Momordica charantia) cultivars grown in Texas and levels of various phytonutrients. *Journal of the Science of Food and Agriculture*, 99(1): 379–390.

Prakash, D., Gupta, C., & Sharma, G. 2012. Importance of phytochemicals in nutraceuticals. *Journal of Chinese Medicine Research and Development*, 1: 70–78.

Qin, X. X., Zhang, M. Y., Han, Y. Y., Hao, J. H., Liu, C. J., & Fan, S. X. 2018. Beneficial phytochemicals with anti-tumor potential revealed through metabolic profiling of new red pigmented lettuces (Lactuca sativa L.). *International Journal of Molecular Sciences*, 19(4): 1165.

Rachner, T. D., Khosla, S., & Hofbauer, L. C. 2011. Osteoporosis: Now and the future. *The Lancet*, 377(9773): 1276–1287.

Ramos, P., Herrera, R., & Moya-leon, M. A. 2014. Anthocyanins: Food sources and benefits to consumer's health. In Leah M. Warner (Ed.) *Handbook of Anthocyanins*. New York: Nova Science Publishers Inc., 1–22.

Ramya, V., & Patel, P. 2019. Health benefits of vegetables. *International Journal of Chemical Studies*, 7(2): 82–87.

Rivero, A. G., Keutgen, A. J., & Pawelzik, E. 2022. Antioxidant properties of tomato fruit (Lycopersicon esculentum Mill.) as affected by cultivar and processing method. *Horticulturae*, 8(6): 547.

Römer, S., Fraser, P. D., Kiano, J. W., Shipton, C. A., Misawa, N., Schuch, W., & Bramley, P. M. 2000. Elevation of the provitamin A content of transgenic tomato plants. *Nature Biotechnology*, 18(6): 666–669.

Salamatullah, A. M., Hayat, K., Alkaltham, M. S., Ahmed, M. A., Arzoo, S., Husain, F. M., … Al-Harbi, L. N. 2021. Bioactive and antimicrobial properties of oven-dried beetroot (pulp and peel) using different solvents. *Processes*, 9(4): 588.

Scripsema, N. K., Hu, D. N., & Rosen, R. B. 2015. Lutein, zeaxanthin, and meso-zeaxanthin in the clinical management of eye disease. *Journal of Ophthalmology*, 13. 10.1155/2015/865179.

Sharma, S., Katoch, V., Kumar, S., & Chatterjee, S. 2021. Functional relationship of vegetable colors and bioactive compounds: Implications in human health. *The Journal of Nutritional Biochemistry*, 92. 10.1016/j.jnutbio.2021.108615.

Singh, R., Singh, M. K., Chandra, L. R., Arora, M. S., Naiwal, T. K., & Pande, V. 2012. In vitro Antioxidant and free radical scavenging activity of Dolichos biflorus (Gahat dal) from Kumauni region. *Journal of Fundamental and Applied Sciences*, 1: 9–11.

Sinha, N. K., Hui, Y. H. Evranuz, E. Ö., Siddiq, M., & Ahmed, J. 2010. Handbook of Vegetables and Vegetable Processing. John Wiley & Sons. 92, 352. ISBN 978-0-470-95844-5.

Soares, G. R., de Moura, C. F. G., Silva, M. J. D., Vilegas, W., Santamarina, A. B., Pisani, L. P., & Ribeiro, D. A. 2018. Protective effects of purple carrot extract (Daucus carota) against rat tongue carcinogenesis induced by 4-nitroquinoline 1-oxide. *Medical Oncology*, 35(4): 1–14.

Suparmi, S., Fasitasari, M., Martosupono, M., & Mangimbulude, J. C. 2016. Comparisons of curative effects of chlorophyll from *Sauropus androgynus* (L) Merr leaf extract and cu-chlorophyllin on sodium nitrate-induced oxidative stress in rats. *Journal of Toxicology*: 1–7. 10.1155/2016/8515089.

Steinmetz, K. A., & Potter, J. D. 1991. Vegetables, fruit, and cancer. I. Epidemiology. *Cancer Causes and Control*, 2(5): 325–357.

Sun, C., Zhao, C., Guven, E. C., Paoli, P., Simal-Gandara, J., Ramkumar, K. M., & Xiao, J. 2020. Dietary polyphenols as antidiabetic agents: Advances and opportunities. *Food Frontiers*, 1(1): 18–44.

Svendsen, L., Rattan, S. I., & Clark, B. F. 1994. Testing garlic for possible anti-ageing effects on long-term growth characteristics, morphology and macromolecular synthesis of human fibroblasts in culture. *Journal of Ethnopharmacology*, 43(2): 125–133.

Terry, L. 2011. Health-Promoting Properties of Fruits and Vegetables. CABI, 2–4. ISBN 978-1-84593-529-0.

UC vegetable research information centre frequently asked questions. http://vric.ucdavis.edu/main/faqs.htm.

Vainio, H., & Bianchini, F. 2003. Fruits and Vegetables. IARC, 2. ISBN 978-92-832-3008-3.

Veiga, A. A., Irioda, A. C., Mogharbel, B. F., Bonatto, S. J., & Souza, L. M. 2022. Quercetin-rich extracts from onions (Allium cepa) play potent cytotoxicity on adrenocortical carcinoma cell lines, and quercetin induces important anticancer properties. *Pharmaceuticals*, 15(6): 754.

Venkataraman, S., Kumaran, S., & Jayapalan, S. 2018. Phytochemical constituents and pharmacological activities of Lagenaria siceraria: A comprehensive review. *Drug Invention Today*, 10: 1431–1436.

Wani, S. A., Naik, H. R., Wagay, J., Ganie, N. A., Mulla, M. Z., & Dar, B. N. 2022. Mentha: A review on its bioactive compounds and potential health benefits. *Quality Assurance and Safety of Crops and Foods*, 14(4): 154–168.

Wargovich, M. J. 2000. Anticancer properties of fruits and vegetables. *Hortscience*, 35(4): 573–575.

Weil, M. J., Zhang, Y., & Nair, M. G. 2005. Tumor cell proliferation and cyclooxygenase inhibitory constituents in horseradish (*Armoracia rusticana*) and Wasabi (*Wasabia japonica*). *Journal of Agricultural and Food Chemistry*, 53(5): 1440–1444.

Zeng, L., Zhang, G., Liao, Y., & Gong, D. 2016. Inhibitory mechanism of morin on α-glucosidase and its anti-glycation properties. *Food and Function*, 7(9): 3953–3963.

CHAPTER 5

Cereals as a Functional Food for Health

Nusrat Jan*, **Shakeel Ahmad Sofi**, **Gousia**,
Sajad Mohd Wani and **Syed Zameer Hussain**

5.1 INTRODUCTION

Cereals are described as grains or edible seeds produced for their highly nutritious edible seed and are members of the grass family (Bender and Bender 1999). Since the beginning of civilisation, several kinds of cereal have served as staple foods for both human consumption as well as livestock feed (Gani et al. 2012). The most popular cereals consumed globally include maize, rice, wheat, oats, barley, millet, rye and sorghum. In addition to being a crucial component of the diet, these grains are abundant in several health-promoting substances (Slavin 2003). The world's population relies heavily on cereals as a primary source of proteins, carbohydrates, minerals and B vitamins. A variety of substances are present in cereals that may have health-promoting properties, which are known as plant bioactive components or phytochemicals (Goldberg 2003).

Bioactive substances are extra nutritional components that often appear in foods in trace amounts. Although these compounds are useful to human health, they are not necessary for the human body (Kris-Etherton et al. 2002). The bran/germ portion of cereal grains contains the majority of the bioactive components found in whole grains. Therefore, consuming these whole-grain fractions may help lower the chance of developing chronic disorders. However, compared to fruits and vegetables, the bioactive substances present in cereals have not received great interest. Regular consumption of whole grains and whole-grain products has been linked to lower risks for a variety of chronic diseases, including type 2 diabetes (Meyer et al. 2000; Liu et al. 2000; Tapola et al. 2005), cardiovascular diseases (Anderson 2003; Okarter and Liu 2010) and some cancers (Kasum et al. 2002; Haas et al. 2009). Whole grains or foods derived from whole grains contain endosperm, bran, and germ as opposed

* Corresponding author E-mail: nusratj353@gmail.com

DOI: 10.1201/9781003315100-5

to refined grains in which the germ and bran have been removed during the polishing process. Plant-based foods such as whole grains are rich sources of bioactive components including vitamins, fibre, minerals and phytochemicals, which may have positive health effects in reducing the risk of chronic disorders (Liu 2004; Slavin 2000). In this regard, this chapter focuses on bioactive components in cereal grains, their nutraceutical characteristics and the development of functional foods from cereals.

5.2 BIOACTIVE COMPOUNDS IN CEREALS

Whole grains consist of specific bioactive substances that complement those in fruits and vegetables when consumed together. Dietary fibres (mostly beta-glucan), phytosterols, γ-oryzanols, ferulic acid, cinnamic acid, phytic acid, lignans, avenanthramides and inositols are the principal bioactive components of whole-grain cereals (Gani et al. 2012; Slavin 2003; Jones et al. 2004). Several bioactive components are unique to certain cereals, such as avenanthramides and saponins in oats, γ-oryzanol in rice, alkylresorcinol in rye and beta-glucans in oats and barley. These substances are also present in other cereals like wheat, although they are present in relatively smaller quantities. The following sections address the vital bioactive substances found in whole-grain cereals.

5.2.1 Phenolic Compounds

Grains contain various phenolic substances, such as flavonoids, phenolic acids, lignans and tannins, which are not evenly distributed throughout the grain (Kandil et al. 2005). Phenolic substances are secondary metabolites of plants extensively present in foods with a plant origin (Wani et al. 2022). They participate in protective mechanisms against pathogen aggressiveness or ultraviolet radiation (Rocha, Monteiro, and Teodoro 2012; Manach et al. 2004). The most common phenolic components in grains are phenolic acids, which tend to deposit in crusts mostly in the pericarp, followed by testa and aleurone (Kandil et al. 2005). These components can accumulate in various plant tissues and cells during ontogenesis and under the influence of environmental factors. As a result, distinct phenolics might be found in specific locations throughout the plant (Weidner et al. 2000).

Hydroxybenzoic acid derivatives and hydroxycinnamic acid derivatives are the two classes of phenolic acids (Karamac et al. 2005; Mattila, Pihlava, and Hellstrom 2005). Derivatives of hydroxybenzoic acid include chlorogenic acid, caffeic acid, p-coumaric acid, ferulic acid and sinapic acid. These acids broadly occur in bound forms because they are bound to structural components of the cell wall (Wang et al. 2013; Adom and Liu 2002). According to researchers, 62% of phenolic acids are in the bound form in rice, 75% in wheat and oat, and 85% in maize (Adom and Liu 2002). Ferulic acid can be found free, dimerised or esterified with proteins and polysaccharides in the cell wall (Fazary and Ju 2007). Significant amounts of ferulic acid and its oxidatively linked derivatives are found in cereal brans (called diferulic acids or ferulic acid dehydrodimers). Scientists have also shown that diferulic acids can be produced by the microbial esterase digestion of ferulic acid, or their reduced dimers can be produced by colonic microbiota hydrogenation reactions of diferulic acids (Belobrajdic and Bird 2013). Diferulic acids, which come in a variety of forms including 5-5-, 8-O-4-, 8-5- and 8-8-diferulic acid, are strong antioxidants that are ester-linked to polymers of the cell wall (Andreasan et al. 2001). Rye and wheat bran contain significant amounts of these chemicals (Bondia-Pons et al. 2009). Among cereals, popcorn has the highest levels of diferulic acid (between 250 and 475 g/g of flour), followed by maize, rye, barley, oats and wheat (Jilek and Bunzel 2013).

Cinnamic acids have been classified as bioactive constituents of the diet (Teixeria et al. 2013; Shahidi and Chandrasekara 2010). The hydroxyl derivatives of cinnamic acid are coumaric acids. Coumaric acids come in three different forms: p-coumaric acid, o-coumaric acid and m-coumaric acid. The three forms vary in the position of the hydroxyl substitution of the phenolic group (Garrait et al. 2006). p-hydroxycinnamic acid is a synonym for p-coumaric acid because it is a hydroxyl derivative of cinnamic acid (Madhujith, Izydorczyk and Shahidi 2006). The concentration of p-coumaric acid is lowest in the centre of the grain kernel and rises as it moves outward (Awika and Rooney 2004).

5.2.2 Flavonoids

Flavonoids are compounds having a C6-C3-C6 skeleton made up of two aromatic rings linked by a three-carbon bond; they include flavonols, flavanones, flavones and anthocyanins. In nature, more than 5000 flavonoids have been identified (Yao et al. 2004). All cereals contain flavonoids in the pericarp. The greatest variety of flavonoids has been found in sorghum. Cereals have been observed to contain the flavanones found in fruits and vegetables. For example, millet, oat and sorghum have also been identified to contain flavanone apigenin, a substance found in celery and barley (Rice-Evans, Miller and Paganga 1997). Cereals only contain trace amounts of flavonoids, with the exception of barley, which also contains some di and tri procyanidins and significant amounts of catechin (McMurough and Baert 1994).

5.2.3 Avenanthramides

Avenanthramides (AVAs) are particular polyphenols found in oats. There are at least 25 different types of these substituted cinnamic acid amides of anthranilic acids. The three primary AVAs discovered in oats are AVA-1, 3, and 4, also known as AVAs B, C, and A (Collins 1989). Oats have significant phenolic compounds called AVAs. Although they are present in nearly all milling fractions and are constitutively expressed in the kernels, they are most prevalent in the bran and outer kernel layers (Peterson, Hahn, and Emmons 2002; Jastrebova et al. 2006; Meydani 2009).

5.2.4 Alkylresorcinols

Alkylresorcinols are phenolic lipids generated from plants that are most frequently noticed in whole-grain cereals. Rye grains have the maximum concentration of alkylresorcinols ranging from 360 to 3200 mg/g db. Wheat and barley grains have lower concentrations of alkylresorcinols, which can range from 317 to 1430 and 41 to 210 mg/g db, respectively (Zarnowski et al. 2002; Ross et al. 2003c; Kulawinek and Kozubek 2007). The majority of homologues found in rye grains are hydrocarbon chains with 15–27 carbon atoms, with C21:0 being the most prevalent. Their most prevalent analogues in wheat grains have 15–25 carbon atoms (90–95%), with C19:0 being the most common. Alkylresorcinols typically have between 15 and 20% of unsaturated hydrocarbon chains, ketone and hydroxyl groups (Suzuki, Esami and Yamaguchi 1999; Kozubek and Tyman 1999).

5.2.5 Carotenoids

The majority of the carotenoid profile in cereals is made up of xanthophylls, with lutein being the most prevalent. Zeaxanthin and α-cryptoxanthin are next in abundance, while tiny levels of α- and β-carotene are also present. Although the embryo only makes up 3–5% of the seed's weight, it contains the highest concentration of carotenoids. On the other hand, the endosperm, which makes up around 80–85% of cereal grains, plays the biggest role in the grain's overall carotenoid content. Since zeaxanthin, α- and β-carotene are concentrated in the germ and bran, the distribution of this carotenoid profile appears to differ between genotypes of the same species of cereal (Siebenhandl et al. 2007) and within the same grain, while lutein is distributed more uniformly (Borrelli et al. 2008; Konopka, Kozirok and Rotkiewicz 2004; Ndolo and Beta 2013).

Cereals have lower amounts of carotenoids than fruits and vegetables; however, yellow-genotype corn (*Zea mays*) may be accounted as the grain with the highest concentrations (up to 63 lg/g) of these phytochemicals. Most cultivars exhibit zeaxanthin as the predominant pigment, with minor levels of β-cryptoxanthin, and α- and β-carotene also present (Moros et al. 2002; Panfili, Fratianni and Distaam 2004). It is evident that the coloured varieties of rice (*Oryza sativa*) are the ones that draw some attention from a quantitative point of view. The carotenoid content of conventional rice grains is almost entirely concentrated in the bran, therefore common processing techniques like milling and stripping the grain significantly limit the amount of this component in the final product (Tan et al. 2005). In crops with dark or black pigmentation, lutein is the main pigment (Belefant-Miller and Grace 2010), followed by zeaxanthin and β-carotene (Frei and Becker 2005; Kim et al. 2010). One of the grains with the lowest amount of carotenoids is oats (*Avena sativa*), which have lutein and carotenes (α- and β-carotene) concentrations of about 0.20 mg/kg and 0.01 mg/kg, respectively (Panfili, Fratianni and Distaam 2004). Wheat (genus *Triticum*), rice and corn are the most prominent grains due to their widespread consumption. Einkorn wheat (*Triticum monococcum*), with the diploid genome (AA), is renowned for having the highest concentrations of carotenoids, typically containing two to four times more of these beneficial compounds compared to other wheat varieties. Additionally, Einkorn stands out as having the highest antioxidant capacity (Lavelli et al. 2009). This is followed by tetraploid wheat (AABB), which has been discovered to have the greatest carotenoid levels (about 5–6 lg/g) among durum wheat cultivars

(*Triticum turgidum* conv. durum). However, other cultivars have been reported, such as the ancestor Emmer (*T. turgidum* subsp. dicoccoides), with similar levels (about 3–5 lg/g). The common bread wheat (*T. aestivum*), which displays an average amount of about 2 lg/g, is regarded as having the lowest carotenoid concentration among wheat types with the hexaploid genome (AABBDD) (Hidalgo et al. 2006; Abdel-Aal et al. 2007).

5.2.6 Gamma-oryzanol

γ-Oryzanol is a constituent of rice-bran oil and was initially believed to be a single component. Subsequently, it was revealed that it contains at least ten pyhtosterylferulates (e.g. methyl sterols esterified to ferulic acid) along with sterols and ferulic acid. Its concentration in whole-grain rice is 18–63 mg/100g (DW) (Britz et al. 2007). The amount in rice-bran ranges from 185–421 mg/100g depending on the milling time, rice variety, extraction methods and stabilisation process (Yu et al. 2007).

5.2.7 Beta-glucan

β-glucan are polysaccharides that are primarily present in the cell walls of the aleurone layer and endosperm in oat and barley kernels. They are more prevalent in the endosperm of barley than they are in the aleurone layer of oats (Bhatty 1993). These are the linear polymers of glucose molecules joined by 30% of β-(1-3) and 70% of β-(1-4)- linkages. Barley (3–11%) and oats (3–7%) contain the highest concentrations of β-glucan, with rye (1–2%) and wheat (< 1%) containing the lowest concentrations. Only minimum concentrations have been recorded in sorghum, rice, corn and other important cereals used as food (Bartlomiej 2011). In recent years, oat-based breakfast cereals have also received renewed interest because of their high content of β-glucan which has been considered a bioactive constituent (Behall et al. 2006)

5.2.8 Tocopherols and Tocotrienols

Tocopherols and tocotrienols, often known as tocols, are chemical compounds known as vitamin E. They rank among foods that possess antioxidant activity and are lipid soluble. This group includes eight congeners: four tocopherols (α-T, β-T, γ-T and δ-T) and four tocotrienols (α-T3, β-T3, γ-T3 and δ-T3) that differ in the number of methyl groups connected to the ring of 6-chromanol. Only photosynthetic plants can produce tocols; hence, they must be consumed through diet. Cereal grains with a higher concentration of tocotrienols than tocopherols are the richest sources of vitamin E in the human diet. The reports of Zielinski, Kozlowska and Lewczuk (2001) found that wheat and rye grains contained 27.81 and 27.78 mg db, respectively, were the highest sources of tocols. Oat and barley grains had lower concentrations of these substances with 11.59 and 18.73 mg db, respectively. These grains varied not just in terms of their contents but also in terms of their tocol profiles. The richest sources of α-tocotrienols, α-tocopherol and β-tocotrienol were barley, rye and wheat, respectively.

5.2.9 Phytosterols

Plant sterols, also known as phytosterols, are secondary metabolites that are mostly found in the seeds of cereals and oilseed crops. Sitosterol, stigmasterol, campesterol and brassicasterol are the sterols that are most prevalent in cereal grains. The two main stanols: sitostanol and campestanol found in cereal grains account for 10–30% of all cereal sterols (Piironen, Toivo and Lampi 2002; Nystrom et al. 2007). Rye grains contain the highest concentrations of phytosterols, with an average of 80–90 mg/100 g. Their concentrations in grains of wheat, barley and oat are lower, with respective concentrations of 70, 70–80 and 45–50 mg/100 g. (Piironen, Toivo and Lampi 2002).

5.3 NUTRACEUTICAL PROPERTIES AND POTENTIAL HEALTH BENEFITS OF CEREALS

Strong epidemiological studies have examined clearly that whole-grain cereals protect the body from age-related disorders like diabetes, cardiovascular disease and various cancers (Das, Raychaudhuri and Chakraborty 2012; Venn and Mann 2004). Table 5.1 presents the health-beneficial properties of various cereal grains.

5.3.1 Phenolic Compounds

The influence of phenolic amides from plant-based foods on human health has gained great attention in recent years. These phenolic amides exhibit potential therapeutic and antioxidant properties including anti-inflammatory, anti-proliferative and anti-genotoxic effects. Hydroxycinnamic acid is described as an ester with cell wall polysaccharides in the colon. It can exist in free form due to the action of an esterase produced by bacteria. Bacteria in the colon have the ability to both absorb these acids and transform them into other substances. Of the body's total feruloyl groups, 95% travel through the colon in free form, where they play a significant role in preventing cancer due to their antimutagenic activity (Kroon et al. 1997; Ferguson et al. 2003). Ferulic and coumaric acids are the two principal phenolic chemicals found in wheat grains. Ferulic acid has a great antioxidant impact and has a high radical scavenger effect for free radicals such as superoxide, nitrogen dioxide, hydrogen peroxide and hydroxyl radicals. At a concentration of 250 mg/1, it quenches 92.5% of hydroxyl radicals (Ou and Kwok 2004).

A study conducted by Mori et al. (1999) on oral cancer in rats revealed that ferulic acid significantly prevents oral cancer due to its chemopreventive activity after feeding rats with ferulic acid at a dose of 0.5 g/kg after exposing them to 4-nitroquinoline-1-oxide in drinking water for five weeks at a dose of 0.02 g/kg. Additionally, Kuenzig et al. (1984) claimed that phenolic compounds are an essential component

Table 5.1 Health-beneficial characteristics of various whole-grain cereals

S. No.	Cereals	Nutraceutical properties	References
1.	Rice	Rice protein has a good amount of lysine and is hypoallergenic. As a result, it might serve as a viable element for baby food formulations while adding variety to the restricted diets of children with food allergies. The amino acid profile of rice protein is superior compared to soy protein isolate and casein in fulfilling the amino acid requirements for 2–5-year-old children.	Burks and Helm (1994); Gurpreet and Sogi (2007); Wang et al. (1999).
2.	Wheat	Whole-wheat and wheat bran-based cereals are a significant source of dietary antioxidants. Free and esterified phenolic acids in wheat have the greatest potential to be beneficial to health.	Baublis et al. (2000).
3.	Maize	Maize consists of many nutrients such as dietary fibre, folate, thiamin (Vitamin-B1), pantothenic acid (Vitamin-B5), vitamin C, manganese and phosphorous. A natural carotenoid pigment found in maize called beta-cryptoxanthin has been shown to lower the incidence of lung cancer by 27% when consumed regularly.	Chaturvedi et al. (2011).
4.	Barley	β-glucan is the major component in barley and is a significant source of soluble fibre and has been linked to reducing the risk of hypoglycaemia, hypercholesterolemia and a decreased risk of chemically induced colon cancer in experimental animals. Barley has also been investigated for lowering adipose tissue in mice fed a high-fat diet.	Bhatty (1993); Jan et al. (2021).
5.	Oat	Oats have recently drawn attention mostly due to their high β-glucan content and compounds with antioxidant capacity. Oat β-glucan consumption has been suggested to be beneficial for healthy subjects and patients with type 2 diabetes. At a daily consumption level of 10 g oat β-glucan, the health benefits of β-glucan are linked to enhanced gastrointestinal function, cholesterol reduction and glucose metabolism. Soluble oat fibres in balance with insoluble fibres play pivotal functions in the digestive tract and are crucial for maintaining a healthy colon wall and colonic environment because they are available for fermentation by "good bacteria" (probiotic microbes), and as a result, they have prebiotic effects.	Malkki et al. (2004); Jan et al. (2021). Malkki and Virtanen (2001); Macfarlane, Macfarlane and Cummings (2006).
6.	Millet	Various health benefits associated with regular intake of millet foods include hypocholesterolemic, hypoglycaemic and anti-ulcerative activities.	McDonough, Rooney and Earp (1986).
7.	Buckwheat	Buckwheat is regarded as an important source of protein, and its amino acid composition is nutritionally better than cereal grains. The significant levels of catechins, rutin and other polyphenols as well as their potential antioxidant capacity are also of significance to the dietary value. Buckwheat grains are used in the prevention of diabetes and obesity because they are a significant source of soluble and total dietary fibre.	Pomeranz and Robbins (1972); Oomah and Mazza (1996); Wanatabe (1998); Brennan (2005).
8.	Brown rice	Brown rice is loaded with health-promoting components particularly dietary fibre and bioactive substances such as polyphenols and γ-aminobutyric acid, which can help in combating various cardiovascular diseases including type 2 diabetes and hypercholesterolemia.	Ye et al. (2016).
9.	Psyllium	The plant-based laxative psyllium, which is present in Metamucil, promotes regularity and has antidiarrheal properties, and lowers the risk of heart disease and obesity. Psyllium seed improves the water content and weight of the stool due to its high-fibre content.	Winston (2000).

of the body's protective mechanism for stopping the N-nitroso forms of caffeine and ferulic acid. Diferulic acids may have a significant effect on dietary fibre by altering the chemical structure of dietary fibre components (Bunzel et al. 2001). Studies have revealed that intestinal enzymes can liberate esterified diferulates from cereals and that free diferulic acids can be absorbed and enter the circulatory system (Andreasan et al. 2001). Furthermore, bound antioxidative phenolic acids in cereals may protect the colon after being released by microbial fermentation (Shahidi and Chandrasekara 2013). Additionally, various research showed that ferulic acids are less effective antioxidants than diferulic acids (Andreasen et al. 2001). Researchers have revealed that p-coumaric acids have the potential to scavenge free radicals and are believed to have an antioxidant effect. In addition, it has been shown that coumaric acid possesses anticancer properties against malignant tumours in humans.

5.3.2 Flavonoids

Clinical studies have suggested that flavonoids, a class of bioactive compounds present in fruit, vegetables and cereal grains, may be used to treat a variety of diseases, including coronary heart disease and cancer, when foods containing them are consumed (Gani et al. 2012).

5.3.3 Avenanthramides

In general, oats and oat products are regarded as healthy, and eating oat bran is claimed to reduce LDL cholesterol because of the presence of essential phenolic components called AVAs (Singh and Belkheir 2013; Liu et al. 2004). AVAs prevent free radicals from harming LDL cholesterol (Singh, De and Belkheir 2013), and in vitro studies revealed that an oat extract rich in AVAs mixed with vitamin C reduces LDL oxidation synergistically (Meydani 2009). Oat antioxidants have been shown to reduce cardiovascular risks by lowering blood cholesterol and decreasing LDL cholesterol oxidation and peroxidation in both human clinical trials and animal research (Inglett and Chen 2012). Consuming oats and oat products are therefore advocated as being essential for reducing the risk of cardiovascular disease (Chen et al. 2004; Singh, De and Belkheir 2013). In a study, consuming oat and oat bran may lessen the incidence of colon cancer owing to its high-fibre content as well as AVAs (Guo et al. 2010). Rats fed a meal enriched with oat extract containing AVAs at a dose of 100 mg/kg were shown to have higher superoxide dismutase activity in their skeletal muscles, liver and kidneys, as well as higher glutathione peroxidase activity in their hearts and skeletal muscles (Meydani 2009). According to Liu et al. (2004), AVAs present in oats offer another possible defence mechanism by which oat consumption may help to lower the risk of atherosclerosis by suppressing the growth of vascular smooth muscle cells.

5.3.4 Alkylresorcinols

Alkylresorcinols found in cereal grains may have biological functions that are yet to be fully understood. Due to the fact that they are only found in the bran layer, it is believed that they are defence molecules protecting cereal seeds (especially during their germination) against parasites like fungus and other microbes (Miche et al. 2003; Deszcz and Kozubek 2000). Alkylresorcinol consumption is advantageous because it lowers cholesterol absorption, controls triacylglycerol metabolism, and affects levels of lipid-soluble vitamins (Kulawinek and Kozubek 2007). Alkylresorcinols seem to be suitable biomarkers of health-promoting whole-grain foods because they are only identified in the bran layer and are absorbed and metabolised by humans (Ross et al. 2003a, b) and can be determined in the native form in blood serum and in urine to evaluate their metabolites (Ross, Aman, and Kamal-Eldin 2004a). Alkylresorcinols can provide insights into variations in the absorption of various grains in the digestive tract of humans because of having different structures from cereal species (Ross et al. 2003c).

Recently, the influence of alkylresorcinols on human health has been investigated. Their low levels in food products result in poorly discernible physiological effects. Alkylresorcinols have a phenolic ring due to which they have antioxidant properties. However, in vitro studies revealed that their antioxidant potential was significantly lower than that of other bioactive substances found in cereal grains, such as tocols (Ross, Kamal-Eldin and Aman 2004b). Alkylresorcinols also exhibit antifungal and antibacterial properties (Kulawinek and Kozubek 2007; Ross, Kamal-Eldin and Aman 2004b). Alkylresorcinols readily react with proteins, especially enzymes, because they include hydrophobic alkyl chains, preventing their catalytic activity. Alkylresorcinols significantly reduce the activities of human digestive enzymes (α-glucosidases, proteases and aldose reductases) even at low concentrations (such as 900 mM), which decreases the absorption of some nutrients (Toyomizu et al. 1993; Ross, Kamal-Eldin and Aman 2004b; Kozubek and Tyman 1999). Alkylresorcinols prevent the oxidation of phospholipids and fatty acids in liposomal membranes, which is triggered by the peroxidation of triacylglycerols and fatty acids by Fe2+ ions (Kozubek and Tyman 1999).

5.3.5 Carotenoids

Carotenoids are one of the phytochemicals found in cereals and are present in varying levels in cereal grains (Liu 2007; Ndolo and Beta 2013; Mellado-Ortega and Hornero-Mendez 2015). Despite being found in relatively small amounts in cereals, carotenoids play a pivotal role in the human diet because of widespread cereal consumption as a staple food (Hussain et al. 2015). The carotenoids lutein and zeaxanthin are particularly abundant in cereals (Zhou et al. 2005; Moore et al. 2005; Leenhardt et al. 2006; Konopka, Czaplicki and Rotkiewicz 2006), and prior research has shown that wheat can provide 40% of the daily requirement of lutein (Hussain et al. 2015). Both of the aforementioned substances have been shown to be beneficial to human health, as they lower the risk of cardiovascular illnesses and increase the health of the skin and eyes (Granado, Olmedilla and Blanco 2003; Ma and Lin 2010).

5.3.6 Gamma-oryzanol

It has been shown that γ-oryzanol exhibits antioxidant activity both *in vivo* and *in vitro* (Suh et al. 2005; Juliano et al. 2005). It is related to lowering plasma cholesterol

(Yoshino et al. 1989) and reduces serum cholesterol. It is also linked to reducing platelet aggregation (Seetharamaiah, Krishnakantha and Chandrasekhara 1990) and reducing cholesterol absorption (Gerhardt and Gallo 1998). γ-Oryzanol has also been employed to treat menopause-related illnesses (Murase and Iishima 1963) and hyperlipidemia (Nakayama et al. 1987), and to improve muscular growth (Bonner 1990).

5.3.7 Beta-glucan

Dietary fibres are essential constituents of a healthy diet. The majority of dietary fibre components, including b-glucans, increase intestinal peristalsis and specifically encourage the formation of potentially healthy colonic microbiota. Their calorie content is decreased by using them as fat substitutes in the diet (Hallfrisch and Behall 1997). Similar to arabinoxylans, large intestine bacteria can also ferment b-glucans, which can encourage the production of short-chain fatty acids. The high water-binding ability of b-glucans helps to reduce the incidence of colorectal cancer by accelerating the removal of detrimental carcinogenic chemicals from colons (Bhatty 1993; Klopfenstein 1988). B-glucans have been found to lower the risk of type 2 diabetes and obesity. Additionally, they decrease the risk of coronary heart disease by lowering blood pressure, lipid and LDL cholesterol levels in blood serum, and enhancing insulin sensitivity, which makes it possible to control blood glucose levels (McKeown et al. 2002; Klopfenstein 1988; Liu 2002). It is claimed that a 1% reduction in blood serum cholesterol levels results in a 2% reduction in the risk of coronary heart disease. B-glucans and other soluble dietary fibre fractions reduce blood serum cholesterol levels by 10–20%, which presumably lowers the risk of heart disease by 20–40%.

5.3.8 Tocopherols and Tocotrienols

Recent reports have shown a range of protective effects for tocotrienols. For example, by preventing cholesterol biosynthesis, they may have a beneficial effect by reducing LDL cholesterol (Theriault et al. 1999; Qureshi et al. 1995). Tocols have also demonstrated a beneficial impact on coronary artery disorders in the last decade (Qureshi et al. 1995). Additionally, studies have demonstrated that the high consumption of α- tocopherols reduces platelet aggregation and lipid peroxidation, and acts as an effective anti-inflammatory agent (Jialal and Devaraj 2005).

5.3.9 Phytosterols

The most prevalent sterols in cereal grains are stigmasterol, campesterol, sitosterol and brassicasterol. Phytosterols reduce the absorption of cholesterol due to the similarity in their structures. It was studied that phytosterol molecules interact with intestinal cells to prevent LDL cholesterol from being absorbed which is primarily released in the faeces. In the intestines, only 0.1–5% of phytosterols are only minimally absorbed, and the vast majority of them are ejected. Epidemiological evidence on representative volunteer groups found that consuming 1–3 g of plant sterols daily reduced blood serum levels of LDL cholesterol by 10% on average while having no effect on levels of "good" high-density lipoprotein cholesterol or triacylglycerols (Lagarda, Garcia-Llatas and Farre 2006; Katan et al. 2003).

5.4 RECENT ADVANCEMENTS IN THE DEVELOPMENT OF FUNCTIONAL CEREALS

The food industry is working to develop new products in the rapidly expanding sector of functional foods, mostly to meet customer demand for healthier eating options. For this reason, cereals like barley, oat and wheat as well as their components provide countless potential for the development of functional foods. These cereals can be used to make a variety of functional food which include;

The baking sector is focusing its research and development (R&D) initiatives on functional foods and functional food constituents in response to the continuously growing demand for healthier foods. Cereals present a wonderful solution for the development of different functional foods including cookies, buns, breakfast cereals, bread and snack bars. Cereals can act as fermentation substrates for probiotic microbial growth (Charalampopoulos et al. 2002). Cereals contain a variety of water-soluble fibres that can act as prebiotic foods, including resistant starch, beta-glucans, oligosaccharides and pentosans. According to the knowledge of the distribution of chemical components in grains, it is possible to separate grain fractions rich in these chemical components using a variety of processing techniques, such as milling, sieving, air classification, debranning or pearling. In cereal grains, some of the constituents, particularly starch (and its derivatives, such as modified starches, dextrins, etc.), can be employed as encapsulation substances for probiotics in order to increase their viability during passage through the deleterious conditions of the human gastrointestinal tract and to enhance their stability during handling, distribution and storage. Oat-based baked goods not only have a greater satiety value but also have been demonstrated to reduce blood cholesterol levels in human subjects (Berti et al. 2005; Frank et al. 2004). Compared to oat bread, pan bread with additional transglutaminase and gluten has been described as having a softer, springier texture. The texture and microstructure that were assessed instrumentally correlated well with these sensory and textural qualities (Salmenkallio-Marttila et al. 2004).

The husk of psyllium is a good source of soluble fibre and contains more than eight times the soluble fibre of the oat bran, but because of its capacity to gel and absorb water, it is difficult to incorporate the necessary levels of psyllium into a single serving of a food product. Researchers have shown that psyllium dramatically reduces both total cholesterol and LDL in people (Jones et al. 2005). It has been claimed that psyllium should be taken with cereal foods rather than consuming it alone in order to have the most serum cholesterol-lowering effect (Jenkins et al. 2005). The textural and sensorial qualities of Arabic flat bread (pita bread) were improved in the lab and during pilot scale production in a commercial bakery when psyllium (up to a 5 %level, on a dry basis) was added to straight-grade flour (WF) and whole-wheat flour (WWF) (Sidhu et al. 2005). As a result, adding psyllium husk to WF or WWF produces high-fibre pita bread that is not only lighter in colour and softer in texture but also more acceptable to consumers. Because of this reason, psyllium is used in a variety of cereal-based products like breakfast cereals, cookies, pasta, cakes, bread and so on (Sidhu, Kabir and Huffman 2007). Ahluwalia, Kaur and Sidhu (1995) have incorporated psyllium at different levels to develop cookies, cake and bread. They noticed that bread had a lower specific volume and higher water absorption ability. The testers did not notice any variations in the sensory qualities between samples of baked goods containing psyllium and control. It has been investigated that the preparation of sourdough-based baked products improved nutritional content as well as palatability. According to Lopez et al. (2001), the usage of sourdough resulted in increased phytate breakdown (almost 90%), which also enhanced phosphorus and magnesium solubility through acidification by the sourdough process. It has also been demonstrated that whole-wheat bread with citric acid and phytase supplements increased iron dialysability and phytate-phosphorus release (Porres et al. 2001).

Barley is another essential cereal grain that can be employed in the preparation of functional baked products. It has been demonstrated that barley bread produced with different levels of whole-wheat flour reduces LDL cholesterol, improves glucose tolerance and enhances sensory attributes (Sidhu, Kabir and Huffman 2007). A range of functional foods including pan bread, flatbread, breakfast cereals, extruded snacks and cookies can be prepared with hull-less barley. Consuming whole-grain oat cereal has been proven to lower LDL and total cholesterol, improve fasting blood glucose levels, and lessen the need for antihypertensive medication (Keenan et al. 2002).

5.5 CONCLUDING REMARKS AND FUTURE DIRECTIONS

Cereal and cereal-based products continue to be an essential part of diets throughout the world. They significantly contribute to the consumption of sodium, selenium, magnesium, zinc, vitamin E, several B vitamins, carbohydrates, protein and fibre. However, it appears that their contribution to health promotion extends beyond merely supplying nutrients; there is strong evidence to suggest that regularly eating cereal products, particularly whole grains, may contribute to the reduction of chronic disorders like diabetes, cardiovascular disorders and colon cancer. These health-promoting functions provided by whole-grain cereals are because of the presence of bioactive components in the whole-grain cereals, which are mostly found in the germ and bran fractions. Further research is required to isolate and characterise these bioactive components that improve health. Since most of these components are bound to the matrix of the grain, extracting them might be difficult. Additionally, the identification of these components becomes more challenging due to the lack of adequate standards. Identification and quantification of cereal bioactive components will help us to select grains with higher concentrations of these health-promoting components. Investigation into their bioavailability, metabolism and impact on human health is also necessary. Compared to refined cereals, whole-grain cereals have a significantly greater variety of chemicals with possible antioxidant effects which include carotenoids, tocols, alkylresorcinols, phytic acid and lignans. However, because a considerable amount of antioxidant chemicals are strongly attached to the fibre fraction of the outer layers of grain, the extraction solvents do not entirely release all antioxidant compounds. As a result, the antioxidant capacity of cereal and cereal products has likely been underestimated. The polyphenolic concentration of cereals (mainly ferulic acid) often correlates with their in vitro antioxidant potential. Investigations on correlations with different kinds of chemicals are still continuing. Reports on the antioxidant capacity of cereal polyphenols are frequently conducted on human plasma, rather than directly in vivo, to examine their ability to prevent LDL oxidation. In order to evaluate the antioxidant effect of whole-grain cereals, a total of 11 investigations have explored the antioxidant activities through postprandial or intervention studies in human subjects. The mechanisms involved are intricate, thus more knowledge of the mechanisms involved is required to generate strong, persuading arguments for people to consume more whole-grain cereal products, provide better knowledge about their health advantages and create future health claims. There are no studies dealing with the influx of whole-grain cereal consumption affecting the induction/repression of genes coding for antioxidant chemicals. More research is required to investigate this new field of study employing the most recent genomic and transcriptomic techniques. Another recent worldwide approach i.e., metabolomics, can also be used to examine the effects of cereal consumption. This new approach will allow us to explore further how complex antioxidant-rich foods like cereal and cereal products can modify general metabolism, and which metabolic pathways are affected by antioxidants. This will offer new information on the advantages of whole-grain cereals for health.

REFERENCES

Abdel-Aal, E. M., J. C. Young, I. Rabalski, P. Hucl, and J. Fregeau-Reid. 2007. "Identification and quantification of seed carotenoids in selected wheat species." *Journal of Cereal Science* 55: 787–794.

Adom, K. K., and R. H. Liu. 2002. "Antioxidant activity of grains." *Journal of Agricultural and Food Chemistry* 50(21): 6182–6187.

Ahluwalia, P., A. Kaur, and J. S. Sidhu. 1995. "Effect of psyllium Mucilloid husk as a source of fiber on baking properties of the flour and acceptability of the baked products." *Chem Mikro Technology Lebensm* 17(3/4): 118–122.

Anderson, J. W. 2003. "Whole grains protect against atherosclerotic cardiovascular disease." *Proceedings of the Nutrition Society* 62(1): 135–142.

Andreasan, M. F., P. A. Kroon, G. Williamson, and M. T. Garcia-Conesa. 2001. "Intestinal release and uptake of phenolic antioxidant diferulic acids." *Free Radical Biology and Medicine* 31(3): 304–314.

Awika, J. M., and L. W. Rooney. 2004. "Sorghum phytochemicals and their potential impact on human health." *Phytochemistry* 65(9): 1199–1221.

Bartlomiej, S., R. K. Justyna, and N. Ewa. 2011. "Bioactive compounds in cereal grains – Occurrence, structure, technological significance and nutritional benefits – A review." *Food Science and Technology International* 18(6): 559–568.

Baublis, A. J., C. Lu, F. M. Clydesdale, and E. A. Decker. 2000. "Potential of wheat-based breakfast cereals as a source of dietary antioxidants." *Journal of the American College of Nutrition* 19(3): 308–311.

Behall, K. M., D. J. Scholfield, J. G. Hallfrisch, and H. G. Liljeberg-Elmstahl. 2006. "Consumption of both resistant starch and beta-glucan improves postprandial plasma glucose and insulin in women." *Diabetes Care* 29(5): 976–981.

Belefant-Miller, H., and S. C. Grace. 2010. "Variations in bran carotenoid levels within and between rice subgroups." *Plant Foods for Human Nutrition* 65(4): 358–363.

Belobrajdic, D. P., and A. Bird. 2013. "The potential role of phytochemicals in wholegrain cereals for the prevention of type-2 diabetes." *Nutrition Journal* 12: 62.

Bender, D. A., and A. E. Bender. 1999. *Dictionary of Nutrition and Food Technology*, 7th edition. Woodhead Publishing, Abington.

Berti, C., P. Riso, A. Brusamolino, and M. Porrini. 2005. "Effect of appetite control of minor cereal and pseudocereal products." *British Journal of Nutrition* 94(5): 850–858.

Bhatty, R. S. 1993. "The potential of hull-less barley." *Cereal Chemistry Journal* 76(5): 589–599.

Bondia-Pons, I., A. M. Aura, S. Vuorela, M. Kolehmainen, H. Mykkanen, and K. Poutanen. 2009. "Rye phenolics in nutrition and health." *Journal of Cereal Science* 49(3): 323–336.

Bonner, B., B. Warren, and L. Bucci. 1990. "Influence of ferulate supplementation on post exercise stress hormone levels after repeated exercise stress." *Journal of Applied Sports Science*: 4–10.

Borrelli, G. M., A. M. De Leonardis, C. Platani, and A. Troccoli. 2008. "Distribution along durum wheat kernel of the components involved in semolina colour." *Journal of Cereal Science* 48(2): 494–502.

Brennan, Ch. S. 2005. "Dietary fibre, glycaemic response and diabetes." *Molecular Nutrition and Food Research* 49(6): 560–570.

Britz, S. J., P. V. Prasad, R. A. Moreau, L. H. Allen Jr., D. F. Kremer, and K. J. Boote. 2007. "Influence of growth temperature on the amounts of tocopherols, tocotrienols, and gamma-oryzanol in brown rice." *Journal of Agricultural and Food Chemistry* 55(18): 7559–7565.

Bunzel, M., J. Ralph, J. M. Marita, R. D. Hatfield, and H. Steinhar. 2001. "Diferulates as structural components in soluble and insoluble cereal dietary fibre." *Journal of the Science of Food and Agriculture* 81(7): 653–660.

Burks, A. W., and R. M. Helm. 1994. "Hypoallerginicity of rice protein." In *Presented at the Annual Meeting of the American Association of Cereal Chemists*. Nashville, TN.

Charalampopoulos, D., R. Wang, S. S. Pandiella, and C. Webb. 2002. "Application of cereals and cereal components in functional foods: A review." *International Journal of Food Microbiology* 79(1–2): 131–141.

Chaturvedi, N., P. Sharma, K. Shukla, R. Singh, and S. Yadav. 2011. "Cereals nutraceuticals, health ennoblement and diseases obviation: A comprehensive review." *Journal of Applied Pharmaceutical Science* 1(7): 6–12.

Chen, C. Y., P. E. Milbury, H. K. Kwak, F. W. Collins, P. Samuel, and J. B. Blumberg. 2004. "Avenanthramides and phenolic acids from oats are bioavailable and act synergistically with vitamin C to enhance hamster and human LDL resistance to oxidation." *Journal of Nutrition* 134(6): 1459–1466.

Collins, F. W. 1989. "Oat phenolics: Avenanthramides, novel substituted N-cinnamoylanthranilate alkaloids from oat groats and hulls." *Journal of Agricultural and Food Chemistry* 37(1): 60–66.

Das, A., U. Raychaudhuri, and R. Chakraborty. 2012. "Cereal based functional food of Indian subcontinent: A review." *Journal of Food Science and Technology* 49(6): 665–672.

Deszcz, L., and A. Kozubek. 2000. "Higher cardol homologs (5-n-alkyloresorcinols) in rye seedlings." *Biochimica et Biophysica Acta* 1483(2): 241–250.

Fazary, A. E., and Y. H. Ju. 2007. "Feruloyl esterases as biotechnological tools: Current and future perspectives." *Acta Biochimica et Biophysica Sinica* 39(11): 811–828.

Ferguson, L. R., I. F. Lim, A. E. Pearson, J. Ralph, and P. J. Harris. 2003. "Bacterial antimutagenesis by hydroxycinnamic acids from plant cell walls." *Mutation Research* 542(1–2): 49–58.

Frank, J., B. Sundberg, A. Kamal-Eldin, B. Vessby, and P. Aman. 2004. "Yeast-leavened oat breads with high or low molecular weight beta-glucans do not differ in their effect on blood concentrations of lipids, insulin, or glucose in humans." *Journal of Nutrition* 134(6): 1384–1388.

Frei, M., and K. Becker. 2005. "Fatty acids and all-trans-b-carotene are correlated in differently colored rice landraces." *Journal of the Science of Food and Agriculture* 85(14): 2380–2384.

Gani, A., S. M. Wani, F. A. Masoodi, and G. Hameed. 2012. "Whole-grain cereal bioactive compounds and their health benefits: A review." *Journal of Food Processing and Technology* 3: 3.

Garrait, G., J. F. Jarrige, S. Blanquet, E. Beyssac, J. M. Cardot, and M. Alric. 2006. "Gastrointestinal absorption and urinary excretion of trans-cinnamic and p-coumaric acids in rats." *Journal of Agricultural and Food Chemistry* 54(8): 2944–2950.

Gerhardt, A. L., and N. B. Gallo. 1998. "Full-fat rice bran and oat bran similarly reduce hypercholesterolemia in humans." *Journal of Nutrition* 128(5): 865–869.

Goldberg, G. 2003. *"Plants: Diet and Health." The Report of the British Nutrition Foundation Task Force*. Blackwell, Oxford.

Granado, F., B. Olmedilla, and I. Blanco. 2003. "Nutritional and clinical relevance of lutein in human health." *British Journal of Nutrition* 90(3): 487–502.

Guo, W., L. Nie, D. Wu, M. L. Wise, F. W. Collins, S. N. Meydani, and M. Meydani. 2010. "Avenanthramides inhibit proliferation of human colon cancer cell lines in vitro." *Nutrition and Cancer* 62(8): 1007–1016.

Gurpreet, K. C., and D. S. Sogi. 2007. "Functional properties of rice bran protein concentrates." *Journal of Food Engineering* 79(2): 592–597.

Haas, P., M. J. Machado, A. A. Anton, A. S. Silva, and A. De Francisco. 2009. "Effectiveness of whole grain consumption in the prevention of colorectal cancer: Metaanalysis of cohort studies." *International Journal of Food Sciences and Nutrition* 21: 1–13.

Hallfrisch, J., and K. M. Behall. 1997. "Evaluation of foods and physiological responses to menus in which fat content was lowered by replacement with Oatrim." *Cereal Foods World* 42(2): 100–103.

Hidalgo, A., A. Brandolini, C. Pompei, and R. Piscozzi. 2006. "Carotenoids and tocols of einkorn wheat (*Triticum monococcum* ssp. monococcum L.)." *Journal of Cereal Science* 44(2): 182–193.

Hussain, A., H. Larsson, R. Kuktaite, M. E. Olsson, and E. Johansson. 2015. "Carotenoid content in organically produced wheat: Relevance for human nutritional health on consumption." *International Journal of Environmental Research and Public Health* 12(11): 14068–14083.

Inglett, G. E., and D. Chen. 2012. "Antioxidant and pasting properties of oat β-glucan hydrocolloids." *Food and Nutrition Sciences* 3: 827–835.

Jan, N., T. A. Wani, F. A. Masoodi, A. Gani, and H. R. Naik. 2021. Beta-Glucans. In *Food Biopolymers: Structural, Functional and Nutraceutical Properties*, 93–125.

Jastrebova, J., M. Skoglund, J. Nilson, and L. H. Dimberg. 2006. "Selective and sensitive LC-MS determination of avenanthramides in oats." *Chromatographia* 63(9–10): 419–423.

Jenkins, D. J. A., C. W. C. Kendall, A. Marchie, D. A. Faulkner, J. M. W. Wong, R. de Souza, A. Emam, T. L. Parker, E. Vidgen, E. A. Trautwein, K. G. Lapsley, R. G. Josse, L. A. Leiter, W. Singer, and P. W. Connelly. 2005. "Direct comparison of a dietary portfolio of cholesterol-lowering foods with a statin in hypercholesterolemic participants." *American Journal of Clinical Nutrition* 81(2): 380–387.

Jialal, I., and S. Devaraj. 2005. "Scientific evidence to support a vitamin E and heart disease health claim: Research needs." *Journal of Nutrition* 135(2): 348–353.

Jilek, M. L., and M. Bunzel. 2013. "Dehydrotriferulic and dehydrodiferulic acid profiles of cereal and pseudocereal flours." *Cereal Chemistry Journal* 90(5): 507–514.

Jones, J. M., M. Reicks, J. Adams, G. Fulcher, and L. Marquart. 2004. "Becoming proactive with the whole-grains message." *Nutrition Today* 39(1): 10–17.

Jones, P. J., M. Raeini-Sarjaz, D. J. A. Jenkins, C. W. C. Kendall, E. Vidgen, E. A. Trautwein, K. G. Lapsley, A. Marchie, S. C. Cunnane, and P. W. Connelly. 2005. "Effects of a diet high in plant sterols, vegetable proteins, and viscous fibers (dietary portfolio) on circulating sterol levels and red cell fragility in hypercholesterolemic subjects." *Lipids* 40(2): 169–174.

Juliano, C., M. Cossu, M. C. Alamanni, and L. Piu. 2005. "Antioxidant activity of gamma-oryzanol: Mechanism of action and its effect on oxidative stability of pharmaceutical oils." *International Journal of Pharmaceutics* 299(1–2): 146–154.

Kandil, A., J. Li, T. Vasanthan, and D. C. Bressler. 2005. "Phenolic acids in some cereal grains and their inhibitory effect on starch liquefaction and saccharification." *Journal of Agricultural and Food Chemistry* 60(34): 8444–8449.

Karamac, M., A. Bucinski, R. B. Pegg, and R. Amarowicz. 2005. "Antioxidant and antiradical activity of ferulates." *Czech Journal of Food Sciences* 23(2): 64–68.

Kasum, C. M., D. R. Jr. Jacobs, K. Nicodemus, and A. R. Folsom. 2002. "Dietary risk factors for upper aerodigestive tract cancers." *International Journal of Cancer* 99(2): 267–272.

Katan, M. B., S. M. Grundy, P. Jones, M. Law, T. Miettinen, and R. Paoletti. 2003. "Efficacy and safety of plant stanols and sterols in the management of blond cholesterol levels." *Mayo Clinic Proceedings* 78(8): 965–978.

Keenan, J. M., J. J. Pins, D. Geleva, C. Franzel, P. J. O'Connor, and L. M. Chemey. 2002. "Whole grain oat cereal consumption reduces antihypertensive medication need: A cost analysis." *Preventive Medicine in Managed Care* 3(1): 9–17.

Kim, J. K., S. Y. Lee, S. M. Chu, S. H. Lim, S. C. Suh, Y. T. Lee, H. S. Cho, and H. S. Ha. 2010. "Variation and correlation analysis of flavonoids and carotenoids in Korean pigmented rice (*Oryza sativa* L.) cultivars." *Journal of Agricultural and Food Chemistry* 58(24): 12804–12809.

Klopfenstein, C. F. 1988. "The role of cereal beta-glucans in nutrition and health." *Cereal Foods World* 33(10): 865–869.

Konopka, I., W. Kozirok, and D. Rotkiewicz. 2004. "Lipids and carotenoids of wheat grain and flour and attempt of correlating them with digital image analysis of kernel surface and cross-sections." *Food Research International* 37(5): 429–438.

Konopka, I., S. Czaplicki, and D. Rotkiewicz. 2006. "Differences in content and composition of free lipids and carotenoids in flour of spring and winter wheat cultivated in Poland." *Food Chemistry* 95(2): 290–300.

Kozubek, A., and J. H. P. Tyman. 1999. "Resorcinolic lipids, the natural non-isoprenoid phenolic amphiphiles and their biological activity." *Chemical Reviews* 99(1): 1–25.

Kris-Etherton, P. M., K. D. Hecker, A. Bonanome, S. M. Coval, A. E. Binkoski, K. F. Hilpert, A. E. Griel, and T. D. Etherton. 2002. "Bioactive compounds in foods: Their role in the prevention of cardiovascular disease and cancer." *American Journal of Medicine* 113(Suppl 9B): 71S–88S.

Kroon, P. A., C. B. Faulds, P. Ryden, J. S. Robertson, and G. Williamson. 1997. "Release of covalently bound ferulic acid from fiber in the human colon." *Journal of Agricultural and Food Chemistry* 45(3): 661–667.

Kuenzig, W., J. Chau, E. Norkus, H. Holowaschenko, W. Mergens, and A. H. Conney. 1984. "Caffeic and ferulic acid as blockers of nitrosamine formation." *Carcinogenesis* 5(3): 309–313.

Kulawinek, M., and A. Kozubek. 2007. "5-n-Alkilorezorcynole ziaren zboz i pełnoziarnistych produktow spozywczych jako biomarkery zdrowej zywnosci." *Postepy Biochemii* 53(3): 287–296.

Lagarda, M. J., G. Garcia-Llatas, and R. Farre. 2006. "Analysis of phytosterols in foods." *Journal of Pharmaceutical and Biomedical Analysis* 41(5): 1486–1496.

Lavelli, V., A. Hidalgo, C. Pompei, and A. Brandolini. 2009. "Radical scavenging activity of einkorn (*Triticum monococcum* L. subsp. monococcum) wholemeal flour and its relationship to soluble phenolic and lipophilic antioxidant content." *Journal of Cereal Science* 49(2): 319–321.

Leenhardt, F., B. Lyan, E. Rock, A. Boussard, J. Potus, E. Chanliaud, and C. Remesy. 2006. "Genetic variability of carotenoid concentration, and lipoxygenase and perioxidase activities among cultivated wheat species and bread wheat varieties." *European Journal of Agronomy* 25(2): 170–176.

Liu, L., L. Zubik, F. W. Collins, M. Marko, and M. Meydani. 2004. "The antiatherogenic potential of oat phenolic compounds." *Atherosclerosis* 175(1): 39–49.

Liu, R. H. 2004. "Potential synergy of phytochemicals in cancer prevention: Mechanism of action." *Journal of Nutrition* 134(12): 3479S–3485S.

Liu, R. H. 2007. "Whole grain phytochemicals and health." *Journal of Cereal Science* 46(3): 2017–2219.

Liu, S. 2002. "Intake of refined carbohydrates and whole grain foods in relation to risk of type 2 diabetes mellitus and coronary heart disease." *Journal of the American College of Nutrition* 21(4): 298–306.

Liu, S., J. E. Manson, M. J. Stampfer, F. B. Hu, E. Giovannucci, G. A. Colditz, C. H. Hennekens, and W. C. Willett. 2000. "A prospective study of whole-grain intake and risk of type 2 diabetes mellitus in US women." *American Journal of Public Health* 90(9): 1409–1415.

Lopez, H. W., V. Krespine, C. Guy, A. Messager, C. Demigne, and C. Remesy. 2001. "Prolonged fermentation of whole wheat sour dough reduces phytates level and increases soluble magnesium." *Journal of Agricultural and Food Chemistry* 49(5): 2657–2662.

Ma, L., and X. M. Lin. 2010. "Effects of lutein and zeaxanthin on aspects of eye health." *Journal of the Science of Food and Agriculture* 90(1): 2–12.

Macfarlane, S., G. T. Macfarlane, and J. J. Cummings. 2006. "Review: Prebiotics in the gastrointestinal tract." *Alimentary Pharmacology and Therapeutics* 24: 701–714.

Madhujith, T., M. Izydorczyk, and F. Shahidi. 2006. "Antioxidant properties of pearled barley fractions." *Journal of Agricultural and Food Chemistry* 54(9): 3283–3289.

Malkki, Y., O. Myllymaki, K. Teinila, and S. Koponen. 2004. "Method for preparing an oat product and a foodstuff enriched in the content of beta-glucan." US Patent 6, 797–307.

Malkki, Y., and E. Virtanen. 2001. "Gastrointestinal effects of oat bran and oat gum: A review." *LWT - Food Science and Technology* 34(6): 337–347.

Manach, C., A. Scalbert, C. Morand, C. Remesy, and L. Jimenez. 2004. "Polyphenols: Food sources and bioavailability." *American Journal of Clinical Nutrition* 79(5): 727–747.

Mattila, P., J. M. Pihlava, and J. Hellstrom. 2005. "Contents of phenolic acids, alkyl- and alkenylresorcinols, and avenanthramides in commercial grain products." *Journal of Agricultural and Food Chemistry* 53(21): 8290–8295.

McDonough, C. M., L. W. Rooney, and C. F. Earp. 1986. "Structural characteristics of Eleusine coracana (finger millet) using scanning electron and fluorescence microscopy." *Food Microstructure* 5: 247–256.

McKeown, N. M., J. B. Meigs, S. Liu, P. W. F. Wilson, and P. F. Jacques. 2002. "Whole-grain intake in favorably associated with metabolic risk factors for type 2 diabetes and cardiovascular disease in the Framingham offspring study." *American Journal of Clinical Nutrition* 76(2): 390–398.

McMurough, I., and T. Baert. 1994. "Identification of proanthocyanidins in beer and their direct measurement with a dual electrode electrochemical detector." *Journal of the Institute of Brewing* 100(6): 409–414.

Mellado-Ortega, E., and D. Hornero-Mendez. 2015. "Carotenoids in cereals: An ancient resource with present and future applications." *Phytochemistry Reviews* 14(6): 873–890.

Meydani, M. 2009. "Potential health benefits of avenanthramides of oats." *Nutrition Reviews* 67(12): 731–735.

Meyer, K. A., L. H. Kushi, D. R. J. Jacob, J. Slavin, T. A. Sellers, and A. R. Folsom. 2000. "Carbohydrates, dietary fiber, incident type 2 diabetes mellitus in older women." *American Journal of Clinical Nutrition* 71(4): 921–930.

Miche, L., S. Belki, R. Rozen, and J. Balandreau. 2003. "Rice seedling whole exudates and extracted alkylresorcinols induce stress-response in Escherichia coli biosensors." *Environmental Microbiology* 5(5): 403–411.

Moore, J., Z. Hao, K. Zhou, M. Luther, J. Costa, and L. Yu. 2005. "Carotenoid, tocopherol, phenolic acid and antioxidant properties of Maryland-grown soft wheat." *Journal of Agricultural and Food Chemistry* 53(17): 6649–6657.

Mori, H., K. Kawabata, N. Yoshimi, T. Tanaka, T. Murakami, T. Okada, and H. Murai. 1999. "Chemopreventive effects of ferulic acid on oral and rice germ on large bowel carcinogenesis." *Anticancer Research* 19(5A): 3775–3778.

Moros, E. E., D. Darnoko, M. Cheryan, E. G. Perkins, and J. Jerrell. 2002. "Analysis of xanthophylls in corn by HPLC." *Journal of Agricultural and Food Chemistry* 50(21): 5787–5790.

Murase, Y., and H. Iishima. 1963. "Clinical Studies of oral administration of gammaoryzanol on climacteric complaints and its syndrome." *Obstet Gynecol Prac* 12: 147–149.

Nakayama, S., A. Manabe, J. Suzuki, K. Sakamoto, and T. Inagaki. 1987. "Comparative effects of two forms of gamma-oryzanol in different sterol compositions on hyperlipidemia induced by cholesterol diet in rats." *Japanese Journal of Pharmacology* 44(2): 135–143.

Ndolo, V. U., and T. Beta. 2013. "Distribution of carotenoids in endosperm, germ, and aleurone fractions of cereal grain kernels." *Food Chemistry* 139(1–4): 663–671.

Nystrom, L., A. Paasonen, A. M. Lampi, and V. Piironen. 2007. "Total plant sterol ferulates and steryl glycosides in milling fractions of wheat and rye." *Journal of Cereal Science* 45(1): 106–115.

Okarter, N., and R. H. Liu. 2010. "Health benefits of whole grain phytochemicals." *Critical Reviews in Food Science and Nutrition* 50(3): 193–208.

Oomah, D. B., and G. Mazza. 1996. "Flavonoids and antioxidative activities in buckwheat." *Journal of Agricultural and Food Chemistry* 44(7): 1746–1750.

Ou, S., and K. C. Kwok. 2004. "Ferulic acid: Pharmaceutical functions, preparation and applications in foods." *Journal of the Science of Food and Agriculture* 84(11): 1261–1269.

Panfili, G., A. Fratianni, and M. Distaam. 2004. "Improved normal phase high-performance liquid chromatography procedure for the determination of carotenoids in cereals." *Journal of Agricultural and Food Chemistry* 52(21): 6373–6377.

Peterson, D. M., M. J. Hahn, and C. L. Emmons. 2002. "Oat avenanthramides exhibit antioxidant activities in vitro." *Food Chemistry* 79(4): 473–478.

Piironen, V., J. Toivo, and A. M. Lampi. 2002. "Plant sterols in cereals and cereal products." *Cereal Chemistry Journal* 79(1): 148–154.

Pomeranz, Y., and G. S. Robbins. 1972. "Amino acid composition of buckwheat." *Journal of Agricultural and Food Chemistry* 20(2): 270–274.

Porres, J. M., P. Etcheverry, D. D. Miller, and X. G. Lei. 2001. "Phytase and citric acid supplementation in whole-wheat bread improves phytates-phosphorus release and iron dialyzability." *Journal of Food Science* 66(4): 614–619.

Qureshi, A. A., B. A. Bradlow, L. Brace, J. Manganello, D. M. Peterson, B. C. Pearce, J. J. Wright, A. Gapor, and C. E. Elson. 1995. "Response of hyperchoHlesterolemic subjects to administration of tocotrienols." *Lipids* 30(12): 1171–1177.

Rice-Evans, C. A., N. J. Miller, and G. Paganga. 1997. "Antioxidant properties of phenolic compounds." *Trends in Plant Science* 2(4): 152–159.

Rocha, L. D., M. C. Monteiro, and A. J. Teodoro. 2012. "Anticancer properties of hydroxycinnamic acids – A review." *Cancer and Clinical Oncology* 1(2): 109–121.

Ross, A. B., P. Aman, and A. Kamal-Eldin. 2004a. "Identification of cereal alkylresorcinol metabolites in human urine – Potential biomarkers of wholegrain wheat and rye intake." *Journal of Chromatography. Part B* 809(1): 125–130.

Ross, A. B., A. Kamal-Eldin, and P. Aman. 2004b. "Dietary alkylresorcinols: Absorption, bioactivities, and possible use as biomarkers of whole-grain wheat- and rye-rich products." *Nutrition Reviews* 62(3): 81–95.

Ross, A. B., A. Kamal-Eldin, E. Lundin, J. X. Zhang, G. Hallmans, and P. Aman. 2003a. "Cereal alkylresorcinols are absorbed by humans." *Journal of Nutrition* 133(7): 2222–2224.

Ross, A. B., M. J. Shepard, K. F. Bach Knudsen, L. V. Glitsø, E. Bowey, J. Phillips, I. Rowland, Z. X. Guo, D. J. Massy, P. Aman, and A. Kamal-Eldin. 2003b. "Absorption of dietary alkylresorcinols in ileal-cannulated pigs and rats." *British Journal of Nutrition* 90(4): 787–794.

Ross, A. B., M. J. Shepard, M. Schupphaus, V. Sinclar, B. Alfaro, A. Kamal-Eldin, and P. Aman. 2003c. "Alkylresorcinols in cereal and cereals products." *Journal of Agricultural and Food Chemistry* 51(14): 4111–4118.

Salmenkallio-Marttila, M., K. Roininen, K. Autio, and L. Lahteenmaki. 2004. "Effects of gluten and transglutaminase on microstructure, sensory characteristics and instrumental texture of oat bread." *Agricultural and Food Science* 13(1–2): 138–150.

Seetharamaiah, G. S., T. P. Krishnakantha, and .N. Chandrasekhara. 1990. "Influence of oryzanol on platelet aggregation in rats." *Journal of Nutritional Science and Vitaminology (Tokyo)* 36(3): 291–297.

Shahidi, F., and A. Chandrasekara. 2010. "Hydroxycinnamates and their in vitro and in vivo antioxidant activities." *Phytochemistry Reviews* 9(1): 147–170.

Shahidi, F., and A. Chandrasekara. 2013. "Millet grain phenolics and their role in disease risk reduction and health promotion: A review." *Journal of Functional Foods* 5(2): 570–581.

Sidhu, J. S., Y. Kabir, and F. G. Huffman. 2007. "Functional foods from cereal grains." *International Journal of Food Properties* 10(2): 231–244.

Sidhu, J. S., S. N. Al-Hooti, J. M. Al-Saqer, H. A. Al-Amiri, A. Al-Othman, and E. Al Salem. 2005. "In Use of Psyllium as a Dietary Fiber Source in Arabic Flat (Pita) Bread." Abstract No. 12, Whole Grains & Health: A Global Summit. University of Minnesota, Minneapolis, MN, May 18–20.

Siebenhandl, S., H. Grausgruber, N. Pellegrini, D. D. Rio, V. Fogliano, R. Pernice, and E. Berghofer. 2007. "Phytochemical profile of main antioxidants in different factions of purple and blue wheat, and black barley." *Journal of Agricultural and Food Chemistry* 55(21): 8541–8547.

Singh, R., S. De, and A. Belkheir. 2013. "Avena sativa (oat), a potential neutraceutical and therapeutic agent: An overview." *Critical Reviews in Food Science and Nutrition* 53(2): 26–144.

Slavin, J. 2003. "Why whole grains are protective: Biological mechanisms." *Proceedings of the Nutrition Society* 62(1): 129–134.

Slavin, J. L. 2000. "Mechanisms for the impact of whole grain foods on cancer risk." *Journal of the American College of Nutrition* 19: 300S–307S.

Suh, M. H., S. H. Yoo, P. S. Chang, and H. G. Lee. 2005. "Antioxidative activity of microencapsulated gamma-oryzanol on high cholesterol-fed rats." *Journal of Agricultural and Food Chemistry* 53(25): 9747–9750.

Suzuki, Y., Y. Esami, and I. Yamaguchi. 1999. "Structures of 5-n-alkyloresorcinol-related analogues in rye." *Phytochemistry* 52(2): 281–289.

Tan, J., N. Baisakh, N. Oliva, V. Parkhi, M. Rai, L. Torrizo, K. Datta, and S. K. Datta. 2005. "The screening of rice germplasm, including those transgenic rice lines which accumulate b-carotene in their polished seeds, for their carotenoid profile." *International Journal of Food Science and Technology* 40(5): 563–569.

Tapola, N., H. Karvonen, L. Niskanen, M. Mikola, and E. Sarkkinen. 2005. "Glycemic responses of oat bran products in type 2 diabetic patients." *Nutrition, Metabolism, and Cardiovascular Diseases* 15(4): 255–261.

Teixeria, J., A. Gaspar, E. M. Garrido, J. Garrido, and F. Borges. 2013. "Hydroxycinnamic acid antioxidants: An electrochemical overview." *BioMed Research International*: Article ID 251754: 11 pages.

Theriault, A., J. T. Chao, Q. Wang, A. Gapor, and K. Adeli. 1999. "Tocotrienol: A review of its therapeutic potential." *Clinical Biochemistry* 32(5): 309–319.

Toyomizu, M., S. Sugiyama, R. L. Jin, and T. Nakatsu. 1993. "α-glucosidase and aldose reductase inhibitors: Constituents of cashew, Anacardium occidentale, nut shell liquids." *Phytotherapy Research* 7(3): 252–254.

Venn, B. J., and J. I. Mann. 2004. "Cereal grains, legumes and diabetes." *European Journal of Clinical Nutrition* 58(11): 1443–1461.

Wanatabe, M. 1998. "Catechins as antioxidant from buckwheat (*Fagopyrum esculentum* Moench) groats." *Journal of Agricultural and Food Chemistry* 46(3): 839–845.

Wang, L., Y. Yao, Z. He, D. Wang, A. Liu, and Y. Zhang. 2013. "Determination of phenolic acid concentrations in wheat flours produced at different extraction rates." *Journal of Cereal Science* 57(1): 67–72.

Wang, M., N. S. Hettiarachchy, M. Qi, W. Burks, and T. Siebenmogen. 1999. "Preparation and functional properties of rice bran protein isolate." *Journal of Agricultural and Food Chemistry* 47(2): 411–416.

Wani, S. A., H. R. Naik, J. Wagay, N. A. Ganie, M. Z. Mulla, and B. N. Dar. 2022. "Mentha: A review on its bioactive compounds and potential health benefits." *Quality Assurance and Safety of Crops and Foods* 14(4): 154–168.

Winston, J. C. 2000. "Psyllium: Soluble fiber to the rescue". Vibrant Life, 16(6), 40–41.

Weidner, S., R. Amarowicz, M. Karamac, and E. Frtczeka. 2000. "Changes in endogenous phenolic acids during development of Secale cereale caryopses and after dehydration treatment of unripe rye grains." *Plant Physiology and Biochemistry* 38: 595–602.

Yao, L. H., Y. M. Jiang, J. Shi, F. A. Tomas-Barberan, N. Datta, R. Singanusong, and S. S. Chen. 2004. "Flavonoids in food and their health benefits." *Plant Foods for Human Nutrition* 59(3): 113–122.

Ye, L., C. Wang, S. Wang, S. Zhou, and X. Liu. 2016. "Thermal and rheological properties of brown flour from Indica rice." *Journal of Cereal Science* 70: 270–274.

Yoshino, G., T. Kazumi, M. Amano, M. Tateiwa, T. Yamasakin, S. Takashima, M. Iwai, H. Hatanaka, and S. Baba. 1989. "Effects of gamma-oryzanol and Probucol on hyperlipidemia." *Current Therapeutic Research* 45: 975.

Yu, S., Z. T. Nehus, T. M. Badger, and N. Fang. 2007. "Quantification of vitamin-E and gamma-oryzanol components in rice germ and bran." *Journal of Agricultural and Food Chemistry* 55(18): 7308–7313.

Zarnowski, R., Y. Suzuki, I. Yamaguchi, and S. J. Pietr. 2002. "Alkylresorcinols in barley (*Hordeum vulgare* L. distichon) grains." *Zeitschrift fur Naturforschung* 57c(1–2): 57–62.

Zhou, K., J. J. Yin, and L. Yu. 2005. "Phenolic acid, tocopherol and carotenoid compositions, and antioxidant functions of hard red winter wheat bran." *Journal of Agricultural and Food Chemistry* 53(10): 3916–3922.

Zielinski, H., H. Kozlowska, and B. Lewczuk. 2001. "Bioactive compounds in the cereal grains before and after hydrothermal processing." *Innovative Food Science and Emerging Technologies* 2(3): 159–169.

CHAPTER 6

Pulses as a Functional Food for Health

Vinita Sharma, Anu Chandel and **Loveleen Sharma***

6.1 INTRODUCTION

Pulses are dry seeds of annual legume plants of the Leguminosae family (or Fabaceae family), as defined by the Food and Agriculture Organization (FAO). The FAO recognises 11 primary pulses as follows: dry beans (including kidney, mung, navy, pinto, tepary beans, ricebean, azuki, black gram, moth and scarlet runner), dry broad beans (horse, field, broad and bean), pigeon peas, lupins, chickpeas, dry peas, black-eyed peas, lentils, vetch, bambara groundnut and other "minor" pulses (velvet, jack, yam and winged beans).

There are 57 million metric tonnes of pulses produced worldwide, despite the fact that Asia produced only 33 million metric tonnes of pulses in 2004. A regular monsoon season and an increase in government support prices to encourage the growth of pulses could lead India to become one of the world's major producers of pulses, increasing output by 30% of overall production (FAO statistics 2008). The highest source of dietary fibre and complex carbohydrates that contribute to low-GI diets is found in pulses. As leguminous fibres cause hypo-glycosuria because they contain more amylose than amylopectin, pulses help decrease cholesterol and triglycerides. The easy availability of the proteins in pulses is demonstrated by their in vitro digestibility (Derbyshire & Delange 2021). Pulses offer twice as much protein as cereals do, and the proteins from the two types of foods work well together in a daily diet (Curran 2012). In terms of nutrition, pulses are a significant source of proteins that, despite lacking tryptophan and sulfur-containing amino acids, have higher concentrations of lysine, arginine, glutamic acid and aspartic acid amides than cereal grains do. Consuming pulses has lately been linked to preventative or therapeutic effects on chronic health disorders, such as diabetes, cardiovascular disease, obesity, being overweight and cancer, in addition to its obvious nutritional

* Corresponding author E-mail: sharmaloveleen88@gmail.com

DOI: 10.1201/9781003315100-6

benefits (Mohanty & Satyasai 2015; Marinangeli & Jones 2011; Patterson & Maskus 2009; Zhu et al. 2012). Table 6.1 shows the composition of commonly grown pulses in India.

Pulses are excellent sources of bioactive compounds and can be crucial sources of ingredients for use in functional foods and other applications. They also contain a number of bioactive compounds that have metabolic effects on people who regularly consume these foods, and their content and composition change depending on how long the seeds are stored and how long they take to germinate (Randhir, Lin & Shetty 2004).

Additionally, pulse grains are a source of several naturally occurring bioactive compounds, including enzyme inhibitors, lectins, oxalates, phytic acid, oligosaccharides and phenolic compounds. These compounds are vital for the metabolism of people and other animals who regularly consume these amazing seeds (Champ 2002). Scientific research has demonstrated that eating pulses three or four times per week can help avoid coronary heart disease (CHD) and enhance human well-being (Flight & Clifton 2006). Pulses are a rich source of nutrients, particularly iron-rich beans and lentils and zinc (Vadivel & Janardhanan 2000). The major polyphenolic compounds of pulses are tannins, flavonoids and phenolic acids. Dark grains such as black gram, lentil and red kidney beans are highly rich in phenolic content. These are organic compounds that typically have high molecular weights and at least two phenolic groups linked together in complicated structures. These substances are byproducts of plants' secondary metabolism, and free radical scavenging tests using 2,2-diphenyl-1-picrylhydrazyl (DPPH) demonstrate that they have good antioxidant potential (Awika & Duodu 2017; Xu et al. 2007).

The body gets the appropriate quantities of vitamins, lipids, carbohydrates and proteins from functional food to ensure a healthy life. Nutraceuticals are functional foods that assist in the prevention and treatment of illnesses or disorders other than anaemia. Nutrients based on pulses have been found to be great sources of bioactive and useful foods. Like many food groups, pulses represent a diverse set of foods.

6.2 BIOACTIVE COMPOUNDS IN PULSES

6.2.1 Phytochemicals

Phytochemicals are a naturally occurring group of chemicals in plants and plant-derived foods that differ substantially in their biochemistry, source distribution and physiological effects (Tor-Roca et al. 2020). The vital nutrients and phytochemicals found in legumes, such as natural antioxidants, that can improve human health are abundant. Figure 6.1 shows the bioactive compounds in pulses.

6.2.1.1 Phenolic Compounds

All higher plants contain phenolic components, which have been classified as bioactive secondary metabolites primarily produced by the phenylpropanoid, pentose

Table 6.1 Composition of commonly grown pulses in India

Sl. no.	Pulse	Common name/other names	Carbohydrate (per 100g)	Protein (per 100g)	Fat (per 100g)	Ash (per 100g)	Energy (per 100g) kcal KJ	Dietary fibre		
								Soluble	Insoluble	Total
1.	*Green gram, dal (Phaseolus aureus)*		52.59±0.45	23.88±0.61	1.35±0.20	3.04±0.03	1363±10	1.62±0.19	7.75±0.39	9.37±0.38
2.	Bengal gram (Cicer arietinum)	Chickpea Chana	46.72±1.29	21.55±1.45	5.31±0.06	2.10±0.10	1377±10	2.48±0.15	12.67±0.22	15.15±0.17
3.	Moth (Vigna aconitifolius)	Tepary bean	52.09±0.96	19.75±0.38	1.76±0.09	3.14±0.18	1291±16	0.62±0.10	14.50±0.44	15.12±0.49
4.	Horsegram (Dolichos biflorus)	Kulthi	57.24±0.50	21.73±0.29	0.62±0.04	3.24±0.11	1379±9	1.66±0.03	6.22±0.03	7.88±0.02
5.	Red gram (Cajanus cajan)	Tur (arhar or pigeon pea)	55.23±0.83	21.70±0.50	1.56±0.03	3.26±0.03	1384±10	2.39±0.15	6.67±0.23	9.06±0.30
6.	Peas (Pisum sativum)	Matar	48.93±0.45	20.43±0.79	1.89±0.08	2.41±0.09	1269±13	2.47±0.17	14.55±0.73	17.01±0.63
7.	Ricebean (Vigna umbellata)		51.26	19.97	0.74	3.54	1265	3.33	10.04	13.37
8.	Black gram (Phaseolus mungo Roxb.)	Urad	51.00±0.80	23.06±0.59	1.69±0.12	3.17±0.02	1356±9	4.35±0.15	7.58±0.13	11.93±0.26
9.	Cow pea (Vigna unguiculata)		54.62±0.49	20.36±0.59	1.15±0.06	2.90±0.11	1340±7	2.80±0.05	8.75±0.09	11.54±0.13
10.	Kidney bean (black) (Phaseolus vulgaris)		49.59	19.01	1.62	3.35	1247	2.58	15.16	17.74
11.	Kidney bean (brown) (Phaseolus vulgaris)		48.83±0.59	19.50±0.84	1.68±0.07	3.36±0.19	1245±12	2.62±0.16	14.33±0.19	16.95±0.27
12.	Kidney bean (red) (Phaseolus vulgaris)	Rajmah	48.61±0.65	19.91±1.44	1.77±0.04	3.28±0.21	1252±14	2.70±0.20	13.86±0.43	16.57±0.63
13.	Soyabean (whitel) (Glycine max)		10.16	37.80	19.42	4.52	1579	5.59	17.04	22.63
14.	Lentil dal (Lens culinaris)		52.53±1.05	24.35±1.10	0.75±0.04	2.23±0.13	1349±11	1.83±0.23	8.60±0.42	10.43±0.39

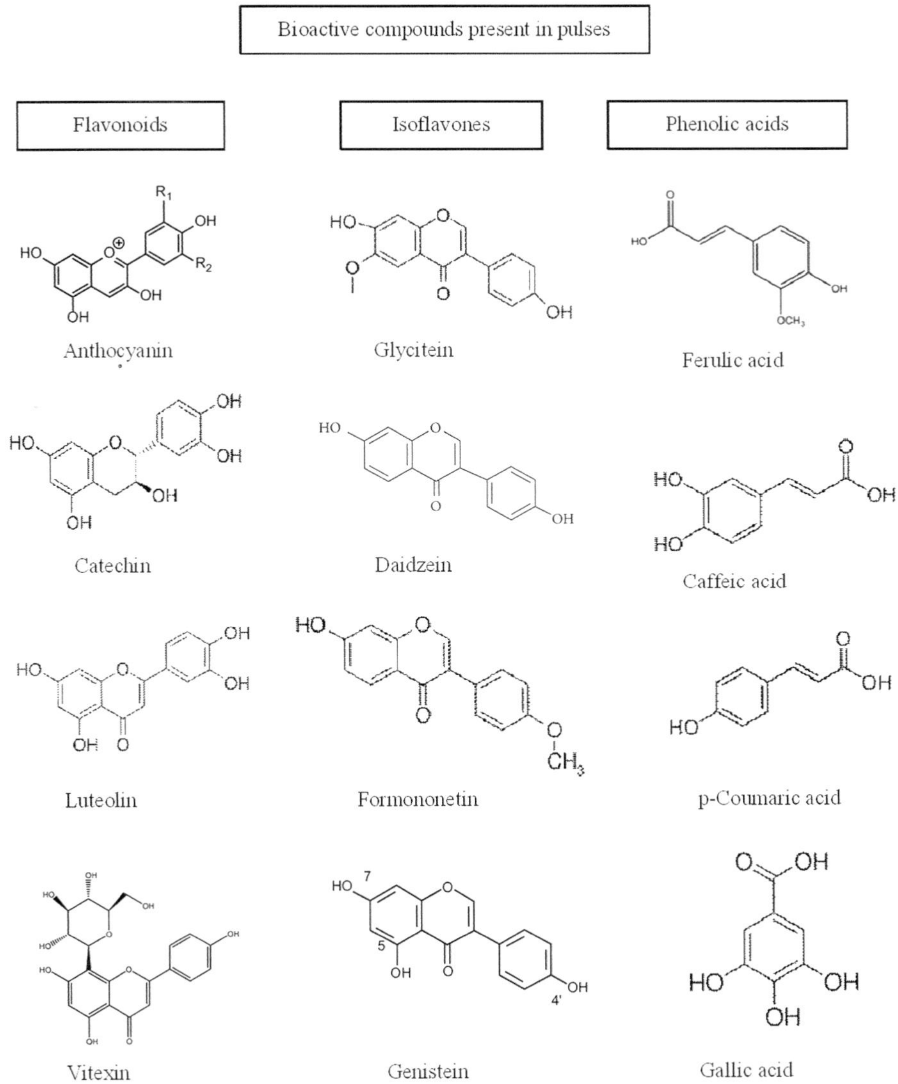

Figure 6.1 Bioactive compounds in pulses

phosphate and shikimic acid pathways (Balasundram et al. 2006). The primary mechanism through which natural phenolics produce positive health effects is their antioxidant activity. Phenolic compounds contribute to the overall antioxidant activities of plant foods. They can be anything from simple molecules to extremely complicated polymers, and structurally they all have one or more hydroxyl groups attached directly to the aromatic ring. (Singh et al. 2017). The synthesis of phenolic compounds takes place from phenylalanine and tyrosine, and phenolic regulates metabolism and lignin synthesis.

Generally, the darkest legume varieties contain higher amounts of phenolic compounds as compared to the light-coloured varieties (Sparvoli, Bollini & Cominelli 2015; Singh, Singh Shevkani, Singh & Haur 2017; Anton, Fulcher & Arntfield 2009; Oomah, Cardadro-Martínez & Loarca-Piña 2005). In general, a large degree of diversity can be found among different varieties of legumes. Various methods were employed for the extraction and quantification of these phenolic compounds due to the high degree of variation in these phenolic compounds (Pedrosa et al. 2015; Amarowicz & Pegg 2008; Oomah, et al., 2005; Amarowicz et al. 2010; Amarowicz & Shahidi 2017; Singh et al., 2017; Yeo & Shahidi 2020). For instance, phenolic compound levels in beans range from 0.3 mg/g in white varieties to 12.6 mg/g in black types; pea and chickpea values range from 0.6 to 2.7 mg/g and 0.6 to 2.7 mg/g, respectively (Pasqualone et al. 2020). Some authors (Amarowicz & Pegg 2008; Singh, Singh, Kaur & Singh 2017) reviewed the phenolic content of various legume seeds and documented contents from 11.2 to 48.3 mg/g in dry beans, from 117.8 to 157.6 mg/g in different faba bean genotypes, from 4.9 to 68 mg/g in lentils and from 0.98 to 183 mg/g in chickpeas.

In a recent study, the simple polyphenols detected in various legumes are mainly phenolic acids and flavonoids (Kalogeropoulos, Chiou, Ioannou, Karathanos, Hassapidou, Nikolaos and Andrikopoulos, 2010). Green split peas and big lentils had levels of simple polyphenols that varied from 321 to 2404 mg/100 g fresh weight, respectively, and lower, in the subsequent order: white bean types, broad beans and split peas come before lentils, chickpeas, pinto beans, lupins and lentils. Secondary metabolites are presently sold as functional foods and nutritional supplements because they are considered antinutrients while also providing health benefits.

6.2.1.2 Flavonoids

Flavonoids are present across the plant kingdom, whereas isoflavonoids are more restricted. Only a small number of papers, meanwhile, have dealt with the identification and measurement of flavonoids in dietary legumes. In the Papilionoideae subfamily of the Leguminosae, isoflavonoids are particularly prevalent. They are also particularly abundant in soybean (including genistein and daidzein) and its products, where they are widely dispersed and serve as preformed or inducible antimicrobial or anti-insect compounds, as inducers of the nodulation genes of symbiotic Rhizobium bacteria or as allelopathic agents (Bouchenak & Lamri-Senhadji 2013). Similar in composition to oestrogens, flavonoids can exhibit both oestrogenic and anti-oestrogenic actions as well as potentially offer protection against heart disease.

Isoflavonoids have a variety of functions in plant-microbe interactions, including defence mechanisms and rhizobia-legume symbiosis. Due to their capacity to trigger the nodulation genes, isoflavones are crucial for nodulation. Another subclass of phenolic phytonutrients is isoflavonoids. Including genistein and daidzein, soybeans are an unusually concentrated source of isoflavones, and soy is the main food supply for these compounds. Soybeans are rich in isoflavones (610–2440 lg/g), lentils do not contain significant amounts of these isoflavones (0.23–0.4 lg/g) while the soy sprout is a potent source of coumestrol, the major coumestan. Chickpea contains

0.04 mg/100 g daidzein, 0.06 mg/100 g genistein, 0.14 mg/100 g formononetin, and 1.7 mg/100 g biochanin A. Soybeans have high levels of daidzein (47 mg/100 g) and genistein (74 mg/100 g) but contain less formononetin and biochanin A compared to chickpeas (0.03 and 0.07 mg/100 g, respectively) (Rochfort & Panozzo 2007).

Isoflavones have drawn a lot of attention because of their ability to prevent CVD, and a number of mechanistic studies show their potential efficacy. Some of these effects derive from the substances' capacity to behave in a manner that is comparable to oestrogens. The TFC of yellow pea varied from 0.18 to 0.32 mg CAE/g; that of green pea from 0.08 to 0.39 mg CAE/g; that of lentil from 0.72 to 2.21 mg CAE/g; that of black soybean from 0.72 to 2.57 mg CAE/g; and that of red kidney beans from 0.85 to 2.93 mg CAE/g.

6.2.1.3 Tannins Content

Due to their capacity to bind and precipitate proteins, condensed tannins are known to have an astringent taste as well as certain anti-nutritional effects. But from the perspective of health, various studies that have been published in the literature describe phenolic compounds as bioactive molecules having antioxidant, antibacterial, anticarcinogenic, immunomodulating, cardioprotective, anti-hypertensive and anti-inflammatory properties.

Lentil, colourful soybean, pea and common bean seeds are all rich in condensed tannins, the main phenolic components in legume seeds (Elias et al. 1979; Vidal-Valverde et al. 1994; Troszynska et al. 1997; Takahata et al. 2001; Troszynska & Ciska 2002; Beninger & Hosfield 2003). Tannins, which are mostly found in the testa, are crucial components of the defence mechanism of seeds that are subject to oxidative damage from a variety of environmental conditions (Troszynska & Ciska 2002).

6.3 NUTRACEUTICAL PROPERTIES AND POTENTIAL HEALTH BENEFITS OF PULSES

The term "nutraceutical" was coined by Stephen DeFelice in 1989. This word has evolved from two words, with "nutrition" and "pharmaceutical," which indicate a food or component of a food that provides health benefits, including the prevention and treatment of disease. Isolated nutrients, nutritional supplements, designer foods, particular diets, herbal items and processed beverages or processed foods might all fall under this category (Prasad & Singh 2015: Carbonaro 2011; Wani et al. 2022). Nutraceuticals and "functional foods" are terms that are frequently used interchangeably. Functional foods are referred to as whole foods or their concentrates, whereas nutraceuticals are used to describe an isolated molecular extract (Carbonaro 2011). Pulses are widely used around the world and have beneficial effects on human health such as preventing cancer and cardiovascular diseases and lowering blood pressure (Figure 6.2). Pulses are also used to prevent gene and cell damage, and to help in controlling diabetes. They are also rich in dietary fibres which prevent heart

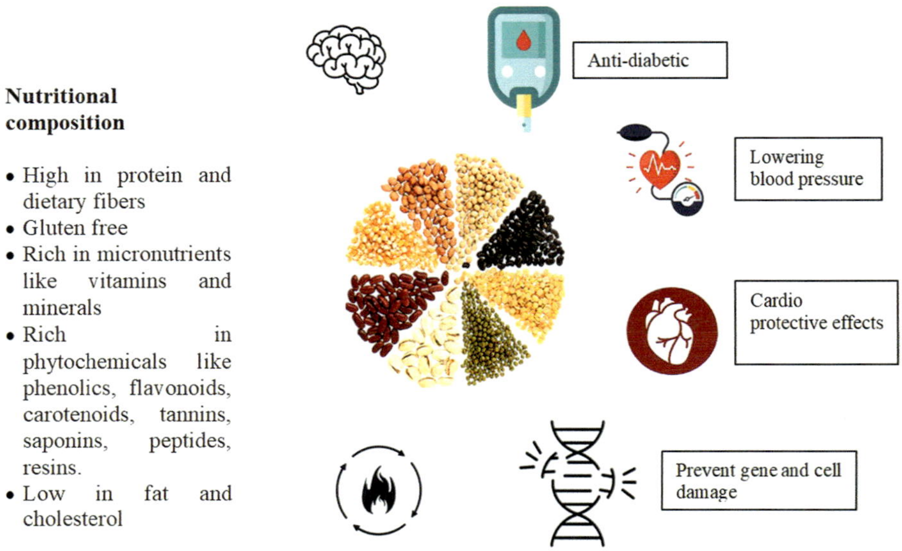

Figure 6.2 Health benefits of pulses

diseases, lower cholesterol and control diabetes (due to their low-glycemic index, and can decrease the risk of colon cancer. Pulses also show that the prebiotic properties are linked with the beneficial effects in improving glycemic control, modulating lipid metabolism, showing anticancer properties, and having hypocholesterolemic effects and other anti-inflammatory effects. Pulses help to prevent cardiovascular disease by lowering blood pressure and reducing serum homocysteine concentration, as they are rich sources of potassium and magnesium. Another factor for reduction of CVD is due to their low methionine content, which reduces serum homocysteine concentration. Table 6.2 shows bioactive compounds present in pulses and their nutraceutical properties.

6.3.1 Health Benefits

6.3.1.1 Cardiovascular Diseases

The leading cause of death in adults is cardiovascular disease (CVD), and concomitant conditions like diabetes raise the likelihood of complications and mortality from CVD. Dietary changes are a crucial part of CVD and diabetes prevention and management strategies that reduce CVD risk. In the last few decades, pulses have drawn attention for their superior nutritional profile as high-fibre, low-glycemic-index foods and have been investigated for their potential to lower the risk of CVD and diabetes (Lukus et al. 2020). The reduction in CVD risk is due to its nutritional components such as fibre, protein, minerals, vitamins, starch, phospholipids, oligosaccharides,

Table 6.2 Bioactive compounds present in pulses and their nutraceutical properties

Sl. no	Pulse name	Components present	Nutraceutical properties	Reference
1.	Soya bean	Isoflavones, bioactive peptides, saponins, protease inhibitor	Antiproliferative, anticarcinogenic, anti-diabetic, anti-atherosclerosis, neuroprotection, anti-fibrosis, cardioprotective effects	Dukariya et al. 2020; Larkin et al. 2008; Asif and Acharya et al. 2013
2.	Red gram or pigeon pea	Alkaloids, Flavonoids Terpenes, Saponins, Tannins	Hypercholesterolaemia, hyperglycaemia, antioxidant, anticancer, anti-inflammatory, cardiac depressant properties	Oke 2014
3.	Bengal gram	Hydroxybenzoics p-Hydroxybenzoic acid, Syringic acid, Gentisic acid, Isoflavones saponins	Antioxidative, inhibits tumour-cell proliferation, anti-inflammatory, hypogycemic	Wang et al . 2021
4.	Horse gram	phytosterols, phenolics, terpenoids, alkaloids, carotenoids, prebiotics, fatty acids,, saponins, and flavonoids, tannins, insoluble and soluble dietary fibres	Reduction of a variety of diseases, including diabetes, obesity and cardiovascular disease	Prasad & Singh 2015
5.	Black gram	Polyphenolic compounds like vanillic, gallic, gentisic, p-coumaric protocatechuic acid, caffeic acid, syringic acid, ferulic acid, vitexin, isovitexin, kaempferol 3-rutinoside and kaempferol 7-rhamnoside	Acts as antioxidant, anticancer agent, helps in diabetes management, useful for inhibition of platelet aggregation, protects against anti-inflammatory responses and heart diseases, reducing obesity risk	Singh & Kaur 2021
6.	Peas	Oligosaccharides, oxalates, flavonoids, phytic acid, lectins, saponins, tannins and phenolic compounds	Main gastrointestinal, metabolic and cardiovascular health Shows anticarcinogenic and hypocholesterolaemic activity	Boye & Ma 2015, Dahl et al. 2012
7.	Kidney beans	Unsaturated fatty acids, peptides, tocopherols, polyphenols, flavonoids	Acts as an antioxidant, shows antihyperglycemic, antimicrobial and anticancer effects Protects from neurodegenerative disorders and cardiovascular diseases	Singh et al. 2021
8.	Green gram	Galactooligosaccharides, oligosaccharides, tannins, saponins	Shows anticancer, anti-inflammatory, antioxidant, hypolipidemic activity activities	Mekkara & Bukkan 2021

antioxidants and isoflavones. All probably contribute to the cardioprotective effects of pulses on glycaemia and blood pressure (Anderson & Major 2022).

6.3.1.2 Diabetes

Diabetes is one of the major health issues around the world. It mainly occurs due to the lack of physical activity, poor diet and excessive weight gain (obesity). Diabetes can be managed by controlling the glycemic index. Pulses like chickpeas, lentils and beans are rich in fibre and also have a low-glycemic-index which helps in decreasing the risk of diabetes (type 2) (Ramdath et al. 2016, Mudryj et al. 2014).

6.3.1.3 Cancer

The United States Food and Drug Administration, United States Food and Drug Administration World Cancer Research Fund (WCRF) and the Canadian Cancer Society recommended the consumption of pulses to reduce the risk of cancer. The high non-digestible fibre content exhibits anti-proliferative activity against colon cancer (Mudryj et al. 2014). Protective components like saponins, lectins, phytates, oligosaccharides, resistant starch and non-starch polysaccharides are provided by pulses, which help to prevent cancer when consumed in adequate amounts (Mathers 2002).

6.3.1.4 Protection against other diseases

Pulses are also widely used for weight management by controlling obesity (Singh et al. 2017).

6.4 RECENT ADVANCEMENTS IN THE DEVELOPMENT OF FUNCTIONAL PULSES

Pulses are being used for developing products having functional properties by utilising the functional bioactive compounds in the pulses. All around the world, many issues like diabetes, cardiovascular disease and obesity have reached epidemic levels. Nowadays, consumers aspire to have healthier food alternatives, which has created a platform for both food industries and food technologists to work and develop food products rich in taste as well as nutritional quality. A lot of research has been conducted that shows that legumes like lentils, chickpeas and dry peas can be used for developing functional foods as they are rich in bioactive compounds (Figure 6.3 shows the process of developing nutraceutical compounds). Many products like pasta, noodles, fortified ready-to-drink beverages, rolls and biscuits, power soup, sauce, nutrition bars and extruded snacks are developed all over the world.

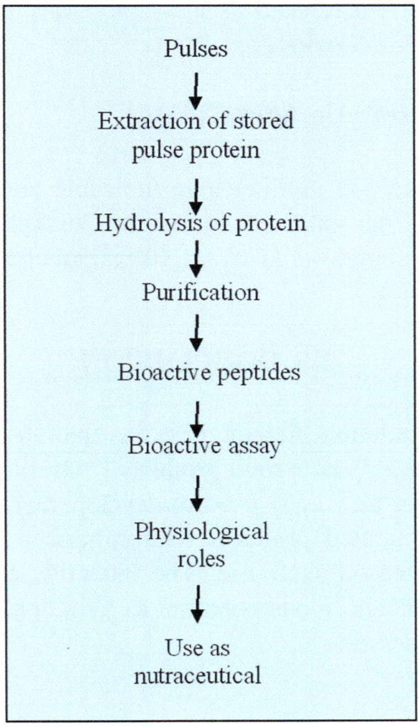

Figure 6.3 Development of nutraceutical compounds (Tak et al., 2021)

6.4.1 Pulse-Based Snacks as Functional Foods

The consumption of pulses has not risen over the past three decades despite its high nutritional content and possible health advantages. To promote pulse consumption, a number of tactics have been used, including its use in baked goods, breakfast cereals and snacks (Escobedo & Mojica 2021).

6.4.2 Pulse-Based Fermented Products

In developing nations, grinding, soaking and fermentation practices have been in use for 100 years. Recently, the fermentation of pulse-based products and their development has gained attention. Fermentation can be defined as the use of microorganisms to develop new products rich in nutrients, with enhanced shelf life, and better taste and flavour. Examples of fermented products are sepu bari, dhokla, tempeh, vadai and wadi (Adebo et al. 2017).

6.4.3 Bioactive Peptides and Hydrolysates as Functional Ingredients

The amino acid sequence derived from the food protein source (e.g. lunasin from soy) is widely used as a functional ingredient to develop nutraceutical products, as pulses are good sources of proteins (20%–40% dry weight basis). Bioactive peptides present in the pulses are widely used in developing various products like chewing gum and toothpaste (López-Barrios et al. 2014).

6.4.4 Use of γ-aminobutyric Acid (GABA)

GABA is a proteinogenic and non-nutritive molecule and shows potential health benefits and a number of physiological functions. Chickpeas, lentils, kidney beans and faba beans are good sources of GABA (Nikmarsm et al. 2017).

6.5 CONCLUSION

With the increasing population and increase in malnutrition and hunger around the world, it's time to develop new food products from different sources. As pulses are rich sources, they can be widely used to develop various functional foods like biscuits, snacks and fermented products with enhanced nutritional value, better flavour, texture and increased shelf life. The bioactive compounds or molecules present in the pulses have enormous potential to fight against various diseases and prevent many serious illnesses.

REFERENCES

Adebo, O. A., Njobeh, P. B., Adebiyi, J. A., Gbashi, S., Phoku, J. Z., & Kayitesi, E. 2017. "Fermented pulse-based food products in developing nations as functional foods and ingredients". In M. C. Hueda (Ed.), *Functional Food—Improve Health through Adequate Food* (pp. 77–109). https://doi.org/10.5772/intechopen.69170.

Amarowicz, R., Troszyńska, A., Baryłko-Pikielna, N. I. N. A., & Shahidi, F. 2004. "Polyphenolics extracts from legume seeds: Correlations between total antioxidant activity, total phenolics content, tannins content and astringency". *Journal of Food Lipids*, *11*(4), 278–286.

Amarowicz, R., Estrella, I., Hernández, T., Robredo, S., Troszyńska, A., Kosińska, A., & Pegg, R. B. 2010. "Free radical-scavenging capacity, antioxidant activity, and phenolic composition of green lentil (*Lens culinaris*)". *Food Chemistry*, *121*(3), 705–711.

Amarowicz, R., & Pegg, R. B. 2008. "Legumes as a source of natural antioxidants". *European Journal of Lipid Science and Technology*, *110*(10), 865–878.

Amarowicz, R., & Shahidi, F. 2017. "Antioxidant activity of broad bean seed extract and its phenolic composition". *Journal of Functional Foods*, *38*, 656–662.

Anderson, J. W., & Major, A. W. 2002. "Pulses and lipaemia, short-and long-term effect: Potential in the prevention of cardiovascular disease". *British Journal of Nutrition*, *88*(S3), 263–271. https://doi.org/10.1079/BJN2002716.

Anton, A., Fulcher, R., & Arntfield, S. 2009. "Physical and nutritional impact of fortification of corn starch based extruded snacks with common bean (Phaseolus vulgaris L.) flour". Effects of bean addition and extrusion cooking. *Food Chemistry*, *133*, 989–996.

Asif, M., & Acharya, M. 2013. "Phytochemicals and nutritional health benefits of soy plant". *International Journal of Nutrition, Pharmacology, Neurological Diseases*, *3*(1), 64. https://doi.org/10.4103/2231-0738.106998.

Awika, J. M., & Duodu, K. G. 2017. "Bioactive polyphenols and peptides in cowpea (Vigna unguiculata) and their health promoting properties: A review". *Journal of Functional Foods*, *38*, 686–697.

Balasundram, N., Sundram, K., & Samman, S. (2006). "Phenolic compounds in plants and agri-industrial by-products: Antioxidant activity, occurrence, and potential uses". *Food Chemistry*, *99*(1), 191–203.

Beninger, C. W., Hosfield, G. L., & Bassett, M. J. (1999). "Flavonoid composition of three genotypes of dry bean (Phaseolus vulgaris) differing in seedcoat color". *Journal of the American Society for Horticultural Science*, *124*(5), 514–518.

Bouchenak, M., & Lamri-Senhadji, M. (2013). "Nutritional quality of legumes, and their role in cardiometabolic risk prevention: A review". *Journal of Medicinal Food*, *16*(3), 185–198.

Boye, J. I., & Ma, Z. 2015. "Impact of processing on bioactive compounds of field peas". In *Processing and Impact on Active Components in Food* (pp. 63–70). Academic Press. https://doi.org/10.1016/B978-0-12-404699-3.00008-1.

Carbonaro, M. 2011. "Role of pulses in nutraceuticals". In *Pulse Foods: Processing, Quality and Nutraceutical Applications* (pp. 385–418). https://doi.org/10.1016/j.foodres.2009.09.003.

Curran, J. 2012. "The nutritional value and health benefits of pulses in relation to obesity, diabetes, heart disease and cancer". *British Journal of Nutrition*, *108*(S1), S1–S2.

Dahl, W. J., Foster, L. M., & Tyler, R. T. 2012. "Review of the health benefits of peas (Pisum sativum L.)". *British Journal of Nutrition*, *108*(S1), S3–S10. https://doi.org/10.1017/S0007114512000852.

Derbyshire, E., & Delange, J. 2021. "The nutritional value of whole pulses and pulse fractions". In B.K. Tiwari, A. Gowen & B. McKenna (Eds.) *Pulse Foods* (pp. 9–29). Academic Press.

Dukariya, G., Shah, S., Singh, G., & Kumar, A. 2020. "Soybean and its products: Nutritional and health benefits". *J Nut Sci Heal Diet*, *1*(2), 22–29.

Elias, L. G., De Fernandez, D. G., & Bressani, R. (1979). "Possible effects of seed coat polyphenolics on the nutritional quality of bean protein". *Journal of Food Science*, *44*(2), 524–527.

Escobedo, A., & Mojica, L. 2021. "Pulse-based snacks as functional foods: Processing challenges and biological potential". *Comprehensive Reviews in Food Science and Food Safety*, *20*(5), 4678–4702. https://doi.org/10.1111/1541-4337.12809.

Flight, I., & Clifton, P. 2006. "Cereal grains and legumes in the prevention of coronary heart disease and stroke: A review of the literature". *European Journal of Clinical Nutrition*, *60*(10), 1145–1159.

Kalogeropoulos, N., Chiou, A., Ioannou, M., Karathanos, V. T., Hassapidou, M., Nikolaos, K., & Andrikopoulos, N. K. 2010. "Nutritional evaluation and bioactive microconstituents (phytosterols, tocopherols, polyphenols, triterpenic acids) in cooked dry legumes usually consumed in the Mediterranean countries". *Food Chemistry*, *121*(3), 682–690.

Larkin, T., Price, W. E., & Astheimer, L. 2008. "The key importance of soy isoflavone bioavailability to understanding health benefits". *Critical Reviews in Food Science and Nutrition*, *48*(6), 538–552. https://doi.org/10.1080/10408390701542716.

López-Barrios, L., Gutiérrez-Uribe, J. A., & Serna-Saldívar, S. O. 2014. "Bioactive peptides and hydrolysates from pulses and their potential use as functional ingredients". *Journal of Food Science*, *79*(3), R273–R283. https://doi.org/10.1111/1750-3841.12365.

Lukus, P. K., Doma, K. M., & Duncan, A. M. 2020. "The role of pulses in cardiovascular disease risk for adults with diabetes". *American Journal of Lifestyle Medicine*, *14*(6), 571–584. https://doi.org/10.1177/1559827620916698.

Mathers, J. C. 2002. "Pulses and carcinogenesis: Potential for the prevention of colon, breast and other cancers". *British Journal of Nutrition, 88*(S3), 273–279. https://doi.org/10.1079/BJN2002717.

Marinangeli, C. P., & Jones, P. J. 2011. "Whole and fractionated yellow pea flours reduce fasting insulin and insulin resistance in hypercholesterolaemic and overweight human subjects". *British journal of Nutrition, 105*(1), 110–117.

Mekkara nikarthil Sudhakaran, S., & Bukkan, D. S. 2021. "A review on nutritional composition, antinutritional components and health benefits of green gram (Vigna radiata (L.) Wilczek)". *Journal of Food Biochemistry, 45*(6), e13743. https://doi.org/10.1111/jfbc.13743.

Mohanty, S., & Satyasai, K. J. 2015. "Feeling the pulse, Indian pulses sector". *NABARD Rural Pulse, 10,* 1–4. https://www.nabard.org/auth/writereaddata/tender/2009164501Rural-Pulse-July-August-2015.pdf.

Mudryj, A. N., Yu, N., & Aukema, H. M. 2014. "Nutritional and health benefits of pulses". *Applied Physiology, Nutrition, and Metabolism, 39*(11), 1197–1204. https://doi.org/10.1139/apnm-2013-0557.

Oke, D. G. 2014. "Proximate and phytochemical analysis of Cajanus cajan (pigeon pea) leaves". *Chemical Science Transactions, 3*(3), 1172–1178. https://doi.org/10.7598/cst2014.785.

Oomah, B. D., Cardadro-Martínez, A., & Loarca-Piña, G. 2005. "Phenolics and antioxidative activities in common beans (Phaseolus vulgaris L.)". *Journal of the Science of Food and Agriculture, 85*(6), 935–942.

Prasad, S. K., & Singh, M. K. 2015. "Horse gram-an underutilized nutraceutical pulse crop: A review". *Journal of Food Science and Technology, 52*(5), 2489–2499.

Pedrosa, M. M., Cuadrado, C., Burbano, C., Muzquiz, M., Cabellos, B., Olmedilla-Alonso, B., & Asensio-Vegas, C. 2015. "Effects of industrial canning on the proximate composition, bioactive compounds contents and nutritional profile of two Spanish common dry beans (Phaseolus vulgaris L.)". *Food Chemistry, 166*, 68–75.

Patterson, C. A., Maskus, H., & Dupasquier, C. 2009. "Pulse crops for health". *Cereal Foods World (CFW), 54*(3), 108.

Pasqualone, A., Costantini, M., Coldea, T. E., & Summo, C. 2020. "Use of legumes in extrusion cooking". *A Review Foods, 9*, 958.

Ramdath, D., Renwick, S., & Duncan, A. M. 2016. "The role of pulses in the dietary management of diabetes". *Canadian Journal of Diabetes, 40*(4), 355–363. https://doi.org/10.1016/j.jcjd.2016.05.015.

Randhir, R., Lin, Y. T., Shetty, K., & Lin, Y. T. 2004. "Phenolics, their antioxidant and antimicrobial activity in dark germinated fenugreek sprouts in response to peptide and phytochemical elicitors". *Asia Pacific Journal of Clinical Nutrition, 13*(3), 295–307.

Rochfort, S., & Panozzo, J. 2007. "Phytochemicals for health, the role of pulses". *Journal of Agricultural and Food Chemistry, 55*(20), 7981–7994.

Sparvoli, F., Bollini, R., & Cominelli, E. 2015. "Nutritional value". In A. M. D. Ron (Ed.), *Grain Legumes* (pp. 291–326). New York: Springer.

Singh, A., Gupta, A., & Sharma, S. 2021. "Kidney beans: Nutritional properties, biofunctional components, and health benefits". In Sneh Punia Bangar, Anil Siroha, Manoj Kumar (Eds.). *Handbook of Cereals, Pulses, Roots, and Tubers* (pp. 357–376). CRC Press.

Singh, B., & Kaur, G. 2021. "Black gram: Bioactive components for human health and their functions". In Sneh Punia Bangar, Anil Siroha, Manoj Kumar (Eds.) *Handbook of Cereals, Pulses, Roots, and Tubers* (pp. 377–392). https://doi.org/10.1201/9781003155508-25.

Singh, B., Singh, J. P., Kaur, A., & Singh, N. 2017. "Phenolic composition and antioxidant potential of grain legume seeds: A review". *Food Research International, 101*, 1–16.

Singh, B., Singh, J. P., Shevkani, K., Singh, N., & Kaur, A. 2017. Bioactive constituents in pulses and their health benefits. *Journal of Food Science and Technology, 54*(4), 858–870. https://doi.org/10.1007/s13197-016-2391-9.

Singh, N. 2017. "Pulses: An overview". *Journal of Food Science and Technology, 54*(4), 853–857. https://doi.org/10.1007/s13197-017-2537-4.

Tak, Y., Kaur, M., Amarowicz, R., Bhatia, S., & Gautam, C. 2021. "Pulse derived bioactive peptides as novel nutraceuticals: A review". *International Journal of Peptide Research and Therapeutics, 27*(3), 2057–2068. https://doi.org/10.1007/s10989-021-10234-8.

Takahata, Y., Ohnishi-Kameyama, M., Furuta, S., Takahashi, M., & Suda, I. (2001). "Highly polymerized procyanidins in brown soybean seed coat with a high radical-scavenging activity". *Journal of Agricultural and Food Chemistry, 49*(12), 5843–5847.

Tor-Roca, A., Mayordomo-Cava, J., Andres-Lacueva, C., Serra-Rexach, J. A., & Urpi-Sarda, M. 2023. "Adherence to mediterranean diet and response to an exercise program to prevent hospitalization-associated disability in older adults: A secondary analysis from a randomized controlled trial". *The Journal of Nutrition, Health & Aging, 27*, 500–506.

Troszynska, A., & Ciska, E. (2002). "Phenolic compounds of seed coats of white and coloured varieties of pea (Pisum sativum L.) and their total antioxidant activity". *Czech Journal of Food Sciences, 20*(1), 15–22.

Troszynska, A., Bednarska, A., Latosz, A., & Kozlowska, H. (1995). "Polyphenolic compounds in the seed coat of legume". *Polish Journal of Food and Nutrition Sciences, 6*(3), 37–45.

Vadivel, V., & Janardhanan, K. 2000. "Chemical composition of the underutilized legume Cassia hirsuta L". *Plant Foods for Human Nutrition, 55*, 369–381.

Vidal-Valverde, C., Frias, J., Estrella, I., Gorospe, M. J., Ruiz, R., & Bacon, J. (1994). "Effect of processing on some antinutritional factors of lentils". *Journal of Agricultural and Food Chemistry, 42*(10), 2291–2295.

Wang, X., Warkentin, T. D., Briggs, C. J., Oomah, B. D., Campbell, C. G., & Woods, S. (1998). "Total phenolics and condensed tannins in field pea (Pisum sativum L.) and grass pea (Lathyrus sativus L.)". *Euphytica, 101*, 97–102.

Wang, J., Li, Y., Li, A., Liu, R. H., Gao, X., Li, D., Kou, X., & Xue, Z. 2021. "Nutritional constituent and health benefits of chickpea (Cicer arietinum L.): A review". *Food Research International, 150*(A), 110790. https://doi.org/10.1016/j.foodres.2021.110790.

Wani, S. A., Naik, H. R., Wagay, J., Ganie, N. A., Mulla, M. Z., & Dar, B. N. 2022. "Mentha: A review on its bioactive compounds and potential health benefits". *Quality Assurance and Safety of Crops and Foods, 14*(4), 154–168.

Xu, B. J., Yuan, S. H., & Chang, S. K. C. 2007. "Comparative analyses of phenolic composition, antioxidant capacity, and color of cool season legumes and other selected food legumes". *Journal of Food Science, 72*(2), S167–S177.

Yeo, J., & Shahidi, F. 2020. "Identification and quantification of soluble and insoluble-bound phenolics in lentil hulls using HPLC-ESIMS/ MS and their antioxidant potential". *Food Chemistry, 315*, 126202.

Zhu, Z., Jiang, W., & Thompson, H. J. 2012. "Edible dry bean consumption (Phaseolus vulgaris L.) modulates cardiovascular risk factors and diet-induced obesity in rats and mice". *British Journal of Nutrition, 108*(S1), S66–S73.

CHAPTER 7

Milk as a Functional Food for Health

Amarjeet Kaur*, **Monika Choudhary**† and **Surbhi Kapoor**‡

7.1 INTRODUCTION

Milk is considered a complete food provided by nature to mankind. It is a unique and complex liquid that is both consumed raw in the form of fluid milk with minimum processing and is a raw material for the preparation of a range of products (Park 2009). Due to its exceptional nutritional properties, the main function of milk is fulfilling the nutritional requirement of neonates (mammalian sp.) in the early phase of life. As the nutritional needs of species vary, the milk secreted by mammary glands of various mammals, therefore, differs in composition to fulfil the needs of that particular species. The conventional characteristic of milk of providing only nutrition to infants has changed with time. It helps the growth of children and is a very good source of nourishment for adults too. (Raikos and Dassios 2014).

Milk and dairy products not only possess good nutritional value but also provide many bioactive components. Various elements present in milk play a diverse but significant part. These are compounds available in foodstuffs that carry physiological activity. Consumption of traditional and functional foods or dietary supplements helps to provide bioactive components of food origin to humans. The interest in the availability of bioactive components in milk and its products is increasing at present. Extensive research is being conducted to understand the health benefits of various food components. Consumers and food manufacturers both have established a better understanding of the association between foodstuffs and their consequences on health. Increased familiarity regarding the functionality of bioactive constituents on human health will lead to the development of new drugs, treatments or food rich in these components in the coming years (Gobbetti and Minervini 2007; Park 2009).

* Corresponding author E-mail: sekhonamarjeet@pau.edu
† Corresponding author E-mail: moni0986@gmail.com
‡ Corresponding author E-mail: surbhikapoor6677@gmail.com

 DOI: 10.1201/9781003315100-7

As today's consumers are very well aware of their health and are quite conscious of making healthier food choices, there is a surge in demand for functional foods globally. Numerous institutions and organisations with a focus on health, including the American Dietetic Association, support the growing need for functional meals in the near future. As was already noted, milk is one of the only naturally occurring, perfectly balanced foods that is also a great source of fat, protein, minerals and vitamins. A particularly good source of calcium that is important for preventing the bone condition osteoporosis is milk. Additionally, it is essential for the growth and development of infants and children (Gulati et al. 2000; Hasler et al. 2004).

As there is a change in the social and eating behaviour of people, the milk available must be of superior value so that it is the preferred choice compared to other dairy products, energy or protein drinks etc. To compete and meet the needs of humans, milk must be designed in a way that results in increased and enhanced properties that fit in the current perspective of people about better and economically viable food choices. In designer or enriched milk, the content is altered from the usual constituents of milk and is designated humanised milk or milk with high therapeutic properties (Singh et al. 2012).

7.2 COMPOSITION OF MILK

Milk is made up of a complex mixture of constituents that include fat, protein (mainly casein), milk sugar (lactose), minerals and water. Its composition is dependent upon many aspects such as the breed or type of mammal it comes from and even on an individual animal. In addition, the lactation stage and feed also play a noteworthy effect on the composition of milk (Walstra et al. 2005).

7.2.1 Lactose

Lactose, in the form of β-D-galactopyranosyl-(1,4)-D-glucopyranose, is the main carbohydrate found in milk. It is a disaccharide made up of d-galactose and d-glucose. In addition to giving the body energy, this disaccharide gives milk its sweet flavour (Walstra et al. 2005). Lactose must first be digested by the enzyme lactase into glucose and galactose in order to be absorbed into the body. Lactose malabsorption occurs when people have insufficient lactose metabolisation as a result of low lactase activity; this condition is eventually considered lactose intolerance.

7.2.2 Lipids

Tocopherols, free fatty acids, phytosterols and other substances that are soluble in non-polar chemical solvents are referred to as lipids. Despite the fact that lard is a complex of triglycerides and solidifies at room temperature, the terms "fat" and "lipids" are sometimes used interchangeably. Lipids in milk are available in the form of fat globules. Milk fat gets emulsified in the aqueous phase and contains approximately 3 to 5% lipids, although this value can vary. About 98% of milk fat is made up of triglycerides (TAG), which are the main form of milk fat. The remaining components are cholesterol, phospholipids, free fatty acids, diglycerides and monoglycerides. (Jensen 2002; Walstra et al. 2005)

7.2.3 Proteins

The primary role of milk protein is to provide newborns with amino acids and nitrogen, but it is also a significant source of protein for adults. The two main types of proteins found in milk are whey and casein. Casein is the primary protein representing 80% of the milk protein. An interaction between casein proteins and calcium phosphate results in the formation of casein micelles (Chandan 2007). These substantial colloidal complexes are spherical in shape. Casein is also divided into fractions such as α_{s1}-, β-, α_{s2}- and -κ fractions (Nagpal et al. 2011). The other protein component whey, which is also known as milk serum, accounts for 20% of milk proteins (Ko 2009). Several enzymes, immunoglobulins, α-lactalbumin, β-lactoglobulin, bovine serum albumin and lactoferrin are all components of milk serum. Whey has a far more organised spherical structure than casein. Whey proteins possess disulphide linkages and become precipitated under heat situations rather than acidic conditions. Proteins and peptides present in milk also show bioactive properties (Chandan 2007).

7.2.4 Vitamins and Minerals

Vitamins that are both fat-soluble and water-soluble are present in milk. Whole milk contains a fatty component that contains vitamin A by nature. Since the low-fat milk fraction contains only minor amounts of vitamin A due to the milk separation process, fortifying milk with vitamin A prior to sale is a common practice. Vitamin D is also present in small amounts in whole milk and hence needs to be fortified. Two principal minerals available from milk are calcium and phosphorus. Calcium is present in the form of calcium caseinate as it gets bound to protein casein, calcium phosphate with phosphorus and also available as calcium citrate. Other minerals found in milk include sulphur, sodium, potassium and magnesium (Chandan 2007; Christian 2008).

7.3 BIOACTIVE COMPONENTS FROM MILK AND MILK PRODUCTS

7.3.1 Bioactive Components from Milk

Bioactive components are those constituents present in minor amounts in foods but have valuable physiological functions along with nutritional properties (Table 7.1). These bioactive compounds are a focus area of research studies, as they may be responsible for the prevention of many ailments and promote good health. Hydrolysis by digestive enzymes in the gastrointestinal tract and during transit can produce bioactive components. These enzymes are naturally found in milk and can add

Table 7.1 Bioactive components of milk

Components	Bioactive compounds	Bioactivities	References
α-,β-caseins	Casomorphins	Opioid against (increase electrolytes uptake and amino acids; decrease gastric emptying rate, gut mobility)	Schanacher et al. 1998; Clare and Swaisgood 2000
	Immunopepides Casomorphins Casokinins	Immunomodulatory (increase phagocytic activity and immune response)	
	Phosphopeptides	Mineral binding (increased absorption of calcium, zinc and phosphorus)	
	Casokinins	ACE-inhibitory (blood flow increases to intestinal epithelium)	
α-casein	Casocidin	Antimicrobial	
	Isracidin	Antimicrobial	
κ-casein	Casoxins	Opioid antagonist	
	Casocidin	Antimicrobial	
	Casoplatelins	Antithrombotic	
	κ-casein glycol-macropeptide	Probiotic (bifidobacterial growth in the gastrointestinal tract)	
α -lactalbumin β-lactogloulin Serum albumin	Lactorphins	ACE-inhibitory	
α -lactalbumin β-lactogloulin	Lactorphins	Opioid agonist	
Serum albumin	Serorphin	Opioid agonist	
Immunoglobulins	IgG, IgA	Antimicrobial (infection and viral attachment to cells decreases and Fe-dependent bacterial inhibition increases), immunomodulatory (passive immunity, killer cell activity, immune response, immunological development of thymocyte trafficking, interleukins-6 increases, while tumour necrosis factor-α decreases) Probiotic activity (bifidobacterial growth in gastrointestinal tract increases)	
Oligosaccharides	Oligosaccharides	Probiotic (bifidobacterial growth in GI tract increases)	
Lactoferrin	Lactoferrroxins	Opioid antagonist	
Prolactin	Prolactin	Immunomodulatory (immune development, thymocyte trafficking and lymphocyte increases)	
Parathromone-P	PTHrP	Uptake and metabolism of calcium increase	
Growth factors	TGF- α, TGF- β, IGF-1, EGF	Organ functions and development	

bioactivity to these products. Additionally, enzymes produced by microbes or plants may release bioactive peptides (Park 2009).

7.3.1.1 Calcium

One of the most prevalent minerals in our bodies, calcium is a necessary element. The body needs calcium for a variety of biological processes to take place, with bone mineralisation being one of the most important. As a result, calcium lost through our skin, nails, hair, urine and faeces needs to be replaced through our diet. The primary sources of calcium in the human diet are milk and its products. Casein is responsible for the majority of the calcium in milk, with smaller quantities going to citrate, phosphorus and other milk proteins. Additionally, a little amount of free or freed calcium exists (Park 2009; Peacock 2010).

An increase in milk consumption by adolescent girls showed enhanced bone mineral acquisition and may also alter levels of peak bone mass. In young women, a high intake of calcium is also related to reduced fat mass (Cadogan et al. 1997; Lin et al. 2000).

7.3.1.2 Conjugated Linoleic Acid (CLA)

A class of polyunsaturated fatty acids known as conjugated linoleic acid (CLA) has 18 carbons and two conjugated double bonds in either the trans or cis position. The cis-9 and trans-11 biological isomers of CLA, which are present in milk and other dairy products, are considered as most active forms. It also goes by the name of rumenic acid. Cis-9 and trans-11 make up between 75 and 90% of the total CLA in milk fat (Aro et al. 2000; Collomb et al. 2006).

Linoleic acid (conjugated) is present in milk, and studies on animals have demonstrated its positive effects on health. In rats, the CLA isomer cis-9, trans-11 has an anti-carcinogenic action. Studies on people also showed that CLA has a lot of advantageous impacts on health (Roche et al. 2001; Corl et al. 2003). There is a substantial association between consuming bovine milk fat and the amount of CLA identified in the subjects' serum, according to a study on the relationship between milk fat and the appearance of favourable effects in human adipose tissue (Dhiman et al. 1999; Jiang et al. 1999).

7.3.1.3 Glycomacropeptide (GMP)

GMP is a portion of the C-terminal f (106- 169) from κ-casein released by the action of the enzyme protease chymosin. It is a tiny molecule and a hydrophilic peptide with an approximate molecular weight of 8000 daltons (Keogh and Clifton 2008; Park 2009). When rennet is applied during the cheese-making process, GMP is released. Chymosin is the primary ingredient in rennet. Through this procedure, glycomacropeptide is kept in the whey fraction and is excluded from cheese curd. GMP has an antibacterial effect in vitro because it prevents bacteria from producing toxins (Brody 2000).

A study conducted on Chinese hamster ovary (CHO)-Kl cells, determined that GMP present in bovine milk prevents cholera toxin binding to its receptor. Similarly, GMP present in human milk helps to promote the growth of bifida bacteria and is considered an important bifida factor. Glycomacropeptide present in human milk is considered to have a fourfold greater bifidobacteria growth-promoting action in comparison to cow's milk. It is also suggested that when GMP from human milk is incubated with pepsin, it has the ability to get further hydrolysed (Azuma et al. 1984; Kawasaki et al. 1992).

7.3.1.4 Immunoglobulin

Antibodies known as immunoglobulins, or Ig, are found in the colostrum and milk of breastfeeding mammals and are created in response to stimulation provided by particular antigens. The primary biological purpose of Ig is to protect newborns from infections and viruses. Therefore, until their own immune systems mature, young mammals rely on colostrum as a passive immune system. Based on their characteristics, structure and biological functions, immunoglobulins are divided into various classes. There are five main categories: IgA, IgD, IgE, IgG and IgM (Mehra et al. 2006; Chandan 2007; Korhonen 2009).

Studies on milk have concluded that milk with Ig can play an active role in the prevention of diseases occurring owing to microbial infections. Taking oral bovine Ig concentrate obtained from colostrum is known to have a protective effect against the pathogen *Shigella* (Tacket et al. 1992).

7.3.1.5 α-lactalbumin

A major milk protein found mostly in human and cow milk is α-lactalbumin. The amino acids methionine, cysteine and tryptophan are found in large concentrations in the protein. The sulphur-containing amino acids give rise to the di-sulfides in the molecule. The mammary gland, where α-La is produced, has been shown to play a crucial physiological role in the biosynthesis of lactose. Together with the enzyme uridine-diphosphate galactosyl transferase, which catalyses the transfer of galactose to glucose to produce lactose as an end product, α-La is a component of lactose synthesis. Additionally, α-La functions as a metal-binding protein, particularly for calcium, and may serve a crucial biological purpose (Hiraoka et al. 1980; Lönnerdal and Glazier 1985).

7.3.1.6 Lactoferrin

Milk and other exocrine secretions contain the iron-binding glycoprotein known as lactoferrin (LF). Its structure is a single-chain polypeptide with two globular lobes. It serves numerous physiological functions in the biological system due to its ability to bind iron. According to the kind of milk, the concentration of lactoferrin and other bioactive ingredients varies significantly. Investigations have also been done on lactoferrin's antimicrobial and anticancer properties. According to studies, lactoferrin's capacity to bind iron and its antibacterial properties are related. When given orally to male rats, bovine lactoferrin may work as a chemo preventer for colon cancer (Lönnerdal and Iyer 1995; Sekine et al. 1997; Korhonen 2009).

7.3.1.7 β-lactoglobulin

A main whey protein present in the milk of different mammalian taxa is β-lactoglobulin. A, B, C and D are its four genetic variations. The cysteine residues in the protein cause it to have a high sulphur content. Bioactive components included in β-lactoglobulin have been studied for their impact on ACE-inhibitory activity and blood cholesterol levels. A study looking at caprine and ovine whey revealed that β-Lg has ACE-inhibitory action. With peptides from β-Lg, more activity was seen following the action of microbial enzymes as opposed to digesting enzymes (Chandan 2007).

When β-Lg was given orally to rats, strong ACE-inhibitory action was also observed. Two peptides were administered for study purposes and both demonstrated a reduction in blood pressure following treatment. However, no ACE-inhibitory activity was seen when the peptide undergoes an imitation of gastrointestinal digestion (Hernández Ledasma et al. 2002)

7.3.1.8 Oligosaccharides

Oligosaccharides are considered to be another chief carbohydrate present in milk after lactose. These contribute substantially to the carbohydrate content of human milk as sugars, accounting for around 1.4% of the total carbohydrate content. These consist of 3–10 units of monosaccharide linked with glycosidic bonds (Mehra and Kelly 2006). The major role of oligosaccharides is to provide protection against pathogens (Kunz et al. 2000). Although physiological protection of these is related to their function at the binding location on the surface of the intestinal epithelium, they function as competitive inhibitors. Goat milk oligosaccharides when administered to rats demonstrated their function in the protection of the intestine and in repairing the damage that occurred due to induced colitis (Gopal and Gill 2000; Lara-Villoslada et al. 2006).

7.3.2 Bioactive Components in Milk Products

7.3.2.1 Cheese and Yoghurt

There are many cheese-making procedures that provide cheese with its unique flavour. Fermentation and ripening, the steps involved in cheese-making, result in the formation of many important bioactive components. In order to satisfy the needs and preferences of consumers, many types of cheese are produced utilising a variety of processing procedures. Lactic acid bacteria can be used as a starter culture to create cheeses of a variety of flavours and tastes that cater to varied health demands (Higurashi et al. 2007).

The effect of Gouda-type cheese consumption on biological markers of metabolic syndrome related to the build-up of abdominal adipose executed on rats resulted in rats that were fed with cheese having an extensively lower mass of adipose tissue. ACE-inhibitory peptides are also reported in ripened cheese with increased activity during the maturation process of cheese. However, this activity declined when proteolysis of the peptides exceeded a certain level (Meisel et al. 1997; Ryhänen et al. 2001)

The term "yoghurt" describes fermented milk made by lactic acid fermentation with *Lactobacillus bulgaricus*, *Streptococcus thermophilus* and *Lactobacillus delbreuckii* present. In the West, bovine milk is most commonly used for preparing yoghurt; however, buffalo, caprine and ovine milk are also used (Donkor et al. 2007).

7.4 POTENTIAL HEALTH BENEFITS OF MILK AND MILK PRODUCTS

There are many claims that dairy product consumption results in an increased risk of chronic problems that include obesity, cardiovascular disease, type 2 diabetes, cancer and osteoporosis. This led to a rise in consumer scepticism over the health effects of consuming dairy products, which in turn caused a dramatic spike in the use of plant-based beverages made from soy, oats, rice or almond. Dairy occupies an important place in the food tradition and culture of Nordic countries, therefore the presence of milk and other dairy products in their diet is quite natural. Taking the increased occurrence of chronic diseases into consideration, it becomes imperative to understand the effects on health of milk and dairy products present in our diets.

7.4.1 Type 2 Diabetes and Obesity

Increased incidences of type 2 diabetes are driven by an increase in obesity cases, so, it is crucial to evaluate the role of milk and its products in controlling body weight. High-quality protein can be found in milk and dairy products. Protein is crucial for weight reduction and weight maintenance since it has a high satiety value. This helps prevent consuming too much calories, thereby reducing body fat reserves (Park et al. 2003; Laaksonen et al. 2010). Dairy proteins include sufficient amounts of the essential amino acids needed for the synthesis of muscle protein, preserving the metabolically active muscle mass while promoting weight loss (Astrup et al. 2015).

Many researchers support the idea that dairy products can help people lose weight, reduce their body fat mass, enhance their body composition and maintain their lean body mass while they are under a calorie restriction, especially adults. The risk of childhood obesity is reduced by a diet high in milk and dairy products. Additionally, it aids in reducing the risk of type 2 diabetes development. Yoghurt and cheese are linked to a lower risk of type 2 diabetes (Gilbert et al. 2011; Bendtsen et al. 2013).

7.4.2 Cardiovascular Disease

Low-fat dairy products with a high calcium content are typically thought to reduce blood pressure. High-fat dairy products raise blood levels of HDL, LDL and cholesterol. The risk of cardiovascular disease can be predicted by these levels, but it may also rely on the size of the LDL-cholesterol particles. Because they have a lower affinity for the LDL receptor and are more susceptible to oxidation, small, dense LDL particles have a higher propensity for atherogenesis than their bigger counterparts (Huth and Park 2012; Mora et al. 2011). Certain fatty acids contained in milk and dairy products are linked to LDL particles that are less dense and small. Additionally, micronutrients included in milk and dairy products reduce the LDL response to consuming high-fat dairy products (Sjogren et al. 2004; Lorenzen and Astrup 2011; Soerensen et al. 2014).

Cheese and other high-fat dairy products have a high percentage of saturated fat, which lowers LDL-cholesterol levels (Hjerpsted et al. 2011). A high intake of cheese does not raise LDL when compared to diets with lower total and saturated fat content or diets with reduced total fat content but a high glycaemic index (Throning et al. 2015). There is currently no conclusive evidence linking milk or dairy consumption to cardiovascular disease or stroke (Soedamah-Muthu et al. 2011).

7.4.3 Osteoporosis and Bone Health

Since milk and milk products contain a variety of vital nutrients that are necessary for the development of healthy bones in childhood and the maintenance of those bones in maturity, they are intended to reduce the risk of bone fractures and osteoporosis in later stages of life. The European Commission has suggested protein, calcium, magnesium, manganese, phosphorus, zinc, vitamin D and vitamin K for the preservation of healthy bones (Rizzoli 2014). All other nutrients, except for vitamin D, are present in adequate levels in milk and its products. Osteoporotic fractures are more common in middle-aged and older people, especially women, who consume less milk and other minerals during early life and adolescence (Power et al. 1999; Kalkwarf et al. 2003).

Lactose intolerant patients or consumers who do not wish to take these dairy foods must consume foods (e.g. soymilk) that are able to offer almost all those nutrients usually found in dairy. The focus must always be on meeting nutrient needs through foods instead of opting for supplements. Plant-based beverages contain calcium in an inorganic form, which may result in a rise in cardiovascular risks (Bolland et al. 2013; Weaver 2014).

7.4.4 Cancer

Dairy products contain a variety of bioactive substances that may influence carcinogenesis in both favourable and unfavourable ways. Its beneficial benefits are linked to calcium, lactoferrin content and the production of fermented foods, whereas its detrimental effects are linked to insulin-like growth factor I (IGF-1) levels (Latino-Martel et al. 2016). Because of the calcium included in dairy, there is thought to be a link between the consumption of dairy foods and lowering the risk of colorectal cancer, with a 24% risk reduction if calcium intake through dairy is at or above 900 mg/day. This is due to the fact that when secondary bile acids and ionised fatty acids bind to calcium, their proliferative effects on the colorectal epithelium are lessened. Additionally, calcium will influence numerous intracellular pathways that cause normal cells to differentiate and altered cells to die. Studies have supported that there is a reduction in cell proliferation in the rectum and colon with good consumption of calcium-rich dairy foods (Newmark et al. 1984; Lamprecht & Lipkin 2001; Keum et al. 2014).

According to WCRF reports, consuming milk and milk products provides a shield against many types of cancer such as colorectal, gastric, bladder and breast cancer. The association between dairy consumption and the risk of pancreatic, lung and ovarian cancers has yet not been fully established. Milk and milk products provide significant health benefits, especially in women, thus reducing the risk of colorectal and breast cancers. Similarly in males, the advantage of the protective nature of milk and its products on colorectal cancer is perceived to be outweighed by a possible higher risk of prostate cancer.

7.4.5 Hypertension and Antihypertensive Activity

Hypertension is a condition when systolic and/or diastolic blood pressure is over 140 and/or 90 mg of mercury respectively. High blood pressure is a serious problem and is linked to CVD and can lead to stroke, heart failure or heart attack. Angiotensin-converting enzyme (ACE) has multiple roles and is a chief element of the renin-angiostatin system. This enzyme is a major regulator of blood pressure, as it converts the enzyme angiotensin I to angiotensin II. Angiotensin II is a hormone that increases blood pressure as it contracts the blood vessels. To avoid this, the use of ACE inhibitors in the form of pharmaceutical drugs is recommended for patients suffering from hypertension. Many studies supported that bioactive peptides obtained from milk protein have ACE-inhibitory activity (Nagpal et al. 2011).

7.4.6 Antimicrobial Activity

Milk contains antimicrobial proteins with complex, wide-ranging activity in addition to a rich complement of antibodies highly specific for particular antigens (Liu and Newburg 2013). One of these is lactoferrin, an 80-kDa cationic glycoprotein that is a member of the nonheme iron-binding protein family (Table 7.2). The physiological actions of lactoferrin are dependent on the binding of iron with high affinity, which places microbiostatic pressure on iron-sensitive microbes of the digestive tract. A direct lactoferrin microbicidal action has also been proposed (Brock 2012). Human cytomegalovirus, hepatitis B and C, HIV, rotavirus, herpesvirus, human papillomavirus, adenovirus, hantavirus and influenza A replication have all been shown to be inhibited by lactoferrin in vitro (Berlutti et al. 2011). Infected neonatal rats treated with oral bovine lactoferrin showed a decrease in gut-derived sepsis, according to animal research, proving that milk lactoferrin protects nursing mammals from infection (Edde et al. 2001). In a randomised, double-blind, placebo-controlled study, prophylactic delivery of recombinant human lactoferrin to a large population of children decreased the length and severity of diarrhoea over six months (Ochoa et al. 2012).

Table 7.2 Clinical trials using milk components as antibacterial agents

Component of milk	Organisms	Results	References
Lactoferrin	Helicobacter pylori	Increased rate of infection resolution when added to regular treatment	De Beortoli et al. 2007
	Not defined	Decreased severity and duration of diarrhoea	Ochoa et al. 2012; Laffan et al. 2011
	Not defined	Sepsis caused by bacteria and fungus is less common in preterm babies	Manzoni et al. 2009; Manzoni et al. 2012
	Not defined	Less common respiratory infections	King et al. 2007
	Not defined	Decreased mortality by 46% in those with acute bacteremia	Guntupalli et al. 2013
	Rhinovirus	Decreased symptom duration and severity of diarrhoea	Vitetta et al. 2013
	Hepatitis C virus	Improved response to treatment when combined with conventional therapy	Kaito et al. 2007
3'-sialyllactose	Helicobacter pylori	No advantage	Parente et al. 2003
Lysozyme and lactoferrin	Not defined	Less frequent cases of diarrhoea	Zavaleta et al. 2007
SIgA	Escherichia coli	90% decrease in the frequency of infection	Otto et al. 2011

Recombinant human lactoferrin-enriched newborn formula also decreased the risk of respiratory infection throughout the first year of life as compared to conventional formula (King et al. 2007).

7.4.7 Antiepileptic Activity

One of the most prevalent proteins in human milk is alpha-lactalbumin. The last stage of lactose production is catalysed by a complex containing a-lactalbumin and galactosyltransferase. Animal preclinical research suggests that a-lactalbumin may have uses in the treatment of epilepsy. In genetically predisposed rats, the daily treatment of a-lactalbumin over a period of 12 days dramatically decreased the severity and length of sound stimulus-induced seizures and protected against chemically produced seizures in mice. The improvement in serotonin-mediated functions and the corresponding sharp elevations in serum tryptophan were attributable to this impact (Citraro et al. 2011). In a different animal model of epileptogenesis, prophylactic a-lactalbumin administration prevented chemically treated rats from spontaneously developing seizures. In both rats and mice, bolus treatment of a-lactalbumin during an acute chemically produced seizure reduced seizure activity by up to 90% (Russo et al. 2012).

7.4.8 Antioxidant Activity

The radical scavenging, lipid peroxidation inhibition and metal ion chelation capabilities of bioactive peptides are what give them their antioxidant action. Additionally, it has been suggested that the peptide structure and amino acid sequence may have an impact on the antioxidative characteristics. αs-casein-derived peptides have the ability to scavenge free radicals and prevent both enzymatic and non-enzymatic lipid peroxidation. These dipeptides can encourage the production of glutathione, an essential antioxidant for cellular defence and repair mechanisms (Rival et al. 2001; Sarmadi et al. 2010)

7.5 RECENT ADVANCEMENTS IN THE DEVELOPMENT OF FUNCTIONAL MILK

7.5.1 Protein Modifications k-caseins, β-casein

Genetic engineering can produce higher quantities of β-caseins (8–20%) and k-caseins (twofold) in transgenic cows (Brophy et al. 2003). The most prevalent milk protein, β-casein, is responsible for binding calcium phosphate and regulating milk calcium levels. Smaller micelles, higher heat stability and improved cheese-making qualities are all associated with milk that contains more k-casein. In comparison to non-transgenic control cows, total milk casein increased by 17–35% and total milk protein by 13–20% in transgenic animals. Clearly, this benefits the cheese yield as well as the casein and milk protein concentrate industries. Acid casein is used for paper coatings, cosmetics, button manufacture, paints and textile fabrics, while edible casein is utilised in vitamin pills, quick beverages and newborn formula.

Plasmin, a serine protease that naturally occurs in milk along with plasminogen, is extremely sensitive to caseins and notably the easily digestible β-, αs 1- and αs 2-caseins. Limited proteolysis in milk is caused by plasmin activity. This has the combined drawbacks of reducing the cheese curd output and causing organoleptic flaws and milk gelation when treated at extremely high temperatures. Therefore, adding particular inhibitors of plasmin or a plasminogen activator to milk would benefit the food processing industry (Murthy and Kanawjia 2002).

7.5.2 Fat Modifications

The percentage of mono-saturated fatty acids (MUFA) in the "ideal" milk fat required for human health is 82%. Although this "perfect" milk fat content cannot be obtained, it can still be improved by altering the feeds and feeding habits of dairy cows as well as by implementing genetic interventions (O'Donnell 1989).

7.5.3 Lowering Milk's Saturated Fat Content

The degree of unsaturation of serum lipids, tissue fat and milk fat increases immediately after feeding protected or encapsulated unsaturated fats. Milk fat decreased when highly unsaturated oils (like soybean oil) were fed to cows, but the ratio of USFA to SFA increased. In contrast to feeding oil seeds, feeding canola oil in encapsulated form (to prevent biohydrogenation by rumen bacteria) increases the concentration of linolenic (18:3) and linoleic (18:2) acids. Unsaturated fatty acid-containing milk fat has a lower melting point, which greatly increases the spreadability of butter made from such milk. The spreadability of butter was doubled as a result of feeding a canola and soybean meal blend in a protected form to cows. This butter was as easily spreadable as margarine when removed from the refrigerator at 5°C without losing its sensory properties (Ashes et al. 1997).

7.5.4 Increasing Milk Fat Concentrations of Conjugated Linoleic Acid

Since CLA is produced in the rumen during the biohydrogenation of linoleic acid, dairy products are a significant source of CLA. Conjugated linoleic acid levels in milk fat increase twofold with linoleic acid-rich diets. The inclusion of conjugated linoleic acid with soy oil in the diets or feeds of cows resulted in escalated levels of CLA in milk and also reduced saturated fatty acid content in milk fat. Another point to take into consideration is that the milk from cows that are fed on grass contains more amounts of CLA as compared to cows fed with grains (Hanuš et al., 2018).

7.5.5 Enhancement of Omega Fatty Acids

In comparison to the recommended intake of no more than 4% of total calories (approximate proportions of 2% omega-3 linolenic acid and 2% omega-6 linoleic acid from the same), it is found that the content of polyunsaturated fatty acids in modern diets is too high, providing about 30% of the total calories needed. An optimal balance of essential fatty acids is present in the milk produced by cows that are fed on pasture (EFAs) (Hanuš et al., 2018).

7.5.6 Reducing Fat Content in Milk as a Variation to Altering the Fat Composition

The cost of feed per kilogramme of milk would decrease by around 22% per tonne if an animal's genetic makeup could be changed to enable it to produce 2% more milk. The amount of fat in milk will noticeably decrease as the enzyme Acetyl CoA carboxylase, which controls the rate of fat synthesis in the mammary glands, declines. The amount of energy an animal needs to create milk will also decrease (Wall et al. 1997; Ntambi et al. 1999).

7.5.7 Lactose Modification

As opposed to glucose and galactose (monosaccharide), which are produced as a result of the enzymatic breakdown of lactose, lactose or milk sugar cannot be directly delivered to the bloodstream. For many people, the amount of the enzyme lactase, also known as galactosidase (β-gal), starts to decline early in life and eventually disappears completely by adolescence. Lactose can remain undigested and poorly absorbed in the gut when consumed, causing water retention as a result of its osmotic impact. Intestinal discomfort and dehydration are caused by this water retention and the bacterial generation of high amounts of carbon dioxide. As milk is one of the primary components of the human diet, lactose intolerance confines the utilisation of valuable nutrients for many people (Vesa 1999).

Additionally, as milk is one of the richest sources of calcium and helps in maintaining bone health, lactose intolerance may lead to osteopenia. Osteopenia is a loss of bone mineral density (BMD) and leads to weak bones. It is a more common issue in people older than 50 years of age, especially women. Lactose intolerance can be managed by dietary modification, lactase supplementation and treating an underlying condition in people with secondary lactase deficiency (occurring due to infections, inflammation, injury to intestinal mucosa or other diseases). Modifications in diet alone can play a significant role in the improvement of signs and symptoms of lactose intolerance in most patients (Corazza et al. 1995).

7.5.8 Techniques for Reducing Lactose Before Harvest

One of the most significant milk proteins, α-lactalbumin (α-LA), is found in practically all mammalian milk. It interacts with β-1,4-UDP-galactosyltransferase (UDP-gal) to alter the enzyme's preference for its substrate, effectively producing an exclusive binding site for glucose and triggering the production of lactose (Vilotte 2002). Pre-harvest techniques for reducing lactose either entail removing α-LA and gene "knockout" techniques or adding lactase enzyme via mammary gland-specific expression. Although these effective methods offer useful tools for addressing milk physiology, they also reduce milk's overall sugar content, making the milk much more viscous. Studies on mice have shown that lactose reduction via α-LA deletion was ineffective because it disrupted the regulation of milk volume. Such mice produced milk that was extremely viscous, heavy in fat and protein but lacking in α-LA and lactose. Knocking down the UDP-gal gene in mice likewise resulted in milk with high viscosity but no lactose, according to Karatzas and Turner (1997).

7.5.9 Humanisation of Bovine Milk

It is well-recognised that breast milk is the ultimate naturally available designer food for babies. However, for many reasons, formula milk based on the composition of bovine milk is fed to many infants. The composition of these formulas can be effectively improved by incorporating constituents that have a close resemblance with human milk so as to suit and meet the needs of the infant; this can be achieved by "humanising" the bovine milk (Jost et al. 1999).

Human milk contains around 1 g/l of lactoferrin, while human colostrum has significantly more, about 7 g/l. However, LF levels in cow's milk are only around one-tenth of those in human milk, which has encouraged scientists to work on formulations to replace human milk to fulfil these demands. Pharming, NV (Leiden, the Netherlands) created the first transgenic bull in the late 1980s. A line of transgenic cows was also created to generate a variety of proteins, including human LF (Subramanian 2004).

7.5.10 Eliminating β-lactoglobulin from Milk

β-lg is present in cows' milk and is a common reason for cow milk allergies, especially in children. β-lg is absent in human milk. It is unlikely that the removal of this protein by the knockout of the β-lg gene from cow's milk would have any negative effects on either cow or human formula, and it may help to resolve many of the most serious allergic issues connected to cow's milk. All dietary proteins may be allergens, according to studies on the allergenicity of milk proteins. The protein molecule has a large number of these allergenic structures. In order to characterise allergenic components in food, milk can therefore serve as a superior model (Wal 1988). Therefore, technical techniques utilised in the evaluation of milk protein allergenicity can be simply adapted for creating systems for the identification and evaluation of possible allergenicity in novel foods.

7.5.11 Melatonin-Enriched Milk

Sleep-related conditions including insomnia can be treated with milk that has been melatonin-enriched. As a powerful antioxidant, melatonin can serve as a terminal antioxidant and be utilised to prevent lifestyle diseases. Milk acquired before sunrise has a three to four-times higher melatonin level than milk obtained during the day (Kayumov et al. 2005). Plants and animals are both sources of melatonin. Melatonin secretion in individuals varies depending on their gender, age, season and health circumstances (Singh et al. 2011). Melatonin levels decline with ageing, yet older women secrete more of it than older men do. Seasonal differences also affect melatonin concentration, and winters have higher melatonin concentrations than summers do (Midau et al. 2010). Due to the presence of conjugated linoleic acid in various kinds of milk, melatonin content also changes with milk type (Singh et al. 2011).

7.5.12 Milk with Human Therapeutic Proteins

Production of high-value, low-volume therapeutic proteins in domestic animal milk has garnered a lot of industrial interest. Several human proteins have already been successfully expressed in this setting. More than 60 therapeutic proteins, including plasma proteins, monoclonal antibodies and vaccines, are produced by GTC

Biotherapeutics using both goats and cows. Recombinant human antithrombin III, an anti-coagulant protein found in blood, is one product that is nearing the end of testing (it is made from goat's milk) (Yang et al. 2018). GTC is also engaged in a study to create a goat-milk-based malaria vaccine. To address pancreatic insufficiency in breaking down dietary lipids, PPL Therapeutics (Edinburgh, UK and Blacksburg, VA) is experimenting with rabbits and lambs to manufacture a-1-antitrypsin, fibrinogen and a lipase. From the milk of transgenic cows, lambs or goats, products like insulin and growth hormone have also been produced. Additionally, transgenic animals can leak proteins into their milk, such as blood clotting factors required by people with haemophilia (Suraokar and Bradley 2000).

7.6 CONCLUDING REMARKS AND FUTURE DIRECTIONS

Important elements for a robust and healthy life are found in milk and dairy products. Although a very small percentage of the population has an unpleasant reaction to one or more of the components in milk, this should not discourage or cause others to steer clear of these healthy nutrients. The numerous health advantages of milk and dairy products should be known and understood by consumers. The food sector could also introduce goods to meet the needs of consumers who avoid drinking milk and other dairy products. One such product may be low lactose milk made utilising current technologies like ultrafiltration. Another strategy is to use hypoallergenic items to lessen the allergic reaction to milk protein. In addition to nutrition, milk has a number of physiological advantages, particularly because of the bioactive peptides that are hidden inside the protein and released during hydrolysis and digestion. These peptides have exhibited a wide range of biological actions, including antioxidant, hypotensive, hypocholesterolemic, anticancer and immunomodulatory effects, among many others. These peptides could become more widely available to consumers if they are extracted from protein digest. They could then be used in functional food production or as supplements. These bioactive substances may be shielded from harmful environmental factors and have a wider range of applications if they are encapsulated. Altering the proportion of saturated fatty acids to unsaturated fatty acids is another area of animal husbandry to research further in order to increase the attractiveness of milk and dairy products.

REFERENCES

Aro, A., S. Männistö, I. Salminen, M. L. Ovaskainen, V. Kataja, and M. Uusitupa. 2000. "Inverse association between dietary and serum conjugated linoleic acid and risk of breast cancer in postmenopausal women". *Nutrition and Cancer 38*(2): 151–157.

Ashes, J. R., S. K. Gulati, and T. W. Scott. 1997. "Potential to alter the content and composition of milk fat through nutrition". *Journal of Dairy Science 80*(9): 2204–2212.

Astrup, A., A. Raben, and N. Geiker. 2015. "The role of higher protein diets in weight control and obesity-related comorbidities". *International Journal of Obesity 39*(5): 721–726.

Azuma, N., K. Yamauchi, and T. Mitsuoka. 1984. "Bifidus growth-promoting activity of a glycomacropeptide derived from human K-casein". *Agricultural and Biological Chemistry 48*(8): 2159–2162.

Bendtsen, L. Q., J. K. Lorenzen, N. T. Bendsen, C. Rasmussen, and A. Astrup. 2013. "Effect of dairy proteins on appetite, energy expenditure, body weight, and composition: A review of the evidence from controlled clinical trials". *Advances in Nutrition 4*(4): 418–438.

Berlutti, F., F. Pantanella, T. Natalizi, A. Frioni, R. Paesano, A. Polimeni, and P. Valenti. 2011. "Antiviral properties of lactoferrin—A natural immunity molecule". *Molecules 16*(8): 6992–7018.

Bolland, M. J., A. Grey, and I. R. Reid. 2013. "Calcium supplements and cardiovascular risk: 5 years on". *Therapeutic Advances in Drug Safety 4*(5): 199–210.

Brock, J. H. 2012. "Lactoferrin–50 years on". *Biochemistry and Cell Biology 90*(3): 245–251.

Brody, E. P. 2000. "Biological activities of bovine glycomacropeptide". *British Journal of Nutrition 84*: 39–46.

Brophy, B., G. Smolenski, T. Wheeler, D. Wells, P. L'Huillier, and G. Laible. 2003. "Cloned transgenic cattle produce milk with higher levels of β-casein and κ-casein". *Nature Biotechnology 21*(2): 157–162.

Cadogan, J., R. Eastell, N. Jones, and M. E. Barker. 1997. "Milk intake and bone mineral acquisition in adolescent girls: Randomised, controlled intervention trial". *BMJ: British Medical Journal 315*(7118): 1255–1260.

Chandan, C. R. 2007. "Milk composition, physical and processing characteristics". In: Hui, Y. H. (ed.), *Handbook of food products manufacturing 2*. Hoboken, NJ: John Wiley & Sons, 347.

Christian, E. W. V., and A. Vickie 2008. *Essential of food science*. 3rd ed. San Marcos, CA: Springer.

Citraro, R., F. Scicchitano, S. De Fazio, R. Raggio, P. Mainardi, E. Perucca, G. De Sarro, and E. Russo. 2011. "Preclinical activity profile of α-lactoalbumin, a whey protein rich in tryptophan, in rodent models of seizures and epilepsy". *Epilepsy Research 95*(1–2): 60–69.

Clare, D. A., and H. E. Swaisgood. 2000. "Bioactive milk peptides: A prospectus". *Journal of Dairy Science 83*(6): 1187–1195.

Collomb, M., A. Schmid, R. Sieber, D. Wechsler, and E. L. Ryhänen. 2006. "Conjugated linoleic acids in milk fat: Variation and physiological effects". *International Dairy Journal 16*(11): 1347–1361.

Corazza, G. R., G. Benati, A. Di Sario, C. Tarozzi, A. Strocchi, M. Passeri, and G. Gasbarrini. 1995. "Lactose intolerance and bone mass in postmenopausal Italian women". *British Journal of Nutrition 73*(3): 479–487.

Corl, B. A., D. M. Barbano, D. E. Bauman, and C. Ip. 2003. "cis-9, trans-11 CLA derived endogenously from trans-11 18: 1 reduces cancer risk in rats". *The Journal of Nutrition 133*(9): 2893–2900.

De Bortoli, N., G. Leonardi, E. Ciancia, A. Merlo, M. Bellini, F. Costa, M. G. Mumolo, A. Ricchiuti, F. Cristiani, S. Santi, and M. Rossi. 2007. "Helicobacter pylori eradication: A randomized prospective study of triple therapy: Versus: Triple therapy plus lactoferrin and probiotics". *Official Journal of the American College of Gastroenterology| ACG 102*: 951–956.

Dhiman, T. R., G. R. Anand, L. D. Satter, and M. W. Pariza. 1999. "Conjugated linoleic acid content of milk from cows fed different diets". *Journal of Dairy Science 82*: 2146–2156.

Donkor, O. N., A. Henriksson, T. K. Singh, T. Vasiljevic, and N. P. Shah. 2007. "ACE-inhibitory activity of probiotic yoghurt". *International Dairy Journal 17*(11): 1321–1331.

Edde, L., R. B. Hipolito, F. F. Hwang, D. R. Headon, R. A. Shalwitz, and M. P. Sherman. 2001. "Lactoferrin protects neonatal rats from gut-related systemic infection". *American Journal of Physiology-Gastrointestinal and Liver Physiology 281*(5): G1140–G1150.

Gilbert, J. A., N. T. Bendsen, A. Tremblay, and A. Astrup. 2011. "Effect of proteins from different sources on body composition". *Nutrition, Metabolism, and Cardiovascular Diseases 21*: B16–B31.

Gobbetti, M., F. Minervini, and C. G. Rizzello. 2007. "Bioactive peptides in dairy products". *Handbook of Food Products Manufacturing 2*: 489–517.

Gopal, P. K., and H. Gill. 2000. "Oligosaccharides and glycoconjugates in bovine milk and colostrum". *British Journal of Nutrition 84*: S69–S74.

Gulati, S. K., S. M. Kitessa, J. R. Ashes, G. C. Simos, P. C. Wynn, E. Fleck, and T. W. Scott. 2000. "Designing milk fat for the new millennium by dietary strategies". *Asian-Australasian Journal of Animal Sciences 13*: 538–541.

Guntupalli, K., N. Dean, P. E. Morris, V. Bandi, B. Margolis, E. Rivers, M. Levy, R. F. Lodato, P. M. Ismail, A. Reese, and J. P. Schaumberg. 2013. "A phase 2 randomized, double-blind, placebo–controlled study of the safety and efficacy of talactoferrin in patients with severe sepsis". *Critical Care Medicine 41*(3): 706–716.

Hanuš, O., E. Samková , L. Křížová, L. Hasoňová, R. Kala. 2018. "Role of Fatty Acids in Milk Fat and the Influence of Selected Factors on Their Variability-A Review." *Molecules 23*(7):1636.

Hasler, C. M., A. S. Bloch, C. A. Thomson, E. Enrione, and C. Manning. 2004. "Position of the American dietetic association: Functional foods". *Journal of the American Dietetic Association 104*(5): 814–826.

Hernández-Ledesma, B., I. Recio, M. Ramos, and L. Amigo. 2002. "Preparation of ovine and caprine β-lactoglobulin hydrolysates with ACE-inhibitory activity. Identification of active peptides from caprine β-lactoglobulin hydrolysed with thermolysin". *International Dairy Journal 12*(10): 805–812.

Higurashi, S., Y. Kunieda, H. Matsuyama, and H. Kawakami. 2007. "Effect of cheese consumption on the accumulation of abdominal adipose and decrease in serum adiponectin levels in rats fed a calorie dense diet". *International Dairy Journal 17*(10): 1224–1231.

Hiraoka, Y., T. Segawa, K. Kuwajima, S. Sugai, and N. Murai. 1980. "α-lactalbumin: A calcium metalloprotein". *Biochemical and Biophysical Research Communications 95*(3): 1098–1104.

Hjerpsted, J., E. Leedo, and T. Tholstrup. 2011. "Cheese intake in large amounts lowers LDL-cholesterol concentrations compared with butter intake of equal fat content". *The American Journal of Clinical Nutrition 94*(6): 1479–1484.

Huth, P. J., and K. M. Park. 2012. "Influence of dairy product and milk fat consumption on cardiovascular disease risk: A review of the evidence". *Advances in Nutrition 3*(3): 266–285.

Jensen, R. G. 2002. "The composition of bovine milk lipids: January 1995 to December 2000". *Journal of Dairy Science 85*(2): 295–350.

Jiang, J., A. Wolk, and B. Vessby. 1999. "Relation between the intake of milk fat and the occurrence of conjugated linoleic acid in human adipose tissue". *The American Journal of Clinical Nutrition 70*(1): 21–27.

Jost, B., J. L. Vilotte, I. Duluc, J. L. Rodeau, and J. N. Freund. 1999. "Production of low-lactose milk by ectopic expression of intestinal lactase in the mouse mammary gland". *Nature Biotechnology 17*(2): 160–164.

Kaito, M., M. Iwasa, N. Fujita, Y. Kobayashi, Y. Kojima, J. Ikoma, I. Imoto, Y. Adachi, H. Hamano, and K. Yamauchi. 2007. "Effect of lactoferrin in patients with chronic hepatitis C: Combination therapy with interferon and ribavirin". *Journal of Gastroenterology and Hepatology 22*(11): 1894–1897.

Kalkwarf, H. J., J. C. Khoury, and B. P. Lanphear. 2003. "Milk intake during childhood and adolescence, adult bone density, and osteoporotic fractures in US women". *The American Journal of Clinical Nutrition 77*(1): 257–265.

Karatzas, C. N., and J. D. Turner. 1997. "Toward altering milk composition by genetic manipulation: Current status and challenges". *Journal of Dairy Science 80*(9): 2225–2232.

Kawasaki, Y., H. Isoda, M. Tanimoto, S. Dosako, T. Idota, and K. Ahiko. 1992. "Inhibition by lactoferrin and κ-casein glycomacropeptide of binding of cholera toxin to its receptor". *Bioscience, Biotechnology, and Biochemistry 56*(2): 195–198.

Kayumov, L., R. F. Casper, R. J. Hawa, B. Perelman,S. A. Chung, S. Sokalsky, and C. M. Shapiro. 2005. "Blocking low-wavelength light prevents nocturnal melatonin suppression with no adverse effect on performance during simulated shift work". *The Journal of Clinical Endocrinology and Metabolism 90*(5): 2755–2761.

Keogh, J. B., and P. Clifton. 2008. "The effect of meal replacements high in glycomacropeptide on weight loss and markers of cardiovascular disease risk". *The American Journal of Clinical Nutrition 87*(6): 1602–1605.

Keum, N., D. Aune, D. C. Greenwood, W. Ju, and E. L. Giovannucci. 2014. "Calcium intake and colorectal cancer risk: Dose–response meta-analysis of prospective observational studies". *International Journal of Cancer 135*(8): 1940–1948.

King Jr, J. C., G. E. Cummings, N. Guo, L. Trivedi, B. X. Readmond, V. Keane, S. Feigelman, and R. de Waard. 2007. "A double-blind, placebo-controlled, pilot study of bovine lactoferrin supplementation in bottle-fed infants". *Journal of Pediatric Gastroenterology and Nutrition 44*(2): 245–251.

Ko, S., and Hae-Son, K. 2009. *Bioactive components in whey products*. Ames, IA: Wiley-Blackwell, 263.

Korhonen, H. J. 2009. "Bioactive components in bovine milk". In: Park, W. Y. (ed.), *Bioactive components in milk and dairy products*. Ames, IA: Wiley-Blackwell, 18–22.

Kunz, C., S. Rudloff, W. Baier, N. Klein, and S. Strobel. 2000. "Oligosaccharides in human milk: Structural, functional, and metabolic aspects". *Annual Review of Nutrition 20*: 699–722.

Laaksonen, M. A., P. Knekt, H. Rissanen, T. Härkänen, E. Virtala, J. Marniemi, A. Aromaa, M. Heliövaara, and A. Reunanen. 2010. "The relative importance of modifiable potential risk factors of type 2 diabetes: A meta-analysis of two cohorts". *European Journal of Epidemiology 25*(2): 115–124.

Laffan, A. M., R. McKenzie, J. Forti, D. Conklin, R. Marcinko, R. Shrestha, M. Bellantoni, and W. B. Greenough III. 2011. "Lactoferrin for the prevention of post-antibiotic diarrhoea". *Journal of Health, Population, and Nutrition 29*(6): 547.

Lamprecht, S. A., and M. Lipkin. 2001. "Cellular mechanisms of calcium and vitamin D in the inhibition of colorectal carcinogenesis". *Annals of the New York Academy of Sciences 952*: 73–87.

Lara-Villoslada, F., E. Debras, A. Nieto, A. Concha, J. Gálvez, E. López-Huertas, J. Boza, C. Obled, and J. Xaus. 2006. "Oligosaccharides isolated from goat milk reduce intestinal inflammation in a rat model of dextran sodium sulfate-induced colitis". *Clinical Nutrition 25*(3): 477–488.

Latino-Martel, P., V. Cottet, N. Druesne-Pecollo, F. H. Pierre, M. Touillaud, M. Touvier, M. P. Vasson, M. Deschasaux, J. Le Merdy, E. Barrandon, and R. Ancellin. 2016. "Alcoholic beverages, obesity, physical activity and other nutritional factors, and cancer risk: A review of the evidence". *Critical Reviews in Oncology/Hematology 99*: 308–323.

Lin, Y. C., R. M. Lyle, L. D. McCabe, G. P. McCabe, C. M. Weaver, and D. Teegarden. 2000. "Dairy calcium is related to changes in body composition during a two-year exercise intervention in young women". *Journal of the American College of Nutrition 19*(6): 754–760.

Liu, B., and D. S. Newburg. 2013. "Human milk glycoproteins protect infants against human pathogens". *Breastfeeding Medicine 8*(4): 354–362.

Lönnerdal, B., and C. Glazier. 1985. "Calcium binding by α-lactalbumin in human milk and bovine milk". *The Journal of Nutrition 115*(9): 1209–1216.

Lönnerdal, B., and S. Iyer. 1995. "Lactoferrin: Molecular structure and biological function". *Annual Review of Nutrition 15*: 93–110.

Lorenzen, J. K., and A. Astrup. 2011. "Dairy calcium intake modifies responsiveness of fat metabolism and blood lipids to a high-fat diet". *British Journal of Nutrition 105*(12): 1823–1831.

Manzoni, P., M. Rinaldi, S. Cattani, L. Pugni, M. G. Romeo, H. Messner, I. Stolfi, L. Decembrino, N. Laforgia, F. Vagnarelli, and L. Memo. 2009. "Bovine lactoferrin supplementation for prevention of late-onset sepsis in very low-birth-weight neonates: A randomized trial". *JAMA 302*(13): 1421–1428.

Manzoni, P., I. Stolfi, H. Messner, S. Cattani, N. Laforgia, M. G. Romeo, L. Bollani, M. Rinaldi, E. Gallo, M. Quercia, and M. Maule. 2012. "Bovine lactoferrin prevents invasive fungal infections in very low birth weight infants: A randomized controlled trial". *Pediatrics 129*(1): 116–123.

Mehra, R., and P. Kelly. 2006. "Milk oligosaccharides: Structural and technological aspects". *International Dairy Journal 16*(11): 1334–1340.

Meisel, H. 1997. "Overview on milk protein-derived peptides". *International Dairy Journal 8*(5–6): 363–373.

Midau, A., A. Kibon, S. M. Moruppa, and C. Augustine. 2010. "Influence of season on milk yield and milk composition of Red Sokoto goats in Mubi area of Adamawa state, Nigeria". *International Journal of Dairy Science 5*(3): 135–141.

Mora, S., J. E. Buring, P. M. Ridker, and Y. Cui. 2011. "Association of high-density lipoprotein cholesterol with incident cardiovascular events in women, by low-density lipoprotein cholesterol and apolipoprotein B100 levels: A cohort study". *Annals of Internal Medicine 155*(11): 742–750.

Murthy, G. L. N., and S. K. Kanawjia. 2002. "Designer milk: Nutritional and technological significance". *Indian Dairyman 54*: 49–58.

Nagpal, R., P. Behare, R. Rana, A. Kumar, M. Kumar, S. Arora, F. Morotta, S. Jain, and H. Yadav. 2011. "Bioactive peptides derived from milk proteins and their health beneficial potentials: An update". *Food and Function 2*(1): 18–27.

Newmark, H. L., M. J. Wargovich, and W. R. Bruce. 1984. "Colon cancer and dietary fat, phosphate, and calcium: A hypothesis". *Journal of the National Cancer Institute 72*(6): 1323–1325.

Ntambi, J. M., Y. Choi, and Y. C. Kim. 1999. "Regulation of stearoyl-CoA desaturase by conjugated linoleic acid". *Advances in Conjugated Linoleic Acid 1*: 340–347.

O'Donnell, J. A. 1989. "Milk fat technologies and markets: A summary of the Wisconsin milk marketing board 1988 milk fat roundtable". *Journal of Dairy Science 72*(11): 3109–3115.

Ochoa, T. J., E. Chea-Woo, N. Baiocchi, I. Pecho, M. Campos, A. Prada, G. Valdiviezo, A. Lluque, D. Lai, and T. G. Cleary. 2012. "Randomized double-blind controlled trial of bovine lactoferrin for prevention of diarrhea in children". *The Journal of Pediatrics 162*(2): 349–356.

Otto, W., B. Najnigier, T. Stelmasiak, and R. M. Robins-Browne. 2011. "Randomized control trials using a tablet formulation of hyperimmune bovine colostrum to prevent diarrhea caused by enterotoxigenic Escherichia coli in volunteers". *Scandinavian Journal of Gastroenterology 46*(7–8): 862–868.

Parente, F., C. Cucino, A. Anderloni, G. Grandinetti, and G. B. Porro. 2003. "Treatment of Helicobacter pylori infection using a novel antiadhesion compound (3′ sialyllactose sodium salt). A double blind, placebo-controlled clinical study". *Helicobacter 8*(4): 252–256.

Park, Y. W., S. Zhu, L. Palaniappan, S. Heshka, M. R. Carnethon, and S. B. Heymsfield. 2003. "The metabolic syndrome: Prevalence and associated risk factor findings in the US population from the third national health and nutrition examination survey, 1988–1994". *Archives of Internal Medicine 163*(4): 427–436.

Park, Y. W. (Ed.). 2009. *Bioactive components in milk and dairy products*. John Wiley & Sons, Iowa, USA.

Peacock, M. 2010. "Calcium metabolism in health and disease". *Clinical Journal of the American Society of Nephrology 5*: S23–S30.

Power, M. L., R. P. Heaney, H. J. Kalkwarf, R. M. Pitkin, J. T. Repke, R. C. Tsang, and J. Schulkin. 1999. "The role of calcium in health and disease". *American Journal of Obstetrics and Gynecology 181*(6): 1560–1569.

Raikos, V., and T. Dassios. 2014. "Health-promoting properties of bioactive peptides derived from milk proteins in infant food: A review". *Dairy Science and Technology 94*(2): 91–101.

Rival, S. G., C. G. Boeriu, and H. J. Wichers. 2001. "Caseins and casein hydrolysates. 2. Antioxidative properties and relevance to lipoxygenase inhibition". *Journal of Agricultural and Food Chemistry 49*(1): 295–302.

Rizzoli, R. 2014. "Dairy products, yogurts, and bone health". *The American Journal of Clinical Nutrition 99*: 1256S–1262S.

Roche, H. M., E. Noone, and A. N. M. J. Gibney. 2001. "Conjugated linoleic acid: A novel therapeutic nutrient?" *Nutrition Research Reviews 14*(1): 173–188.

Russo, E., F. Scicchitano, R. Citraro, R. Aiello, C. Camastra, P. Mainardi, S. Chimirri, E. Perucca, G. Donato, and G. De Sarro. 2012. "Protective activity of α-lactoalbumin (ALAC), a whey protein rich in tryptophan, in rodent models of epileptogenesis". *Neuroscience 226*: 282–288.

Ryhänen, E.-L., A. Pihlanto-Leppälä, and E. Pahkala. 2001. "A new type of ripened, low-fat cheese with bioactive properties". *International Dairy Journal 11*(4–7): 441–447.

Sarmadi, B. H., and A. Ismail. 2010. "Antioxidative peptides from food proteins: A review". *Peptides 31*(10): 1949–1956.

Schanbacher, F. L., R. S. Talhouk, F. A. Murray, L. I. Gherman, and L. B. Willett. 1998. "Milk-borne bioactive peptides". *International Dairy Journal 8*(5–6): 393–403.

Sekine, K., E. Watanabe, J. Nakamura, N. Takasuka, D. J. Kim, M. Asamoto, V. Krutovskikh, H. Baba-Toriyama, T. Ota, M. A. Moore, M. Masuda, H. Sugimoto, H. Nishino, T. Kakizoe, and H. Tsuda. 1997. "Inhibition of azoxymethane-initiated colon tumor by bovine lactoferrin administration in F344 rats". *Japanese Journal of Cancer Research: Gann 88*(6): 523–526.

Singh, V. P., N. Sachan, and A. K. Verma. 2011. "Melatonin milk; a milk of intrinsic health benefit: A review". *International Journal of Dairy Science 6*(4): 246–252.

Singh, V., V. Pathak, and K. V. Akhilesh. 2012. "Modified or enriched eggs: A smart approach in egg industry". *American Journal of Food Technology 7*(5): 266–277.

Sjogren, P., M. Rosell, C. Skoglund-Andersson, S. Zdravkovic, B. Vessby, U. de Faire, A. Hamsten, M. L. Hellenius, and R. M. Fisher. 2004. "Milk-derived fatty acids are associated with a more favorable LDL particle size distribution in healthy men". *The Journal of Nutrition 134*(7): 1729–1735.

Soedamah-Muthu, S. S., E. L. Ding, W. K. Al-Delaimy, F. B. Hu, M. F. Engberink, W. C. Willett, and J. M. Geleijnse. 2011. "Milk and dairy consumption and incidence of cardiovascular diseases and all-cause mortality: Dose-response meta-analysis of prospective cohort studies". *The American Journal of Clinical Nutrition 93*(1): 158–171.

Soerensen, K. V., T. K. Thorning, A. Astrup, M. Kristensen, and J. K. Lorenzen. 2014. "Effect of dairy calcium from cheese and milk on fecal fat excretion, blood lipids, and appetite in young men". *The American Journal of Clinical Nutrition 99*(5): 984–991.

Suraokar, M., and A. Bradley. 2000. "Targeting sheep". *Nature 405*(6790): 1004–1005.

Tacket, C. O., S. B. Binion, E. Bostwick, G. Losonsky, M. J. Roy, and R. Edelman. 1992. "Efficacy of bovine milk immunoglobulin concentrate in preventing illness after Shigella flexneri challenge". *The American Journal of Tropical Medicine and Hygiene 47*(3): 276–283.

Thorning, T. K., F. Raziani, N. T. Bendsen, A. Astrup, T. Tholstrup, and A. Raben. 2015. "Diets with high-fat cheese, high-fat meat, or carbohydrate on cardiovascular risk markers in overweight postmenopausal women: A randomized crossover trial". *The American Journal of Clinical Nutrition 102*(3): 573–581.

Vesa, T. H. 1999. "Many factors affect symptoms of lactose intolerance". *Food Reviews International 15*(2): 235–247.

Vilotte, J. L. 2002. "Lowering the milk lactose content in vivo: Potential interests, strategies and physiological consequences". *Reproduction, Nutrition, Development 42*(2): 127–132.

Vitetta, L., S. Coulson, S. L. Beck, H. Gramotnev, S. Du, and S. Lewis. 2013. "The clinical efficacy of a bovine lactoferrin/whey protein Ig-rich fraction (Lf/IgF) for the common cold: A double blind randomized study". *Complementary Therapies in Medicine 21*(3): 164–171.

Wal, J. M. 1998. "Strategies for assessment and identification of allergenicity in (novel) foods". *International Dairy Journal 8*(5–6): 413–423.

Wall, R. J., D. E. Kerr, and K. R. Bondioli. 1997. "Transgenic dairy cattle: Genetic engineering on a large scale". *Journal of Dairy Science 80*(9): 2213–2224.

Walstra, P., P. Walstra, J. T. Wouters, and T. J. Geurts. 2005. *Dairy science and technology*. CRC Press, Boca Raton.

Weaver, C. M. 2014. "Calcium supplementation: Is protecting against osteoporosis counter to protecting against cardiovascular disease?" *Current Osteoporosis Reports 12*(2): 211–218.

Yang, Z., R. Jiang, Q. Chen, J. Wang, Y. Duan, X. Pang, S. Jiang, Y. Bi, H. Zhang, B. Lönnerdal, J. Lai, S. Yin. 2018."Concentration of Lactoferrin in Human Milk and Its Variation during Lactation in Different Chinese Populations". *Nutrients 10*(9): 1235.

Zavaleta, N., D. Figueroa, J. Rivera, J. Sánchez, S. Alfaro, and B. Lönnerdal. 2007. "Efficacy of rice-based oral rehydration solution containing recombinant human lactoferrin and lysozyme in Peruvian children with acute diarrhea". *Journal of Pediatric Gastroenterology and Nutrition 44*(2): 258–264.

CHAPTER 8

Meat as a Functional Food for Health

Toiba Majeed, Nusrat Maqbool, Asima Sajad, Tahmeed Aijaz and **Zakir Showkat Khan***

8.1 INTRODUCTION

The edible parts of terrestrial animals that comprise muscle and by-products such as blood and viscera are referred to as meat (Beriain et al. 2018). Meat is considered the main food when planning meals and is an essential part of most people's diets because of the 70% biological availability of its protein in the body. It is a good source of metabolically active iron and vitamin B complex, including vitamin B12, which is not found in plant foods and enhances its absorption from other food sources. The nutrition of many plant foods is supplemented by the composition of various amino acid present in meat (Cobos and Díaz 2015). The changes in the concentration of some of the micronutrients in meat are mainly due to the continuous innovations in feed composition, changes in the slaughtering method, rearing practices and ageing (Lombardi-Boccia et al. 2005).

Various technological treatments, such as drying, heating, grounding, fermenting or curing, produce different varieties of processed or semi-processed meat products. The processing of muscle with fat as one of the main constituents produces meat products. Besides the main ingredients, various non-meat substances are used to obtain processed meat products (Beriain et al. 2018). Consumers are more concerned about the safety of meat and meat products rather than any other food. The enhancement of meat and meat products has been analysed by Arihara (2006), as well as the bioactive compounds that are present naturally in meat, including carnosine, anserine, taurine, creatine, L-carnitine and glutathione, and their enhancement effects on human physiology. Nutritional and medical fields have pointed out that the risk of chronic cardiovascular diseases in humans is related to the consumption of meat and meat products. The presence of nitrites,

* Corresponding author E-mail: zakirkhan@uumail.in

DOI: 10.1201/9781003315100-8

salts and saturated fatty acids in meat and meat products can lead to major health problems (Lücke 2000).

Thus, products made from meat and poultry are a food class that has both good and bad nutritional qualities. According to Jiménez-Colmenero et al. (2001), ruminants and muscle foods are the main providers of several bioactive substances, including iron, zinc, conjugated linoleic acid and B vitamins. However, meats and processed meats also have nutritional profiles that are frequently regarded negatively, such as in relation to the high amounts of saturated fatty acids, cholesterol, sodium, fat and calories (Whitney and Rolfes 2002). People are now expecting functional beef products with altered fatty acid profiles, less salt, lower nitrite and nitrate concentrations, and lower cholesterol and fat content. The demand from consumers for meat products that include beneficial additives has also increased significantly in the past decade (Zhang et al. 2010). Lean meat cuts, removal of adipose fat, dietary manipulation to change the composition of fatty acids and good portion control to reduce fat consumption and calorie intake can all help to reduce some of the undesirable components in meat. Additionally, the addition of substances that may promote health could enhance the nutritional profile of meat products (Decker and Park 2010). The functional component is added to animal products in order to lower the chance of developing chronic illnesses. The growing idea that food may be used for certain health advantages in addition to sustenance presents new opportunities for the meat sector (Kausar et al. 2019).

8.2 MEAT-BASED BIOACTIVE COMPOUNDS

Our understanding of food has changed as a result of the advancement of nutrition research, and it is now recognised as a way for promoting proper body growth and development. Initially, eating was merely seen as a way to provide the body with the nutrition and energy it needs. Food is now thought to play a significant role in the prevention of various diet-related disorders (Pogorzelska-Nowicka et al. 2018). Consumers today seek more nutritious foods with more health-promoting ingredients, as they are more worried about health issues. Functional foods are those that have a positive impact on health and disease prevention by supplying more nourishment than is necessary for the body's systems. These are classified as functional foods based on how they affect health, therefore both unmodified and modified foods fall under this category. These functional food ingredients perform tertiary functions, which are beneficial in controlling various physiological systems and reducing the chances of certain diseases. Initially, it was believed that functional components came mostly from plants, but subsequent study has revealed that animal products can also include these chemicals (Nisa et al. 2017). Along with chicken, eggs and fish, meat is included in the food pyramid in the category of protein-rich foods. Certainly, meat is a significant source of biologically valuable dietary proteins in many nations. Most people agree that meat and its products are significant sources of nourishment. They are essential for the body's development, upkeep and repair. In addition to being a substantial source of high biological value protein, vitamins

(vitamins B2, B6, vitamin B12, pantothenic acid and niacin), minerals (such as iron, zinc, selenium and phosphorus) and other bioactive compounds (such as conjugated linoleic acid and omega-3 polyunsaturated fats), red meat is also a source of endogenous antioxidants such as coenzyme Q10, glutathione, lipoic acid, and other bioactive constituents such as conjugated linoleic acid and omega-3 fatty acids (Arihara 2006; Arihara and Ohata 2008; Vongsawasdi and Noomhorm 2014). In addition to the basic nutrients, meat-based bioactive substances such as conjugated linoleic acid have received a lot of attention in recent times. Through the adjustment of animal feed, the constitution of animal products can be enhanced. Numerous studies have demonstrated that animal feeding practices have an impact on the levels of bioactive elements in animal products, including conjugated linoleic acid and L-carnitine. Such initiatives might result in the manufacture of meat and meat products that are distinctive (Arihara 2006).

8.2.1 Minerals and Vitamins

An essential part of human nutrition, iron is crucial for several physiological processes. It is essential for numerous metabolic activities in the body, including oxygen delivery and storage as well as electron transfer. It is also required for the healthy growth of the central nervous system, cell division, synthesis and genetic material repair (Pogorzelska-Nowicka et al. 2018; Buzala et al. 2016).

8.2.2 Iron

Myoglobin (17 kDa) is the primary protein in the sarcoplasm of muscle tissue, which has a central heme that is a combination of proto-porphyrin IX and iron (II). Heme is shielded from oxidation to Fe (III) in the hydrophobic pocket of the globin molecule. The primary binding site for oxygen, as well as nitric oxide and carbon oxide, is the sixth coordination position of the iron atom on the opposite side of the heme. Additionally, the existence of three physiological forms of myoglobin is made possible by variations in the occupancy of the sixth coordination point. (Buzala et al. 2016). Breed, age and muscular activity can all have an impact on an animal's myoglobin content in the skeletal muscles. Higher iron levels in meat are also a result of the extended lives of animals (Wadhwani et al. 2011). In terms of myoglobin per gram of fresh meat, beef has the greatest concentration (15 mg), followed by mutton (10 mg), pork (5 mg), poultry, and rabbit meat (5 mg) (Valenzuela et al. 2011). Red meat contains a tenfold greater quantity of heme than white meat, which is responsible for the meat's deep red colour (Buzala et al. 2016). The physical activity of animals may result in more heme iron, particularly in their muscles, with elevated oxidative activity. More physical fitness increases the number of mitochondria in the white fibres, turning them into red fibres and boosting the muscle's oxidative capacity (Castellini et al. 2002). Two forms of iron found in the human diet are heme iron (II), which is quickly absorbed by the body, and non-heme iron (III), which is much more easily absorbed by the body (Schönfeldt and Hall 2011). Humans should obtain heme iron from muscle protein to avoid an iron shortage, which can cause anaemia.

Heme iron, the primary source of which is red meat, is two to three times more readily absorbed by the human body (50–87%) than non-heme iron. In addition, heme iron also helps to give flesh its desirable bright red colour and its unfavourable brown hue. The bulk of dietary iron (about 60%) is non-heme iron, which is primarily found in vegetables and dairy products, but its absorption is low (2–20%) (Buzala et al. 2016). In general, some meats may be viewed as functional foods because they contain such high levels of iron. For example, emu meat comprises 5 mg of Fe/100 g of wet muscle tissue, while rhea meat has 3.2 mg (in contrast, chicken has 0.4 mg and pork has 0.36 mg) (Ramos et al. 2009). While beef and mutton contain 8–220 mg and 33 mg of iron per kg of muscle tissue, respectively. Veal contains 5.5–23 mg, poultry 5.9–20 mg, goat meat 21–44 mg, lamb 9.6–26 mg, pork 5.5–17 mg, and fish 1.2–5 mg of iron per kg of muscles (Buzala et al. 2016). It has been shown that iron in a diet that is rich in cereal items is less bioavailable due to the substantial fibre content compared to a diet incorporating meat, despite other factors influencing iron absorption in the human body (Buzala et al. 2016).

8.2.3 Selenium

Selenium (Se) is an essential trace mineral for both human as well as animal health because it is a component of selenoproteins and enzymes and it controls numerous metabolic processes. The primary mechanism by which selenium regulates redox is by neutralising and eliminating free radicals. It is an essential part of the body's defence against free radicals and plays a role in the immune system's regulation, thyroid hormone metabolism, male reproduction, avoiding the risk of preeclampsia, diabetes mellitus, cardiovascular disease (CVD) and cancers (Riaz and Mehmood 2012). The recommended daily allowance (RDA) for adult men and women from 19 to 50 years of age is 55 mg daily while lactating and pregnant women require about 60 and 70 mg daily, respectively. Meat has an average selenium concentration that ranges from 13.9 mg (duck) to 36.1 mg (pork), 28–33 mg (beef) and 35 mg (chicken breast) per 100 g of edible portion (Peter 2017; www.hsph.harvard.edu/nutritionsource/selenium/). Most seafood contains high levels of selenium, with tuna and white fish containing about 92 mg of Se, salmon with about 40 mg and sardines with about 45 mg per 85 g (3 ounces) (www.hsph.harvard.edu/nutritionsource/selenium/; www.globalseafood.org/advocate/salmon-selenium-1-healthy-serving/).

With the supplementation of selenium in animal feed, especially with organic selenium, its concentration in the meat of different animals rises (Lagin et al. 2008; Poławska et al. 2016). In Korea, in response to consumer demand, premium chicken and pork brands with ten times more selenium content than standard pigs were developed under the names "Selen chicken" and "Selen pork." The addition of selenium to animal feeds increases lipid oxidation stability, delays microbial growth (Kim et al. 2010) and increases the firmness and chewiness of chicken meat (Miezeliene et al. 2011). Supplementation of pork meat with organic selenium reduces drip loss, pale meat frequency and watery meat syndrome (Lagin et al. 2008). Additionally, meat items can be directly infused with selenium (Pogorzelska-Nowicka et al. 2018).

8.2.4 Calcium and Magnesium

The calcium cation plays a variety of important roles, including giving strength to the skeletal system and serving as a secondary messenger to control a variety of intracellular processes by its significant concentration gradient throughout the membranes. Many prevalent and chronic ailments, including osteoporosis, osteoarthritis, CVDs (hypertension and stroke), diabetes, being overweight or obese, and cancer, have been linked to calcium deficiency (Vongsawasdi and Noomhorm 2014). Magnesium is a vital nutrient and the fourth most commonly occurring cation in the human body and has been recognised as an enzymatic cofactor of more than 300 enzymatic reactions connected to the entire corporeal metabolism. Magnesium deficiency may contribute to the development of a broad range of illnesses, including ischaemic cardiopathy, high blood pressure, osteoporosis, diabetes and heart attacks (Jodral-Segado et al. 2003). Calcium has an RDA of 1000–1200 mg/day and magnesium of 320–420 mg/day, but only about 700–800 mg/day of calcium and 260–320 mg/day of magnesium are ingested from food, respectively (Moshfegh et al. 2009). Products made from meat and poultry are not regarded as being good suppliers of either of these minerals. The average magnesium concentration varies between 19 mg (beef) and 25 mg in turkey, while the average calcium level in various species ranges from 7 mg (beef) to 17 mg (pork) (Peter 2017). While the average content of magnesium in raw pork is around 23 mg/100 g, ham and sausage contain 22 and 24 mg/100 g, respectively. Chicken and lamb meats have calcium and magnesium content of about 15–17 and 23 mg/100 g, respectively. Research has shown that the concentration of these micronutrients in meat can be regulated by supplementation of animal feed or by direct addition as salts to meats (Dikeman et al. 2003). The introduction of calcium and magnesium salts has a beneficial impact on the tenderness and emulsion stability, elasticity, cohesiveness, and cooking yield while also having a positive impact on protein denaturation and calpain enzyme activation (Cruzen et al. 2015; Horita et al. 2011).

8.2.5 Vitamins

Vitamins, on the basis of their solubility, are commonly divided into two groups: fat-soluble vitamins and water-soluble vitamins. The B-complex vitamins and vitamin C are examples of water-soluble vitamins. The relevance of meat in terms of nutrition is influenced by its fat-soluble vitamins, such as vitamin A, vitamin D and vitamin K (Ahmad et al. 2018).

Vitamin A is a fat-soluble vitamin that is required for the maintenance of tissue health and normal eyesight or vision (Ross et al. 2020; Carazo et al. 2021). Vitamin A is essential for proper growth and development, immune functioning, cell signalling, and male and female reproduction. It fosters the healthy development and upkeep of the heart, lungs, eyes and other organs while also assisting with cell growth and differentiation. The light-sensitive protein rhodopsin, which reacts to light entering the eye, is a component of vitamin A, which is vital for vision. The human diet includes two sources for vitamin A: preformed vitamin A (retinol and retinyl esters) and provitamin A carotenoids (Blaner 2020). Foods derived from animal sources, such as dairy goods, eggs, seafood and organ meats, contain preformed vitamin A (Blaner 2020; Ross et al. 2020). Provitamin A carotenoids are plant pigments that are changed by the body into vitamin A in the gut. Beta-carotene, alpha-carotene and beta-cryptoxanthin are the three major provitamin A carotenoids found in human diets (https://nap.nationalacademies.org/read/9810/chapter/1). RDA for vitamin A in both adult females and males is in the range of 700–900 mcg RAE (retinol activity equivalents). Preformed vitamin A can be found in animal products such as meat, fish, liver, poultry and dairy products. Most dietary precursors to vitamin A, also known as provitamin A, come from leafy green vegetables, orange and yellow vegetables, tomato products, fruits and some vegetable oils (Blaner 2020; Solomons and Orozco 2003). Vitamin A levels in the liver vary from 6.5 to 18.9 mg RE/100 g in pigs, 1.1 to 6.7 mg RE/100 g in cattle, 2.7 to 21.5 mg RE/100 g in turkeys and 1.6 to 16.6 mg RE/100 g in chickens (Majchrzak et al. 2006).

The 1,25(OH)2-D-3 Vitamin D metabolite regulates calcium and phosphorus metabolism as a hormone with the goal of preserving regular serum calcium and phosphorus levels and maintaining proper bone mineralisation (Holick 2007). The effect of sunlight on human skin is the primary source of vitamin D for humans. Sun exposure, however, is frequently insufficient for adequate vitamin D production due to the modern Western lifestyle, where the majority of everyday activities are performed indoors. Due to this reason, dietary consumption is very important. The primary forms of naturally occurring cholecalciferol (vitamin D-3) are animal products (such as fish, meat, offal, eggs and dairy). The RDA for vitamin D for both genders between the ages of 9 and 70 years in the United States is 15 g/d (600 IU/d), and the Endocrine Society also recommends this amount for adults between the ages of 19 and 50 years to raise the blood level of 25(OH)-D-3 above 30 g/L and for this, at least 37.5–50.0 g/d (1500–2000 IU/d) of vitamin D may be needed (Schmid and Walther 2013). According to the available data, fish, particularly fish liver, has the greatest levels of vitamin D, but offal also contains significant amounts of the vitamin D. Muscle meat generally contains lower contents of the vitamin, e.g. beef has 0.0–9.0 μg/kg, pork has 1.0–23.0 μg/kg, lamb has 1.0–61.0 μg/kg, veal has 0.0–50.0 μg/kg, poultry has 0.0–14.0 μg/kg and other meat products have 0.0–23.0 μg/kg (FOPH:www.bag.admin.ch/themen/ernaehrung_bewegung/05191/index.html?lang=en). Extended literature studies found the amount of vitamin D3 and 25(OH)-D-3of meat and offal as follows: cow liver (2.7–5.3 μg/kg), cow muscle (1.5–3.4 μg/kg), kidney (5.1–9.8 μg/kg); beef mince 5.8 μg/kg, beef lion 2.7 μg/kg; lamb roast 12 μg/kg, leg steak 10.4 μg/kg, shoulder chop 8.4 μg/kg (Purchas et al. 2007); pork fillet <0.6 μg/kg, pork liver 4.4 μg/kg (Clausen et al. 2003); and chicken leg and fillets <2.0 μg/kg (Mattila et al. 2011).

The natural form of vitamin E exists in eight chemical forms: alpha-, beta-, gamma- and delta-tocopherol, and alpha-, beta-, gamma- and delta-tocotrienol, which have varying levels of biological activity. Each form has a different degree of biological activity. The only type of tocopherol acknowledged to satisfy human needs is alpha- (or α-) tocopherol. Vitamin E plays an important role in maintaining a healthy immune system as well as healthy vision and skin. Foods like cereal and dairy products, as well as the fat in meat, poultry and seafood, are all excellent sources of vitamin E. Due to its antioxidant qualities, vitamin E, a necessary non-toxic fat-soluble micronutrient, has a positive impact on the efficiency of livestock. Although it is only necessary in small amounts in the diet, it is essential for improving performance. It has been discovered that providing animals with vitamin E (-tocopherol) supplements can reduce lipid oxidation and enhance the oxidative stability of foods like meat and cheese (Gallardo et al. 2015). RDA for vitamin E (α-tocopherol) in both males and females above 14 years of age is 15 mg/D (https://ods.od.nih.gov/factsheets/VitaminE-HealthProfessional/). In a study, the average vitamin E concentration in raw and cooked beef was found to be 133 g/100 g. However, cooking reduced the amount of vitamin E in the meat by 33–44% (Bennink and Ono 1982). Beef liver and red meats are found to contain 0.4 mg/100g and 1.7 mg/100g of vitamin E. Numerous studies have shown an association between fortification of feed with vitamin E and the rise in vitamin E concentration in muscle cells over the course of supplementation. Reduced lipid oxidation with the fortification of vitamin Ewas observed in poultry (Rymer and Ian Givens 2010), cow (Gobert et al. 2010), pork (Jasińska and Kurek 2017) and rabbit meat. In addition, vitamin E increases tenderness (improved collagen solubility) (Maiorano et al. 2007), reduces toughness (lowering shear force) (Carnagey et al. 2008), increases water retaining capacity and reduces rigour mortis (inactivation of phospholipase A2) (Chen et al. 2010), and has an influence on volatile compounds in meat (Brodowska et al. 2016; Wojtasik-Kalinowska et al. 2016).

8.2.6 B-Group Vitamins

The B vitamins comprise a set of eight water-soluble vitamins [thiamine (B_1), riboflavin (B_2), niacin (B_3), pantothenic acid (B_5), vitamin B_6, biotin (B_7), folate (B_9) and cyanocobalamine (B_{12})] that play vital, closely inter-connected roles in cellular operation by serving as co-enzymes in a wide range of catabolic and anabolic enzymatic processes. Their combined impact is especially noticeable in various aspects of brain function, such as generating energy, repairing and synthesising DNA/RNA, genomic and non-genomic methylation, and creating different neurochemicals and signalling molecules (Kennedy 2016). B vitamins are produced by plants in their chloroplasts, mitochondria and cytosol, and they perform similar cellular functions in plants and animals. However, vitamin B12 is an exception because it is produced by bacteria and is usually not found in animal-derived foods. Since B vitamins dissolve in water, any excess amounts are usually expelled from the body through urine. This means that B vitamins are generally safe to consume in larger amounts than the recommended daily allowance, but they require more consistent intake than fat-soluble vitamins (Rucker et al. 2007; Green and Miller 2007). In the course of a human lifetime, the necessary intake of B vitamins changes depending on the specific needs of each stage of life. Adequate intake of micronutrients and balanced nutrition is crucial for the health and survival of both the mother and foetus during the critical period of pregnancy and lactation, as well as throughout their lives. A diet lacking essential nutrients and micronutrients can lead to depression and stress, negatively impacting the health of a mother and subsequently affecting the well-being of the infant (Rucker et al. 2007). This highlights the importance of consuming a nutrient-rich diet and having a strong social support system (Ali et al. 2022). Meat has a high nutrient content and is a good source of B-complex vitamins.

In a study, the level of B vitamins (B1, B2, B6 and B12) in three pork tissues (Longissimus dorsi, loin; Biceps femoris, gammon; and Triceps brachii, shoulder) along with other cooked, pickled and cured meat preparations was identified. The results showed that the estimation of the daily contribution of B (B1, B2, B6 and B12) vitamins provided by fresh pork represents over 25% of the RDA (Esteve et al. 2002). B vitamin levels in raw and processed meat cuts were also reported by some researchers. The highest concentration of thiamine was found in pork meat cuts, which ranged from 0.6 to 0.9 mg per 100 grams. Horse and ostrich meats had a thiamine concentration of over 0.1 mg per 100 grams (0.18 and 0.16 mg/100 g, respectively), while veal had a thiamine concentration of 0.11 mg per 100 grams. The concentration of thiamine was found to be very low in breast meat samples of chicken and turkey. The riboflavin content in beef meat cuts varied between 0.09 and 0.17 mg per 100 grams, with fillet having the highest concentration. Similar riboflavin concentrations to beef were found in lamb, ostrich, pork and rabbit meats. Horse meat had the highest concentration of riboflavin (0.20 mg/100 g) among all samples studied, whereas poultry breast had the lowest concentration (0.03 mg/100 g). The niacin content was relatively consistent across the species and meat cuts studied, with the highest amounts detected in horse meat and poultry. The impact of cooking on the amount of B vitamins in meat was further reported, which revealed that thermal procedures reduced B vitamin concentration. Among the vitamins, thiamine was certainly most prone to thermal breakdown, while riboflavin was usually more heat stable (Lombardi-Boccia et al. 2005). Dobreva et al. (2018) determined and compared the average daily vitamin intake of various kinds of fish and shellfish from Black Sea waters, as well as their vitamin B_{12} contents. The vitamin B_{12} content was examined in 12 Black Sea specimens (ten fish and two molluscs). With the exemption of two fish species (sprat and garfish), all other fish (demersal fish, rapana and mussel) provided a significantly higher content of cyanocobalamin. When the RDI is taken into account, certain fish such as bluefish, grey mullet, bonito and horse mackerel provide more than 500% of the required daily amount of vitamin B_{12} (Dobreva et al. 2018). Although these fish contain less cyanocobalamin compared to others, they still offer more than 100% RDI. The USDA food composition database lists the amount of vitamin B_{12} in eel, mackerel and turbot as 3.00 $\mu g.100\ g^{-1}$ww, 13.67 $\mu g.100\ g^{-1}$ww and 2.20 $\mu g.100\ g^{-1}$ ww, respectively.

8.3 MEAT-BASED FUNCTIONAL FOODS

According to Ashwell (Ashwell 2002),

> A food may be classified as "functional" when it has a beneficial impact on one or more specific biological functions that are significant for promoting good health and reducing the risk of disease, alongside its nutritional benefits that are considered acceptable.

A functional food must be a food product that is consumed as part of a balanced diet, and its beneficial effects must be demonstrated at levels that are consistent with typical consumption patterns(Fernández Ginés et al. 2005; Dominguez 2013). The term "neutraceutical" is commonly used to refer to a concentrated natural substance that is added to a non-food matrix, such as pills, capsules or powder, and is intended to provide nutritional benefits as a complement or supplement (Kalra 2003). Functional foods can be created through various methods, including increasing the amount of a beneficial constituent already found in the food, adding a beneficial substance that is not naturally present in the food, removing a harmful component from the food, substituting an unhealthy component with a healthier one, improving the absorption of the added ingredient or bioactive substance, or combining any of these techniques. There are various approaches to producing functional meat, including altering the fat and fatty acid levels and incorporating functional components like fibre, vegetable proteins, mono- and polyunsaturated fatty acids, vitamins, minerals and phytochemicals. Additionally, whole foods such as seaweed and walnuts can also be used as functional ingredients (Olmedilla-Alonso et al. 2013; Decker and Park 2010; Fernández Ginés et al. 2005; Jiménez et al. 2001; Jiménez-Colmenero et al. 2010; Librelotto et al. 2008; Bastida et al. 2009).

Meat products contain crucial functional nutrients such as vitamins A, E, and C, as well as essential minerals such as magnesium, calcium and potassium. Furthermore, non-essential elements like dietary fibre (DL), antioxidants, bioactive peptides, long-chain omega-3 fatty acids, conjugated linoleic acid, prebiotics and probiotic bacteria can be incorporated into meat products to boost their nutritional properties.

8.3.1 Meat (Beef and Mutton)

Meat, as well as meat-based products, are a naturally occurring source of bioactive substances such as minerals, peptides, fatty acids or vitamins with a beneficial impact on health (Nowicka et al. 2018). Meat and its derivatives are crucial foods that include vital nutrients required for regular physiological and biochemical activities, such as all essential amino acids, omega-3, fatty acids, B complex, high-quality vitamins, proteins, iron, zinc and vitamin B12 (Hoffmann et al. 2010; Kausar et al. 2019). Meat is an excellent source of anticancer functional components (conjugated linoleic acid; CLA), bioactive compounds (L-taurine, choline, creatine, carnitine) and antioxidants (carnosine and anserin) (Hoffmann et al. 2010).

There are numerous studies on methods for improving the bioactive components in functional foods made from meat, but there is a shortage of in vitro studies that can confirm the level of bioavailability of these added substances (Garci-Martin et al. 2012; González et al. 2012; Olivero-David et al. 2011; Vossen et al. 2012). In particular, there is little evidence available regarding evaluations of how well these items operate in humans: their intended end users. To determine the effectiveness of functional food, reliable proof can only be obtained through human research studies that utilise validated methods to assess the properties of the dietary ingredients and biological endpoints (Asp et al. 2003; Aggett et al. 2010). The evaluation of functional foods containing meat primarily focuses on their potential to lower the risk of cardiovascular diseases, as meat is commonly recognised as a significant contributor to this health issue (Pan et al. 2012). Additionally, LDL cholesterol and plasma are employed as biological markers for regular analysis to track cardiovascular disorders.

Functional elements such as plant proteins, dietary fibre, herbs, fruits, cereal by-products, lactic acid bacteria, spices and legumes can be incorporated into meat products to increase their functional and nutritional value (Kausar et al. 2019). The oxidative degradation of lipids, which results in unfavourable changes in flavour, texture and nutritional value, can be stopped by adding natural extracts with antioxidant capabilities (such as sage, rosemary, citrus peel, soy and sesame seeds) to meat products (Grujić et al. 2012). It is possible for one to modify the composition of meat as well as meat products to minimise or eliminate certain hazardous components (Grujić et al. 2012). Functional alterations include changes in the fatty acid content and cholesterol levels

(Bhat and Bhat 2011). A certain element in the meat can be lowered or enhanced by making functional changes to the animal diet. The polyunsaturated fatty acid content of meat can be increased by raising the polyunsaturated fatty acid level in the diet of animals. Adding selenium, vitamin E, magnesium or iron to the diet of the animals boosts the concentration in the meat (Kumar et al. 2014).

Yılmaz and Daglıoglu (2003) prepared meatballs by substituting oat bran for fat and investigated the influence on the composition of fatty acids and physicochemical qualities. They found that the samples with additional oat bran had a higher ratio of total unsaturated fats to total saturated fats and fewer total trans-fatty acids than the control meatballs. Ylmaz (2004) investigated how the addition of rye bran affected the fatty acid composition and quality attributes of low-fat meatballs. They found that the samples with added rye bran had higher ratios of total unsaturated to total saturated fatty acids and less total trans-fatty acids than the control meatballs. In another study by Yılmaz (2005), a functional component called wheat bran was added to low-fat meatballs in order to examine their physicochemical and sensory property. The author discovered that the samples with additional wheat bran had greater ratios of total unsaturated to total saturated fatty acids and lower total trans-fatty acids than the control meatballs. In a study conducted by Eim et al. (2008), the effects of adding carrot dietary fibres on the ripening process of a dry-fermented sausage called sobrassada were investigated. Different amounts of carrot dietary fibre (3%, 6%, 9% and 12% w/w) were added to the sausage, and the physicochemical, microbiological and sensory characteristics were analysed during the maturation process. The results showed that the addition of carrot dietary fibre had a significant impact on the pH of the sobrassada samples. Samples with more than 3% DF showed unsuccessful fermentation.

Verma et al. (2013) created nuggets of sheep meat, incorporating guava (Psidium guajava L.) powder as a source of dietary fibre and as an antioxidant with functional properties. The authors discovered that adding guava powder to fried sheep meat nuggets could prevent them from lipid oxidation during refrigerated storage. As a result, adding guava powder to sheep meat nuggets at a concentration of 1% did not impact the sensory qualities of the product. Therefore, it can be utilised as a dietary fibre source and antioxidant without affecting the overall appeal of the nuggets.

Hur et al. (2004) found that the substitution of fat considerably boosted the concentration of CLA. The concentration of conjugated linoleic acid in beef patties was not significantly changed by storage for 14 days. The colour stability was enhanced by using conjugated linoleic acid sources in place of fat, perhaps by preventing oxidation of the oxymyoglobin and lipids.

Tang et al. (2001) investigated how the addition of tea catechins to cooked red meat and poultry would affect lipid oxidation. They found that adding 300 mg/kg of minced meat significantly decreased the pro-oxidative effect of NaCl and prevented lipid oxidation in cooked meat patties. One possible explanation for this phenomenon is that tea catechins have a strong affinity for muscle lipid bilayers and can effectively scavenge free radicals, thereby preventing lipid oxidation in cooked muscle foods.

By substituting IVOs (60% to 100%) for beef fat (10%) in frankfurters made with cottonseed, olive and palm oils, Vural et al. (2004) discovered that the SFA:PUFA ratio and linoleic acid and oleic content significantly increased without affecting the texture, colour, flavour, appearance or other sensory attributes. According to other studies, introducing high oleic sunflower oil to low-fat frankfurters can provide them with a source of monounsaturated fat.

DF integration in meat products is ideally suited due to its numerous functional qualities, such as texture modification, emulsion stability, flavour alteration and water retention. Dietary fibre is currently employed in meat products in addition to its nutritional benefits to improve quality criteria like rheological qualities, emulsion stability, cooking yield, formulation costs and textural profile. When it comes to clinical points, dietary fibre-enriched meat products outperform typical meat products. One element that could be added to meat products to create low-fat, high-fibre functional meat products is dietary fibre (Cofrades et al. 2000). Due to the presence of phenolic and antioxidant compounds, the use of ADF in meat-based formulations not only prolongs the duration of storage but also improves the sensory qualities, physicochemical properties and texture of meat products (Das et al. 2015; Madane et al. 2019).

According to research conducted by Prakongpan et al. (2002), pineapple cores (PC) can serve as a valuable source of dietary fibre and cellulose that, depending on the type and particle size, can be used as an important component in the preparation of meat and bakery products. A PC with a mesh size of over 170 was found to be less effective in reducing shrinkage and improving the texture of beef burgers, whereas a smaller-sized PC was more suitable for these purposes. The water interaction of fibres, particularly their ability to retain water, appears to have a significant impact on their functional characteristics. The risk of coronary heart disease and cardiovascular illnesses is substantially and independently correlated with dietary fibre consumption, especially water-soluble fibre (Bazzano et al. 2003).

Liu et al. (2003) examined that there is a correlation between the consumption of dietary fibre and grain products and weight changes and the likelihood of developing obesity in middle-aged women. The research found that eating high-fibre whole-grain foods was linked to a decrease in weight gain, whereas consuming refined-grain foods was associated with an increase in weight gain. This showed the significance of differentiating between whole-grain products and refined-grain products to help with weight control. Fibre can be classified as soluble or insoluble depending on how easily it dissolves in water. These two types of fibre have different physiological characteristics. Water-soluble fibre can create a thick solution, can hold a significant amount of water but is incapable of retaining it, and can be easily broken down through fermentation. In contrast, insoluble fibre has a lower viscosity, a lower capacity to hold water but a higher capacity to hold water.

In a study conducted by Sáyago-Ayerdi et al. (2009), the effect of grape antioxidant dietary fibre (GADF) on the lipid oxidation of raw and prepared chicken hamburgers was examined. The results showed that incorporating GADF into chicken breast meat prevented oxidation for a duration of up to 13 days of refrigeration storage, thereby prolonging the food's lipid stability and shelf life. When dietary fibres are added to meat products, the health benefits are increased and the stability is improved without the dish losing its appealing characteristics (Verma et al. 2010). According to Bhat and Hina (2011), meat as well as meat-based items can be adjusted by adding fibre sources to reduce the probability of chronic diseases linked with them. The incorporation of these additives into meat products allows processors to improve the nutritional value and health aspects of their products.

8.3.2 Chicken Meat

Kumari (2016) showed that high-quality chicken flesh cutlets can be made by integrating health-promoting beneficial components such as mushrooms, wheat gluten and sesame seeds at 2.5%, 1.5% and 10% respectively. The organoleptic quality of functional chicken flesh cutlets was shown to be higher than that of control samples. Vacuum-wrapped chicken meat cutlets had superior quality and acceptability than conventional packaged chicken meat cutlets. The addition of functional elements such as mushrooms, wheat gluten and sesame seeds to chicken flesh cutlets boosted their nutritional value through rising fibre and protein content. Furthermore, it improved the physical quality of the formed cutlets by boosting water retention and minimising cooking loss.

According to Kumar et al. (2010), incorporating green banana flour (GBF) and soybean husk flour (SHF) into chicken nuggets enhanced their nutritional value and resulted in beneficial outcomes due to the inclusion of dietary fibres.

Verma et al. (2010) prepared low-fat-based chicken nuggets by substituting apple pulp for sodium chloride and came to the conclusion that this allowed for the production of functional chicken nuggets with superior sensory ratings (6.9–7.4) and, low-fat, high-fibre and low-salt content.

By substituting chickpea (Cicerarietinum L.) husk flour for sodium chloride, Verma et al. (2012) made low-fat chicken nuggets. They found that, aside from flavour and appearance, replacing 40% of conventional salt with a blend had no effect on organoleptic attributes. When chickpea hull flour was added, the amount of dietary fibre significantly increased while the amount of total cholesterol significantly dropped. Oat flour was added to low-fat chicken nuggets by Santhi and Kalaikannan

(2014), and they came to the conclusion that up to 10% oat flour might be used for enhanced sensory characteristics and texture.

Zargar et al. (2014) used chicken meat and then substituted pumpkin as a functional ingredient, and they created relatively inexpensive fibre-enriched chicken sausages. They also investigated how the addition of pumpkin would affect the characteristics and durability of chicken sausages during storage that were packaged aerobically. They discovered that as the amount of pumpkin rose, the crude fibre content increased significantly ($P < 0.05$). The sample that included pumpkin contained more crude fibre (1.81%) than the control sample (0.78%). This increase in crude fibre may be related to pumpkins' high-fibre content.

Conjugated linoleic acids are said to have anti-diabetic and anticancer qualities in addition to having noticeable impacts on body composition. Due to their extreme sensitivity to food manipulation, chicken has been regarded as an excellent model for nutritional studies. Because chicken flesh is low in cholesterol and fat, consumers would greatly benefit from the dietary inclusion of CLA in poultry diets. Additionally, the human population would have acquired CLA, which is essential for health and well-being, through the intake of meat and egg products that are CLA-enriched. Therefore, it can be said that CLA has a lot of potential to be added to poultry meats as a functional food (Kumari et al. 2017).

8.3.3 Rabbit Meat

The rabbit diet manipulation is very effective at raising levels of PUFA, CLA, EPA, DHA, vitamin E and selenium, and lowering the n-6/n-3 ratio, which is crucial for controlling CVD and other chronic diseases, consumers may regularly consume bioactive compounds from rabbit meat. Rabbit meat is higher in protein with necessary amino acids than meat from other animal species, especially the L. dorsi muscle, which also has a higher digestibility value. It also has less cholesterol than meat from other animal species. Rabbit meat is also distinguished by being a large source of vitamin B (vitamins B2, B5, B6, B12), having very low levels of MUFA, high levels of PUFA and high amounts of n-3 FA, as well as having very low-salt and high phosphorus concentrations. While rabbit meat already has high nutritional and dietetic qualities, it can be further strengthened with bioactive substances to produce meat that is termed functional.

Adding high amounts of polyunsaturated fatty acids (PUFA) or long-chain omega-3 fatty acids to the diets of rabbits, such as through linseed (whole or oil) or fish oil, can increase the essential fatty acid and bioactive omega-3 fatty acid content in their meat. The fatty acid profile in the meat is directly affected by the feed. To achieve this enrichment and maintain meat quality, a maximum of 4% linseed in the rabbit diet is recommended. Supplementing the diet with PUFA-rich sources, like linseed, should occur for 2–4 weeks prior to slaughter for effectiveness. To improve the oxidative stability of rabbit meat, antioxidants are necessary. To improve the quality of rabbit meat and create a functional food product, several dietary fortification strategies can be employed. For example, incorporating 200 mg α-tocopheryl acetate/kg feed can protect the meat from oxidation, increase its shelf life and boost its vitamin E levels. Adding 0.5% CLA to the diet can increase the CLA content in the meat, providing a healthy value-added product for human consumption. By supplementing the diet with supra-nutritional levels of selenium (0.5 mg/kg), selenium-fortified rabbit meat can be produced, which can significantly contribute to human selenium intake. Additionally, feeding rabbits with highly digestible fibre diets can enhance the meat's biologically significant branched-chain fatty acids. Fortifying the rabbit diet with PUFAs, long-chain n-3 FAs, vitamin E and selenium can also meet the criteria for rabbit meat to be classified as a functional food (Dalle and Szendro 2011).

Dal Bosco and others (2004) found in investigations of rabbit meat that adding n-3 PUFA to the meat did not raise the level of oxidation. The long-chain polyunsaturated n-3 content of meat increased due to the linolenic acid-vitamin E diet, which also improved the resistance to oxidative damage and nutrient content of the meat.

8.3.4 Pork Meat

Soy protein isolates (2%) obtained by thermal/enzymatic processing were added to pig frankfurters by Feng et al. (2003). They came to the conclusion that heat and enzyme hydrolysis of soy proteins had varied effects on textural qualities, decreasing hardness and improving hardness, breaking strength and cohesiveness in the first case.

In beef, the oxidation of unsaturated fatty acids as time goes on can lead to spoilage and reduced shelf life. However, most studies on pig meat have not found a connection between changes in fatty acid levels and the oxidation of myoglobin, which affects muscle colour (Wood et al. 2004).

According to Fagali and Catala (2008), conjugated linoleic acid is oxidised preferentially, which prevents other FAs from oxidising. Joo et al. (2002) found that dietary CLA reduced the quantity of thiobarbituric acid-reactive compounds (TBARS) in pigs given CLA, which inhibited lipid oxidation and changed the fatty acid component as well as improved the stability of the colour of pork loin when kept refrigerated.

Enser et al. (2000) investigate the effect of feeding linseed to pigs on the levels of n-3 polyunsaturated fatty acids (PUFA) in pork meat. The study found that pork meat can be a valuable source of n-3 PUFA in the human diet and that it is possible to increase the levels of these fatty acids in pork meat by feeding pigs with linseed. In numerous animal models, conjugated linoleic acid has been shown to have anti-carcinogenic and antioxidant characteristics. Olive oil is the vegetable oil that contains the highest amount of monounsaturated fats. This type of oil is considered to be highly beneficial for our health and has been associated with a reduced risk of heart disease and breast cancer. Additionally, olive oil is known to have a high biological value (Pappa et al. 2000)

In low-fat frankfurters and other forms of cooked goods, vegetable oils have also been employed as a partial replacement for the back fat of pork, resulting in commodities with better fatty acid profiles and levels of cholesterol than traditional products (Muguerza et al. 2001).

Studies by Pappa et al. (2000), showed that greater quantities of olive oil had lower overall acceptability, even when colour features were unaffected when olive oil is used to substitute (0% to 100%) back fat of pork for the manufacturing of low-fat frankfurters.

Muguerza et al. (2001), developed a traditional Spanish sausage and replaced varying amounts (0–30%) of pork backfat with pre-emulsified olive oil. The resulting sausages had comparable texture and colour to commercially available ones. However, the sausages with olive oil had higher levels of healthy oleic and linoleic acids and lower levels of cholesterol. This indicates that replacing up to 25% of pork backfat with olive oil can be a beneficial nutritional alternative. Ansorena and Astiasarán (2004) reported that increasing the amount of monounsaturated fatty acids (MUFA) and preventing lipid oxidation during storage by adding olive oil to sausages was more successful than using vacuum-storing techniques.

Muguerza et al. (2002) research on fermented sausages, substituting olive oil for 20% of the pork backfat has no impact on weight reduction and results in lighter, more yellow-coloured sausages. The product has a good smell and flavour, but its look is undesirable due to the heavily wrinkled surfaces and the progression of case hardening. Muguerza et al. (2003) discovered that the oxidation process was greatly slowed and the MUFA content was significantly raised. In their studies into "salami" products, Severini et al. (2003) discovered that with the exception of water activity and stiffness, partial substitution of pork backfat by extra virgin olive oil had no significant effect on the chemical, physical and sensory properties of the products. The addition of extra virgin olive oil, which is high in unsaturated fatty acids, had no effect on lipid oxidation, owing to the antioxidant effect of both polyphenols and tocopherols. Sensory evaluations revealed no variations from the standard formulation. Ansorena and Astiasarán (2004b) discovered that using linseed oil instead of pork backfat in the making of dry-fermented sausages lowered the n-6:n-3 ratio from 14.1 to 2.1 by increasing the amount of linolenic acid present. This change had a notable effect on the sausages' nutritional value while having no significant effect on their flavour or oxidation.

Jo et al. (2003) conducted a study on the potential of using irradiation green tea leaf extract to improve the functional properties of raw and cooked pork patties. Their results indicated that this extract could be beneficial as it did not negatively impact the physical or sensory characteristics of the patties and exhibited favourable biochemical properties. Essentially, the study found that the addition of irradiation green tea extract powder could improve the quality of pork patties without any negative effects. In a study conducted by Jo et al. (2003), fried pork patties were mixed with freeze-dried green tea that had been irradiated. The results showed that adding green tea had no adverse effects on the sensory or physical qualities of the patties and actually helped reduce lipid oxidation, resulting in less cooking loss. The addition of green tea extract also exhibited beneficial biochemical properties. Another extract that is commonly used in meat products is rosemary, which contains high levels of phenolic compounds with antioxidant properties. These include carnosic acid, carnosol, epirosmanol, isorosmanol, rosmanol, rosmarinic acid, rosmariquinone and rosmaridiphenol (Gil et al.2001; Fernández-López et al. 2003).

Yang et al. (2007) adding hydrated oatmeal and tofu to low-fat pork sausages resulted in a slight reduction in fat content, with values of 3.14 g/100 g and 3.43 g/100 g at a 25% addition level, respectively, compared to the control sample's 4.09 g/100 g.

8.4 POTENTIAL HEALTH BENEFITS OF MEAT AND MEAT PRODUCTS

Consumption of meat can be advantageous for our health as it contains essential nutrients such as macronutrients (proteins and carbohydrates) and micronutrients (vitamins and minerals). These nutrients aid in boosting our immune system, promoting overall body development and maintaining haemoglobin levels. Additionally, meat consumption can alleviate skin ailments and provide sustained energy. Overall, meat is associated with various health benefits due to its nutritional value. The meat industry has been exploring the use of fermented meat products to improve sensory qualities and nutritional content by breaking down meat proteins through proteolytic reactions, resulting in new components. This has been a topic of interest in meat science and industry in recent years. During the fermentation of meat, the breakdown of meat proteins into peptides is mainly carried out by enzymes naturally present in the meat itself, including cathepsin B, D, H and L. While some microbial proteolytic enzymes also play a role in this process, lactic acid bacteria, which are commonly found in fermented meat products, have relatively low proteolytic activity. However, lactic acid bacteria do contribute to the breakdown of meat proteins by lowering the pH, which enhances the activity of muscle proteolytic enzymes (Arihara and Ohata 2008).

Conjugated linoleic acid (CLA) is a bioactive compound found in meat, which has been the subject of research due to its potential health benefits. It consists of a blend of positional and geometric isomers of octadecadienoic acid that are produced through bacterial hydrogenation in the rumen (Vongsawasdi and Noomhorm 2014). The compound CLA is a sought-after bioactive ingredient in the development of functional foods. In particular, there has been growing interest in creating fermented meat products with probiotics that contain CLA as a nutraceutical component. Recent studies have shown that probiotic bacteria can enhance the production of CLA in both media and fermented meat products. Additionally, there is evidence from epidemiological studies to suggest that consuming high-fat dairy foods and CLA may be associated with a lower risk of colorectal cancer (Larsson et al. 2005). Conjugated linoleic acid (CLA) has been found to have various positive effects, including but not limited to, its ability to fight against cancer, reduce the risk of atherosclerosis, act as an antioxidant and modulate the immune system. Conjugated linoleic acid (CLA) might have beneficial effects in managing obesity, lowering the likelihood of developing diabetes and regulating bone health (Arihara and Ohata 2008). Taurine is an amino acid that contains sulfur and can be produced in small quantities in the human body through the synthesis of methionine and cysteine. It is primarily found in bile and can also be synthesised in the liver from methionine and cysteine. Animal products such as meat, poultry, fish, clams and shellfish are rich sources of taurine compared to vegetable products. For instance, lamb contains approximately 110 mg of taurine per 100g, pork contains 50–70 mg per 100 g, and beef contains 77 mg per 100 g (Purchas et al. 2004). Taurine plays a crucial role in maintaining the healthy functioning of the nervous system, myocardium and retina of the eye, as well as supporting bile formation in the body. Additionally, it has been found to lower blood cholesterol levels, thus reducing the risk of developing atherosclerosis. Inadequate levels of taurine can result in the degeneration of the myocardium (cardiomyopathy) (Ostaszewski 2018).

Creatine is a compound made up of three amino acids: arginine, glycine and methionine. It is primarily found in the muscles, organs and brain. The liver, kidneys and pancreas are responsible for producing creatine, with the ability to produce around 1 gram per day. The body stores approximately 90% of its creatine in the muscles as free creatine and phosphocreatine, which is utilised for energy (Kaviani et al. 2020). Creatine occurs naturally in animal tissues like meat, fish and poultry. Fresh beef contains a substantial amount of creatine, with 401 mg present per 100 g. On the other hand, lower concentrations of creatine are found in organ tissues such as beef heart and liver. Lamb meat has a creatine content ranging from 278 to 511 mg per 100 g, while pork has a lower concentration, ranging from 247 to 347 mg per 100 g (Ostaszewski 2018). Creatine and its phosphorylated form play an essential role in energy metabolism by regulating the levels of ATP at specific locations during high-intensity exercise. Therefore, having a reserve of phosphocreatine before starting physical activity can improve both performance and recovery (Venderley and Campbell 2006). Creatine kinase has additional roles besides its function as a catalyst for the conversion of creatine to phosphocreatine for energy storage. It also serves as an energy transporter and helps regulate the ATP/ADP ratio in specific areas of the cell where it is linked to processes that consume or produce ATP (Greenhaff 2001).

In a cellular environment, creatine acts as an antioxidant by eliminating radical species (Sestili et al. 2011). It is thought to enhance muscle strength and is proposed as an additional therapy for situations that result in muscle weakness, such as muscular dystrophy, congestive heart failure, Huntington's disease, McArdle's disease (a type of glycogen storage disease or myophosphorylase deficiency), amyotrophic lateral sclerosis, myasthenia gravis, Parkinson's disease and post-injury or post-surgery (Schmid 2009; Gualano et al. 2012).

Dietary fibre can be categorised into two types: soluble and insoluble. These types of fibre offer many health advantages, such as keeping the digestive system healthy, reducing blood cholesterol levels, regulating blood sugar levels and providing a non-caloric bulking agent that can assist in weight loss by substituting high-calorie food components such as fat. To put it differently, both soluble and insoluble fibre play an essential role in promoting overall health by supporting digestion, regulating cholesterol and blood sugar levels, and aiding in weight management. According to the Dietary Guidelines for Americans (2005), insufficient consumption of dietary fibres is a common issue among adults, suggesting that incorporating fibre fortification in meat products could promote better health outcomes. Studies have demonstrated that adding significant levels of fibre (2–3 g/serving) to sausages can be achieved without compromising their taste or sensory attributes (Besbes et al. 2008; Choi et al. 2008; Salazar et al. 2009; Yılmaz and Gecgel 2009). Apart from the advantage of higher fibre intake, meat products containing dietary fibres provide additional benefits like replacing fat, enhancing the ability to retain water and improving resistance to oxidation especially when the fibre source contains phenolic antioxidants (Choi et al. 2008; Sáyago-Ayerdi et al. 2009).

Antioxidants have been found to have potential benefits for immune function, heart disease and cancer. However, many people consume vitamins A, C and E at levels below the recommended dietary intake (Dietary Guidelines for Americans 2005). Adding vitamin C to meats is challenging because it is not very stable at meat pH and can promote lipid oxidation (Haak et al. 2009). Moreover, it is often prohibited due to its ability to stabilise meat colour, which is considered adulteration. While β-carotene is a crucial source of dietary vitamin A, meats are generally not good sources of β-carotene or other carotenoids except for chicken. Increasing β-carotene concentrations can be achieved through dietary supplementation (Bou et al. 2009). However, adding β-carotene to meats is complicated because of its chemical instability, and its exogenous addition or dietary fortification is limited because it alters the muscle colour (Torrissen 2000).

Meats, particularly poultry and ground beef, are significant dietary sources of vitamin E, specifically α-tocopherol (Ma et al. 2000). Meat products can provide higher amounts of vitamin E if they are fortified with α-tocopheryl acetate or if the animals are given supplements containing this compound. When it comes to fortifying foods with vitamin E, α-tocopheryl acetate is the most effective ingredient to use. This is due to the acetate group added to the tocopherol, which shields the molecule from oxidation (McClements and Decker 2008). Meats are sometimes supplemented with mixed tocopherols to prevent lipid oxidation (Mitsumoto 2000). The vitamin E level in meat can be increased through dietary fortification. As a result, α-tocopherol plays a significant role in safeguarding meat flavour (Schaefer et al. 1995). Supplementing livestock with α-tocopheryl acetate can offer significant advantages in terms of meat quality and nutritional value. This is because α-tocopherol helps to protect meat colour by preventing lipid oxidation, which causes discolouration through myoglobin. Therefore, the inclusion of α-tocopherol in livestock diets can lead to improvements in both meat colour and nutritional composition.

Dietary minerals play a vital role in bone health, nerve and muscle function, blood sugar regulation and prevention of diseases such as hypertension, osteoporosis, diabetes and cardiovascular disease (Whitney and Rolfes 2002). However, calcium, potassium and magnesium are frequently deficient in the diet (Dietary Guidelines for Americans 2005). Nevertheless, adding iron to meat can cause discolouration and lipid oxidation, making it problematic (McClements and Decker 2008). On the other hand, fortification of processed meat with magnesium potassium and calcium can be done without significantly altering their taste and texture (Moon et al. 2008; Schoene et al. 2009; Selgas et al. 2009). Fortifying processed meats with calcium, potassium and magnesium offers the additional benefit of reducing sodium levels, leading to an improved nutritional profile (Moon et al. 2008). The addition of salt, such as potassium lactate, can also be advantageous, as it can hinder lipid oxidation and microbial growth (Moon et al. 2008) and activate the protease calpain, ultimately resulting in more tender meat. Studies have demonstrated that infusing calcium into beef cattle can increase muscle calcium concentrations (Dikeman et al. 2003) leading to better beef tenderness after ageing (14 days). However, the presence of these minerals can inhibit certain proteolytic enzymes essential to the development of flavour in dry cured meats (Armenteros et al. 2009) although their presence does not negatively affect salty flavour (Armenteros et al. 2009). There has been a lot of attention given to the idea of adding selenium to livestock feed in order to increase the amount of selenium in meat, which may be good for human health. Studies have shown that feeding livestock with selenium can increase the levels of this mineral in the muscles of beef (Juniper et al. 2008), pork (Morel et al. 2008) and chicken (Pan et al. 2007), which is thought to be linked to the activity of an enzyme called glutathione peroxidase that helps break down certain types of fats. However, while increased levels of this enzyme were observed in beef, it did not seem to improve the stability of the meat (Juniper et al. 2008). Research into the effects of selenium on human health has focused on its potential to reduce the risk of prostate cancer and support proper thyroid function (Lippman and Watkins 2004; Combs et al. 2009).

8.5 RECENT ADVANCEMENTS IN THE DEVELOPMENT OF FUNCTIONAL MEAT

In improving product life, making safe and quality products, and having a good impact on human health, recent advances in meat and meat products have played a major role. Meat contains nutrients such as vitamins, minerals and fats, and it is also a rich source of protein. For improving shelf stability and nutritional value of meat products, some natural antioxidants in the form of extracts and whole ingredients are added. In order to maintain the nutritional value of meat and meat products, novel processing plays a significant role during processing (Khalid et al. 2022). Meat and meat products can be modified by adding ingredients that are healthful or by eliminating or reducing components that are considered harmful, and utilising these ingredients in meat products allows the processors to improve the health and nutritional qualities of the products.

In order to reach client demand all over the world, food companies are producing novel methods and meat products. In meat processing, nanotechnology is used in the form of nanomaterials that helps in financial matters, food handling and quality purpose (Ramachandraiah et al. 2015). In improving the bioavailability, sensory absorption, bioactive chemical delivery and antimicrobial effects of food nanomaterials play a very important role. Besides nanotechnology, ultrasound has been found effective in several fields of food technology to replace or support traditional forms like meat tenderisation ,cutting and degassing (Boateng and Nasiru 2019). According to the latest evidence, a strategy based on the use of short frequencies (700–1100 nm) in the near-infrared region (NIR) of the electromagnetic range can non-invasively gauge boundaries relevant to meat quality in live animals. This approach is able to remove tissues such as fat and lean, measured through the skin. Depending on NIR spectroscopy, the value of meat quality in live animals can be examined by rapid and non-invasive methods (Chapman et al. 2020).

The use of grain processing products in the technology of combined meat products allows producing a product of stable quality. This enhances the nutritional and biological value of products and is a technology that promotes a stable and even distribution of ingredients (Gabdukaeva et al. 2021). In legumes and oilseeds, the protein content is high and thus is proven important in the production of functional foods. It is determined that the addition of collagen and rye bran alters the technological and sensory characteristics of products, and in the technology of processed meat products, it is a good substitute for fat (Hjelm et al. 2019). The meat products are also enriched with the use of fruits, vegetables, berries and products of their processing. For the prevention of various diseases like inflammatory diseases of the digestive system and colorectal cancer in humans, functional meat products enriched with anthocyanins can be an effective strategy (Fernández Fernández et al. 2018).

8.6 CONCLUDING REMARKS AND FUTURE DIRECTIONS

Meat systems have significant potential for providing important nutrients to the diet. The inclusion of bioactive food ingredients or the addition of bioactive compounds into animal diets can alter the nutritional composition of meat products. Novel technologies play a crucial role in retaining the nutritional value of meat products during processing. It is deduced that the recent advances in meat and meat products have played an important role in improving the shelf life of the product, making quality products that are safe to be consumed and having a positive impact on human health. Many strategies have been adopted to produce meat products that are healthier. The producers of meat-based functional foods claim that products having bioactive compounds can promote health and decrease disease risk factors. However, there are not very many studies to evaluate the effect of these compounds on health. We need more advancement in the field of human intervention studies. The incorporation of functional ingredients in meat and meat products is increasing because of its various functional properties such as texture modification, emulsion stability, water retention and neutral flavours for the development of a nutritionally balanced diet. Today, the use of functional ingredients in meat products is increasing, and several studies are required to understand the potential that these components provide so that processing lines can be developed that are able to cope with future changes in formulations and ingredient systems. For the development of low-salt meat products, palatability is the critical barrier; further, immense work is still needed to be done to find a suitable replacement for the multifunctional sodium chloride.

REFERENCES

Aggett, P., Antoine, J.-M., de Vries, J., Gallagher, A., Hendriks, H., Kozianowski, G., Meijet, G., Richardson, D., Rondeau, V., & Tweedie, G. (2010). *Beyond PASSCLAIM–guidance to substantiate health claims on foods.*

Ahmad, R. S., A. Imran, and M. B. Hussain. 2018. "Nutritional composition of meat". *Meat Science and Nutrition*, 61(10.5772), 61–75.

Ali, M. A., H. A. Hafez, M. A. Kamel, H. I. Ghamry, M. Shukry, and M. A. Farag. 2022. "Dietary vitamin B complex: Orchestration in human nutrition throughout life with sex differences". *Nutrients*, 14(19), 3940.

Ansorena, D., & Astiasaran, I. (2004a). Effect of storage and packaging on fatty acid composition and oxidation in dry fermented sausages made with added olive oil and antioxidants. *Meat Science*, *67*(2), 237–244.

Ansorena, D., & Astiasarán, I. (2004b). The use of linseed oil improves nutritional quality of the lipid fraction of dry-fermented sausages. *Food Chemistry*, *87*(1), 69–74.

Arihara, K. 2006. "Strategies for designing novel functional meat products". *Meat Science*, 74(1), 219–229.

Arihara, K., and M. Ohata. 2008. "Bioactive compounds in meat". In *Meat biotechnology* (pp. 231–249). New York, NY: Springer NewYork.

Armenteros, M., Aristoy, M.-C., & Toldrá, F. (2009). Effect of sodium, potassium, calcium and magnesium chloride salts on porcine muscle proteases. *European Food Research and Technology*, *229*(1), 93–98.

Ashwell, M. (2002). Concepts of functional foods. ILSI Europe concise monograph series. *International Life Sciences Institute, Brussels*, 3–45.

Asp, N.-G., Cummings, J. H., Mensink, R. P., Prentice, A., Richardson, D. P., & Saris, W. H. M. (2003). Process for the assessment of scientific support for claims on foods. *European Journal of Nutrition*, *42*(Suppl. 1), 1–3.

Bastida, S., Sánchez-Muniz, F. J., Olivero, R., Pérez-Olleros, L., Ruiz-Roso, B., & Jiménez-Colmenero, F. (2009). Antioxidant activity of Carob fruit extracts in cooked pork meat systems during chilled and frozen storage. *Food Chemistry*, *116*(3), 748–754.

Bazzano, L. A., He, J., Ogden, L. G., Loria, C. M., & Whelton, P. K. (2003). Dietary fiber intake and reduced risk of coronary heart disease in US men and women: The National Health and Nutrition Examination Survey I Epidemiologic Follow-up Study. *Archives of Internal Medicine*, *163*(16), 1897–1904.

Bennink, M. R., and K. Ono. 1982. "Vitamin B12, E and D content of raw and cooked beef". *Journal of Food Science*, 47(6), 1786–1792.

Beriain, M. J., I. Gómez, F. C. Ibáñez, M. V. Sarriés, and A. I. Ordóñez. 2018. "Improvement of the functional and healthy properties of meat products". In *Food quality: Balancing health and disease* (pp. 1–74). Academic Press, Elsevier.

Besbes, S., Attia, H., Deroanne, C., Makni, S., & Blecker, C. (2008). Partial replacement of meat by pea fiber and wheat fiber: Effect on the chemical composition, cooking characteristics and sensory properties of beef burgers. *Journal of Food Quality*, *31*(4), 480–489.

Bhat, Z. F., & Hina, B. (2011a). Fibre-based functional meat products. *Asian Journal of Food and Agro-Industry*, *4*(4), 261–273.

Bhat, Z. F., & Bhat, H. (2011b). Functional meat products: A review. *International Journal of Meat Science*, *1*(1), 1–14.

Blaner, W. S. 2020. "Vitamin A and provitamin A carotenoids". In *Present knowledge in nutrition* (pp. 73–91). Academic Press.

Boateng, E. F., and M. M. Nasiru. 2019. "Applications of ultrasound in meat processing technology: A review". *Food Science and Technology*, 7(2), 11–15.

Bou, R., Codony, R., Tres, A., Decker, E. A., & Guardiola, F. (2009). Dietary strategies to improve nutritional value, oxidative stability, and sensory properties of poultry products. *Critical Reviews in Food Science and Nutrition*, *49*(9), 800–822.

Brodowska, M., D. Guzek, A. Kołota, D. Głąbska, E. Górska-Horczyczak, I. Wojtasik-Kalinowska, and A. Wierzbicka. 2016. "Effect of diet on oxidation and profile of volatile compounds of pork after freezing storage". *Journal of Food and Nutrition Research*, 55(1), 40–47.

Buzala, M., A. Slomka, and B. Janicki. 2016. "Heme iron in meat as the main source of iron in the human diet". *Journal of Elementology*, 21(1), 303–314.

Carazo, A., K. Macáková, K. Matoušová, L. K. Krčmová, M. Protti, and P. Mladěnka. 2021. "Vitamin A update: Forms, sources, kinetics, detection, function, deficiency, therapeutic use and toxicity". *Nutrients*, 13(5), 1703.

Carnagey, K. M., E. J. Huff-Lonergan, A. Trenkle, A. E. Wertz-Lutz, R. L. Horst, and D. C. Beitz. 2008. "Use of 25-hydroxyvitamin D3 and vitamin E to improve tenderness of beef from the longissimus dorsi of heifers". *Journal of Animal Science*, 86(7), 1649–1657.

Castellini, C., C. Mugani, and A. Dal Bosco. 2002. "Effect of organic production system on broiler carcass and meat quality". *Meat Science*, 60(3), 219–225.

Chapman, J., A. Elbourne, V. K. Truong, and D. Cozzolino. 2020. "Shining light into meat–a review on the recent advances in in vivo and carcass applications of near infrared spectroscopy". *International Journal of Food Science and Technology*, 55(3), 935–941.

Chen, T., G. H. Zhou, X. L. Xu, G. M. Zhao, and C. Li. 2010. "Phospholipase A2 and antioxidant enzyme activities in normal and PSE pork". *Meat Science*, 84(1), 143–146.

Choi, Y.-S., Choi, J.-H., Han, D.-J., Kim, H.-Y., Lee, M.-A., Lee, E.-S., Jeong, J.-Y., Paik, H.-D., & Kim, C.-J. (2008). Effects of rice bran fiber on quality of low-fat tteokgalbi. *Food Science and Biotechnology*, *17*(5), 959–964.

Clausen, I., J. Jakobsen, T. Leth, and L. Ovesen. 2003. "Vitamin D3 and 25-hydroxyvitamin D3 in raw and cooked pork cuts". *Journal of Food Composition and Analysis*, 16(5), 575–585.

Cobos, A., and O. Díaz. 2015. "Chemical composition of meat and meat products". In *Handbook of food chemistry*, 1, 1–32.

Cofrades, S., Guerra, M. A., Carballo, J., Fernández-Martín, F., & Colmenero, F. J. (2000). Plasma protein and soy fiber content effect on bologna sausage properties as influenced by fat level. *Journal of Food Science*, *65*(2), 281–287.

Combs Jr., G. F., Midthune, D. N., Patterson, K. Y., Canfield, W. K., Hill, A. D., Levander, O. A., Taylor, P. R., Moler, J. E., & Patterson, B. H. (2009). Effects of selenomethionine supplementation on selenium status and thyroid hormone concentrations in healthy adults. *American Journal of Clinical Nutrition*, *89*(6), 1808–1814.

Cruzen, S. M., Y. H. B. Kim, S. M. Lonergan, J. K. Grubbs, A. N. Fritchen, and E. Huff-Lonergan. 2015. "Effect of early postmortem enhancement of calcium lactate/phosphate on quality attributes of beef round muscles under different packaging systems". *Meat Science*, 101, 63–72.

Dal Bosco, A., Castellini, C., Bianchi, L., & Mugnai, C. (2004). Effect of dietary -linolenic acid and vitamin E on the fatty acid composition, storage stability and sensory traits of rabbit meat. *Meat Science*, *66*(2), 407–413.

Dalle Zotte, A., & Szendrő, Z. (2011). The role of rabbit meat as functional food. *Meat Science*, *88*(3), 319–331.

Das, A. K., Rajkumar, V., & Verma, A. K. (2015). Bael pulp residue as a new source of antioxidant dietary fiber in goat meat nuggets. *Journal of Food Processing and Preservation*, *39*(6), 1626–1635.

Decker, E. A., and Y. Park. 2010. "Healthier meat products as functional foods". *Meat Science*, 86(1), 49–55.

Dietary guidelines for Americans (2005). U.S. Department of Health and Human Services, & U.S. Department of Agriculture.

Dikeman, M. E., M. C. Hunt, P. B. Addis, H. J. Schoenbeck, M. Pullen, E. Katsanidis, and E. J. Yancey. 2003. "Effects of postexsanguination vascular infusion of cattle with a solution of saccharides, sodium chloride, and phosphates or with calcium chloride on quality and sensory traits of steaks and ground beef". *Journal of Animal Science*, 81(1), 156–166.

Dobreva, D. A., A. V. Merdzhanova, M. D. Stancheva, D. I. Terziyski, and V. Z. Panayotova. 2018. "Black Sea fish and shellfish as essential source of vitamin B12". *International Journal of Scientific Reports*, 4(8), 199–203.

Dominguez, H. (2013). *Functional ingredients from algae for foods and nutraceuticals*. Elsevier.

Eim, V. S., Simal, S., Rosselló, C., & Femenia, A. (2008). Effects of addition of carrot dietary fibre on the ripening process of a dry fermented sausage (sobrassada). *Meat Science*, *80*(2), 173–182.

Enser, M., Richardson, R. I., Wood, J. D., Gill, B. P., & Sheard, P. R. (2000). Feeding linseed to increase the n-3 PUFA of pork: Fatty acid composition of muscle, adipose tissue, liver and sausages. *Meat Science*, *55*(2), 201–212.

Esteve, M. J., R. Farré, A. Frıgola, and C. Pilamunga. 2002. "Contents of vitamins B1, B2, B6, and B12 in pork and meat products". *Meat Science*, 62(1), 73–78.

Fagali, N., & Catalá, A. (2008). Antioxidant activity of conjugated linoleic acid isomers, linoleic acid and its methyl ester determined by photoemission and DPPH techniques. *Biophysical Chemistry*, *137*(1), 56–62.

Feng, J., Xiong, Y. L., & Mikel, W. B. (2003). Textural properties of pork frankfurters containing thermally/enzymatically modified soy proteins. *Journal of Food Science*, *68*(4), 1220–1224.

Fernández-Ginés, J. M., Fernández-López, J., Sayas-Barberá, E., & Pérez-Alvarez, J. A. (2005). Meat products as functional foods: A review. *Journal of Food Science*, *70*(2), R37–R43.

Fernández-Ginés, J. M., Fernández-López, J., Sayas-Barberá, E., Sendra, E., & PérezÁlvarez, J. A. (2003). Effect of storage conditions on quality characteristics of bologna sausages made with citrus fiber. *Journal of Food Science*, *68*(2), 710–715.

Fernández Fernández, J., L. García, J. Monte, C. J. Villar Granja, and F. Lombó Brugos. 2018. "Functional anthocyanin-rich sausages diminish colorectal cancer in an animal model and reduce pro-inflammatory bacteria in the intestinal microbiota". *Genes*, 9(3).

Gabdukaeva, L. Z., T. Y. Gumerov, A. R. Nurgalieva, and L. V. Abdullina. 2021. "Current trends in the development of functional meat products to improve the nutritional status of the population". *IOP Conference Series: Earth and Environmental Science*, 624(1), 12196.

Gallardo, B., M. G. Manca, A. R. Mantecón, A. Nudda, and T. Manso. 2015. "Effects of linseed oil and natural or synthetic vitamin E supplementation in lactating ewes' diets on meat fatty acid profile and lipid oxidation from their milk fed lambs". *Meat Science*, 102, 79–89.

Garci-Martin, A., Fonseca, C. S., Bastida, S., Vázquez-Velasco, M., González-Torres, L., Benedí, J., & Sánchez-Muniz, F. J. (2012). Restructured meat containing Himanthalia elongata acts as a competitive α glucosidase inhibitor. *Potential Utilities*, *11*.

Gil, M. D., Bañón, S. J., Cayuela, J. M., Laencina, J., & Garrido, M. D. (2001). Utilización de extractos de plantas como antioxidantes naturales en carne y productos cárnicos: Revisión. *Eurocarne*, *101*, 1–10.

Gobert, M., D. Gruffat, M. Habeanu, E. Parafita, D. Bauchart, and D. Durand. 2010. "Plant extracts combined with vitamin E in PUFA-rich diets of cull cows protect processed beef against lipid oxidation". *Meat Science*, 85(4), 676–683.

González-Torres, L., Churruca, I., Moreira, A. R. S., Bastida, S., Benedí, J., Portillo, M. P., & Sánchez-Muniz, F. J. (2012). Effects of restructured pork containing Himanthalia elongata on adipose tissue lipogenic and lipolytic enzyme expression of normo-and hypercholesterolemic rats. *Lifestyle Genomics*, *5*(3), 158–167.

Green, R., and J. W. Miller. 2007. "Handbook of vitamins". In *Handbook of vitamins*. CRC Press, 1–608

Greenhaff, P. L. (2001). The creatine-phosphocreatine system: There's more than one song in its repertoire. *Journal of Physiology*, *537*(3), 657.

Grujić, R., Grujić, S., & Vujadinović, D. (2012). Funkcionalni proizvodi od mesa. *Hrana u Zdravlju i Bolesti: Znanstveno-Stručni Časopis Za Nutricionizam i Dijetetiku*, *1*(1), 44–54.

Gualano, B., Roschel, H., Lancha-Jr., A. H., Brightbill, C. E., & Rawson, E. S. (2012). In sickness and in health: The widespread application of creatine supplementation. *Amino Acids*, *43*(2), 519–529.

Haak, L., Raes, K., & De Smet, S. (2009). Effect of plant phenolics, tocopherol and ascorbic acid on oxidative stability of pork patties. *Journal of the Science of Food and Agriculture*, *89*(8), 1360–1365.

Hjelm, L., L. A. Mielby, S. Gregersen, N. Eggers, and H. C. Bertram. 2019. "Partial substitution of fat with rye bran fibre in Frankfurter sausages–Bridging technological and sensory attributes through inclusion of collagenous protein". *LWT*, 101, 607–617.

Hoffmann, M., Waszkiewicz-Robak, B., & Świderski, F. (2010). Functional food of animal origin. Meat and meat products. *Nauka Przyroda Technologie*, *4*(5), 63.

Holick, M. F. 2007. "Vitamin D deficiency". *New England Journal of Medicine*, 357(3), 266–281.

Horita, C. N., M. A. Morgano, R. M. S. Celeghini, and M. A. R. Pollonio. 2011. "Physico-chemical and sensory properties of reduced-fat mortadella prepared with blends of calcium, magnesium and potassium chloride as partial substitutes for sodium chloride". *Meat Science*, 89(4), 426–433.

Hur, S. J., Ye, B. W., Lee, J. L., Ha, Y. L., Park, G. B., & Joo, S. T. (2004). Effects of conjugated linoleic acid on color and lipid oxidation of beef patties during cold storage. *Meat Science*, *66*(4), 771–775.

Jasińska, K., and A. Kurek. 2017. "The effect of oil plants supplementation in pig diet on quality and nutritive value of pork meat". *Animal Science Papers and Reports*, 35, 137–146.

Jiménez-Colmenero, F., J. Carballo, and S. Cofrades. 2001. "Healthier meat and meat products: Their role as functional foods". *Meat Science*, 59(1), 5–13.

Jiménez-Colmenero, F., Sánchez-Muniz, F. J., & Olmedilla-Alonso, B. (2010). Design and development of meat-based functional foods with walnut: Technological, nutritional and health impact. *Food Chemistry*, *123*(4), 959–967.

Jodral-Segado, A. M., M. Navarro-Alarcón, H. L. G. de la Serrana, and M. C. López-Martínez. 2003. "Magnesium and calcium contents in foods from SE Spain: Influencing factors and estimation of daily dietary intakes". *Science of the Total Environment*, 312(1–3), 47–58.

Jo, C., Ho Son, J., Bae Son, C., & Woo Byun, M. (2003). Functional properties of raw and cooked pork patties with added irradiated, freeze-dried green tea leaf extract powder during storage at 4°C. *Meat Science*, *64*(1), 13–17.

Joo, S. T., Lee, J. I., Ha, Y. L., & Park, G. B. (2002). Effects of dietary conjugated linoleic acid on fatty acid composition, lipid oxidation, color, and water-holding capacity of pork loin. *Journal of Animal Science, 80(1)*, 108–112.

Juniper, D. T., Phipps, R. H., Ramos-Morales, E., & Bertin, G. (2008). Effect of dietary supplementation with selenium-enriched yeast or sodium selenite on selenium tissue distribution and meat quality in beef cattle. *Journal of Animal Science, 86*(11), 3100–3109.

Kalra, E. K. (2003). Nutraceutical-definition and introduction. *AAPS PharmSci, 5*(3), 27–28.

Kausar, T., E. Hanan, O. Ayob, B. Praween, and Z. Azad. 2019. A review on functional ingredients in red meat products. *Bioinformation*, 15(5), 358.

Kaviani, M., Shaw, K., & Chilibeck, P. D. (2020). Benefits of creatine supplementation for vegetarians compared to omnivorous athletes: A systematic review. *International Journal of Environmental Research and Public Health, 17*(9), 3041.

Kennedy, D. O. 2016. B vitamins and the brain: Mechanisms, dose and efficacy—A review. *Nutrients*, 8(2), 68.

Khalid, W., Z. Maqbool, and M. S. Arshad. 2022. "Recent advances and innovation in meat with reference to processing technologies". In Dr. Muhammad Sajid Arshad and Mr. Waseem Khalid (Eds.) *Health risks of food additives - Recent developments and trends in food sector.* Publisher Infotech. DOI: 10.5772/intechopen.108620

Kim, Y. J., W. Y. Park, and I. H. Choi. 2010. "Effects of dietary α-tocopherol, selenium, and their different combinations on growth performance and meat quality of broiler chickens". *Poultry Science*, 89(3), 603–608.

Kumar, V., Biswas, A. K., Sahoo, J., Chatli, M. K., & Devatkal, S. K. (2010). Quality evaluation of chicken nuggets formulated with green banana and soybean hulls flours. *Indian Journal of Poultry Science, 45*(3), 330–335.

Kumari, K. (2016). Effect of Functional Ingredients Such As Mushrooms, Sesame Seeds And wheat Gluten Incorporation Levels on the Sensory Acceptability of Chicken Meat Cutlets. *Journal of Engineering Research and Application, 6(6)*, 11–16.

Kumari, S., Yong Meng, G., & Ebrahimi, M. (2017). Conjugated linoleic acid as functional food in poultry products: A review. *International Journal of Food Properties, 20*(3), 491–506.

Lagin, L., B. Bobček, J. Mrázová, O. Debrecéni, and M. Adamec. 2008. "The effect of organic selenium on slaughter value, physical-chemical and technological quality characteristic of pork". *Biotechnology in Animal Husbandry*, 24(1–2), 97–107.

Larsson, S. C., Bergkvist, L., & Wolk, A. (2005). High-fat dairy food and conjugated linoleic acid intakes in relation to colorectal cancer incidence in the Swedish Mammography Cohort. *American Journal of Clinical Nutrition, 82*(4), 894–900.

Librelotto, J., Bastida, S., Serrano, A., Cofrades, S., Jiménez-Colmenero, F., & Sánchez-Muniz, F. J. (2008). Changes in fatty acids and polar material of restructured low-fat or walnut-added steaks pan-fried in olive oil. *Meat Science, 80*(2), 431–441.

Lippman, H. E., & Watkins, B. A. (2004). Effects of stearidonic acid (SDA) on MC3T3-E1 osteoblast-like cells. *FASEB Journal, 18*(4), A6–A6.

Liu, S., Willett, W. C., Manson, J. E., Hu, F. B., Rosner, B., & Colditz, G. (2003). Relation between changes in intakes of dietary fiber and grain products and changes in weight and development of obesity among middle-aged women. *American Journal of Clinical Nutrition, 78*(5), 920–927.

Lombardi-Boccia, G., S. Lanzi, and A. Aguzzi. 2005. "Aspects of meat quality: Trace elements and B vitamins in raw and cooked meats". *Journal of Food Composition and Analysis*, 18(1), 39–46.

Lücke, F.K. 2000. "Utilization of microbes to process and preserve meat". *Meat Science*, 56(2), 105–115.

Ma, J., Hampl, J. S., & Betts, N. M. (2000). Antioxidant intakes and smoking status: Data from the continuing survey of food intakes by individuals 1994–1996. *American Journal of Clinical Nutrition*, *71*(3), 774–780.

Madane, P., Das, A. K., Pateiro, M., Nanda, P. K., Bandyopadhyay, S., Jagtap, P., Barba, F. J., Shewalkar, A., Maity, B., & Lorenzo, J. M. (2019). Drumstick (Moringa oleifera) flower as an antioxidant dietary fibre in chicken meat nuggets. *Foods*, *8*(8), 307.

Maiorano, G., C. Cavone, R. J. McCormick, A. Ciarlariello, M. Gambacorta, and A. Manchisi. 2007. "The effect of dietary energy and vitamin E administration on performance and intramuscular collagen properties of lambs". *Meat Science*, 76(1), 182–188.

Majchrzak, D., E. Fabian, and I. Elmadfa. 2006. "Vitamin A content (retinol and retinyl esters) in livers of different animals". *Food Chemistry*, 98(4), 704–710.

Mattila, P. H., E. Valkonen, and J. Valaja. 2011. "Effect of different vitamin D supplementations in poultry feed on vitamin D content of eggs and chicken meat". *Journal of Agricultural and Food Chemistry*, 59(15), 8298–8303.

McClements, D. J., & Decker, E. A. (2008). Lipids. In S. Damodaran, K. L. Parkin & O. R. Fennema (Eds.), *Fennema's food chemistry* (pp. 155–216). CRC Press, Taylor & Francis Group.

Miezeliene, A., G. Alencikiene, R. Gruzauskas, and T. Barstys. 2011. "The effect of dietary selenium supplementation on meat quality of broiler chickens". *Biotechnologie, Agronomie, Société et Environnement*, 15(1), 61–69.

Mitsumoto, M. (2000). Dietary delivery versus exogenous addition of antioxidants. In Edited by E. A. Decker, C. Faustman, and C. J. López-Bote (Eds.), *Antioxidants in muscle foods: Nutritional strategies to improve quality* (pp. 315–343).

Moon, S.-S., Kim, Y.-T., Jin, S.-K., & Kim, I.-S. (2008). Effects of sodium chloride, potassium chloride, potassium lactate and calcium ascorbate on the physico-chemical properties and sensory characteristics of sodium-reduced pork patties. *Food Science of Animal Resources*, *28*(5), 567–573.

Morel, P. C. H., Janz, J. A. M., Zou, M., Purchas, R. W., Hendriks, W. H., & Wilkinson, B. H. P. (2008). The influence of diets supplemented with conjugated linoleic acid, selenium, and vitamin E, with or without animal protein, on the composition of pork from female pigs. *Journal of Animal Science*, *86*(5), 1145–1155.

Moshfegh, A., J. Goldman, J. Ahuja, D. Rhodes, and R. LaComb. 2009. *What we eat in America, NHANES 2005–2006: Usual nutrient intakes from food and water compared to 1997 dietary reference intakes for vitamin D, calcium, phosphorus, and magnesium.* US Department of Agriculture, Agricultural Research Service.

Muguerza, E., Ansorena, D., Bloukas, J. G., & Astiasarán, I. (2003). Effect of fat level and partial replacement of pork backfat with olive oil on the lipid oxidation and volatile compounds of Greek dry fermented sausages. *Journal of Food Science*, *68*(4), 1531–1536.

Muguerza, E., Fista, G., Ansorena, D., Astiasarán, I., & Bloukas, J. G. (2002). Effect of fat level and partial replacement of pork backfat with olive oil on processing and quality characteristics of fermented sausages. *Meat Science*, *61*(4), 397–404.

Muguerza, E., Gimeno, O., Ansorena, D., Bloukas, J. G., & Astiasarán, I. (2001). Effect of replacing pork backfat with pre-emulsified olive oil on lipid fraction and sensory quality of Chorizo de Pamplona—A traditional Spanish fermented sausage. *Meat Science*, *59*(3), 251–258.

Nisa, A. U., N. Zahra, S. Hina, S. Masood, A. Javed, and S. ManzarInam. 2017. "Development of meat-based functional foods: A review". *Journal of Applied Biology and Biotechnology*, 5(3), 86–92.

Olivero-David, R., Schultz-Moreira, A., Vázquez-Velasco, M., González-Torres, L., Bastida, S., Benedí, J., Sanchez-Reus, M. I., González-Muñoz, M. J., & Sánchez-Muniz, F. J. (2011). Effects of Nori-and wakame-enriched meats with or without supplementary cholesterol on arylesterase activity, lipaemia and lipoproteinaemia in growing Wistar rats. *British Journal of Nutrition*, *106*(10), 1476–1486.

Olmedilla-Alonso, B., Jiménez-Colmenero, F., & Sánchez-Muniz, F. J. (2013). Development and assessment of healthy properties of meat and meat products designed as functional foods. *Meat Science*, *95*(4), 919–930.

Ostaszewski, M. (2018). Meat and meat products as functional food. *World Scientific News*, *110*, 147–158.

Pan, A., Sun, Q., Bernstein, A. M., Schulze, M. B., Manson, J. E., Stampfer, M. J., Willett, W. C., & Hu, F. B. (2012). Red meat consumption and mortality: Results from 2 prospective cohort studies. *Archives of Internal Medicine*, *172*(7), 555–563.

Pan, C., Huang, K., Zhao, Y., Qin, S., Chen, F., & Hu, Q. (2007). Effect of selenium source and level in hen's diet on tissue selenium deposition and egg selenium concentrations. *Journal of Agricultural and Food Chemistry*, *55*(3), 1027–1032.

Pappa, I. C., Bloukas, J. G., & Arvanitoyannis, I. S. (2000). Optimization of salt, olive oil and pectin level for low-fat frankfurters produced by replacing pork backfat with olive oil. *Meat Science*, *56*(1), 81–88.

Peter, P. P.2017. *New aspects of meat quality: From genes to ethics*. Woodhead Publishing.

Pogorzelska-Nowicka, E., A. G. Atanasov, J. Horbańczuk, and A. Wierzbicka. 2018. "Bioactive compounds in functional meat products". *Molecules*, 23(2), 307.

Poławska, E., Ż. Zdanowska-Sąsiadek, J. Horbańczuk, J. F. Pomianowski, A. Jóźwik, D. Tolik, K. Raes, and S. De Smet. 2016. "Effect of dietary organic and inorganic selenium supplementation on chemical, mineral and fatty acid composition of ostrich meat". *CyTA-Journal of Food*, 14(1), 84–87.

Prakongpan, T., Nitithamyong, A., & Luangpituksa, P. (2002). Extraction and application of dietary fiber and cellulose from pineapple cores. *Journal of Food Science*, *67*(4), 1308–1313.

Purchas, R. W., Rutherfurd, S. M., Pearce, P. D., Vather, R., & Wilkinson, B. H. P. (2004). Concentrations in beef and lamb of taurine, carnosine, coenzyme Q10, and creatine. *Meat Science*, *66*(3), 629–637.

Purchas, R., M. Zou, P. Pearce, and F. Jackson. 2007. "Concentrations of vitamin D3 and 25-hydroxyvitamin D3 in raw and cooked New Zealand beef and lamb". *Journal of Food Composition and Analysis*, 20(2), 90–98.

Ramachandraiah, K., S. G. Han, and K. B. Chin. 2015. "Nanotechnology in meat processing and packaging: Potential applications-a review". *Asian-Australasian Journal of Animal Sciences*, 28(2), 290–302.

Ramos, A., M. C.Cabrera, M. Del Puerto, and A. Saadoun. 2009. "Minerals, haem and non-haem iron contents of rhea meat". *Meat Science*, 81(1), 116–119.

Riaz, M., and K. T. Mehmood. 2012. "Selenium in human health and disease: A review". *Journal of Postgraduate Medical Institute*, 26(2).

Ross, A. C., B. Caballero, R. J. Cousins, and K. L. Tucker. 2020. *Modern nutrition in health and disease*. Jones & Bartlett Learning.

Rucker, R. B., J. Zempleni, J. W. Suttie, and D. B. McCormick. 2007. *Handbook of vitamins*. CRC Press.

Rymer, C., and D. Ian Givens. 2010. "Effects of vitamin E and fish oil inclusion in broiler diets on meat fatty acid composition and on the flavour of a composite sample of breast meat". *Journal of the Science of Food and Agriculture*, 90(10), 1628–1633.

Salazar, P., García, M. L., & Selgas, M. D. (2009). Short-chain fructooligosaccharides as potential functional ingredient in dry fermented sausages with different fat levels. *International Journal of Food Science and Technology*, *44*(6), 1100–1107.

Santhi, D., & Kalaikannan, A. (2014). The effect of the addition of oat flour in low-fat chicken nuggets. *Journal of Nutrition and Food Sciences*, *4*(1), 1.

Sáyago-Ayerdi, S. G., Brenes, A., & Goñi, I. (2009). Effect of grape antioxidant dietary fiber on the lipid oxidation of raw and cooked chicken hamburgers. *LWT-Food Science and Technology*, *42*(5), 971–976.

Schaefer, D. M., Liu, Q., Faustman, C., & Yin, M.-C. (1995). Supranutritional administration of vitamins E and C improves oxidative stability of beef. *Journal of Nutrition*, *125*(Suppl. 6), 1792S–1798S.

Schmid, A. (2009). Bioactive substances in meat and meat products. *Fleischwirtschaft*, *89*(7), 83–90.

Schmid, A., and B. Walther. 2013. "Natural vitamin D content in animal products". *Advances in Nutrition*, 4(4), 453–462.

Schoene, F., Mnich, K., Jahreis, G., Kinast, C., Greiling, A., Kirmse, R., Hartung, H., & Leiterer, M. (2009). Analysis of meat products, produced with mineral salt. *Fleischwirtschaft*, *89*(2), 149–152.

Schönfeldt, H. C., and N. G. Hall. 2011. "Determining iron bio-availability with a constant heme iron value". *Journal of Food Composition and Analysis*, 24(4–5), 738–740.

Selgas, M. D., Salazar, P., & García, M. L. (2009). Usefulness of calcium lactate, citrate and gluconate for calcium enrichment of dry fermented sausages. *Meat Science*, *82*(4), 478–480.

Sestili, P., Martinelli, C., Colombo, E., Barbieri, E., Potenza, L., Sartini, S., & Fimognari, C. (2011). Creatine as an antioxidant. *Amino Acids*, *40*(5), 1385–1396.

Severini, C., De Pilli, T., & Baiano, A. (2003). Partial substitution of pork backfat with extra-virgin olive oil in 'salami'products: Effects on chemical, physical and sensorial quality. *Meat Science*, *64*(3), 323–331.

Solomons, N. W., and M. Orozco. 2003. "Alleviation of vitamin A deficiency with palm fruit and its products". *Asia Pacific Journal of Clinical Nutrition*, 12(3), 373–384.

Tang, S., Kerry, J. P., Sheehan, D., Buckley, D. J., & Morrissey, P. A. (2001). Antioxidative effect of added tea catechins on susceptibility of cooked red meat, poultry and fish patties to lipid oxidation. *Food Research International*, *34*(8), 651–657.

TJ, N. K., Nayar, R., & Rajagopal, K. (n.d.). *Technological and health impacts of meat based functional foods-A review.*

Torrissen, O. J. (2000). Dietary delivery of carotenoids. In *Antioxidants in muscle foods: Nutritional strategies to improve quality* (pp. 289–313).

Valenzuela, C., D. L. de Romaña, C. Schmiede, M. S. Morales, M. Olivares, and F. Pizarro. 2011. "Total iron, heme iron, zinc, and copper content in rabbit meat and viscera". *Biological Trace Element Research*, 143(3), 1489–1496.

Venderley, A. M., & Campbell, W. W. (2006). Vegetarian diets: Nutritional considerations for athletes. *Sports Medicine*, *36*(4), 293–305.

Verma, A. K., Rajkumar, V., Banerjee, R., Biswas, S., & Das, A. K. (2013). Guava (Psidium guajava L.) powder as an antioxidant dietary fibre in sheep meat nuggets. *Asian-Australasian Journal of Animal Sciences*, *26*(6), 886.

Verma, A. K., Sharma, B. D., & Banerjee, R. (2010). Effect of sodium chloride replacement and apple pulp inclusion on the physico-chemical, textural and sensory properties of low fat chicken nuggets. *LWT-Food Science and Technology*, *43*(4), 715–719.

Vongsawasdi, P., and A. Noomhorm. 2014. "Bioactive compounds in meat and their functions". In *Functional foods and dietary supplements: Processing effects and health benefits* (pp. 113–138). DOI: 10.1002/9781118227800.ch5

Vossen, E., Raes, K., Maertens, L., Vandenberge, V., Haak, L., Chiers, K., Ducatelle, R., & De Smet, S. (2012). Diets containing n-3 fatty acids-enriched pork: Effect on blood lipids, oxidative status and atherosclerosis in rabbits. *Journal of Food Biochemistry, 36*(3), 359–368.

Vural, H., Javidipour, I., & Ozbas, O. O. (2004). Effects of interesterified vegetable oils and sugarbeet fiber on the quality of frankfurters. *Meat Science, 67*(1), 65–72.

Wadhwani, R., D. P. Cornforth, L. K. Murdia, and D. Whittier. 2011. "Animal age, packaging and antioxidant treatment effects on sensory characteristics of beef infraspinatus muscle". *International Journal of Food Science and Technology*, 46(9), 1847–1855.

Whitney, E. N., and S. R. Rolfes. 2002. *Understanding nutrition* (9th ed.). Wadsworth.

Wojtasik-Kalinowska, I., D. Guzek, E. Górska-Horczyczak, D. Głąbska, M. Brodowska, D. W. Sun, and A.Wierzbicka. 2016. "Volatile compounds and fatty acids profile in longissimus dorsi muscle from pigs fed with feed containing bioactive components". *LWT – Food Science and Technology*, 67, 112–117.

Wood, J. D., Richardson, R. I., Nute, G. R., Fisher, A. V., Campo, M. M., Kasapidou, E., Sheard, P. R., & Enser, M. (2004). Effects of fatty acids on meat quality: A review. *Meat Science, 66*(1), 21–32.

Yang, H.-S., Choi, S.-G., Jeon, J.-T., Park, G.-B., & Joo, S.-T. (2007). Textural and sensory properties of low fat pork sausages with added hydrated oatmeal and tofu as texture-modifying agents. *Meat Science, 75*(2), 283–289.

Yılmaz, I. (2004). Effects of rye bran addition on fatty acid composition and quality characteristics of low-fat meatballs. *Meat Science, 67*(2), 245–249.

Yılmaz, I. (2005). Physicochemical and sensory characteristics of low fat meatballs with added wheat bran. *Journal of Food Engineering, 69*(3), 369–373.

Yılmaz, İ., & Dağlıoğlu, O. (2003). The effect of replacing fat with oat bran on fatty acid composition and physicochemical properties of meatballs. *Meat Science, 65*(2), 819–823.

Yılmaz, İ., & Gecgel, U. (2009). Effect of inulin addition on physico-chemical and sensory characteristics of meatballs. *Journal of Food Science and Technology (Mysore), 46*(5), 473–476.

Zargar, F. A., Kumar, S., Bhat, Z. F., & Kumar, P. (2014). Effect of pumpkin on the quality characteristics and storage quality of aerobically packaged chicken sausages. *SpringerPlus, 3*(1), 1–10.

Zhang, W., S. Xiao, H. Samaraweera, E. J. Lee, and D. U. Ahn. 2010. "Improving functional value of meat products". *Meat Science*, 86(1), 15–31.

CHAPTER 9

Nutraceuticals from Seafood: Role in Health and Disease Prevention

Kawkabul Saba, Faisal Rashid, Akhila, Vijay Reddy* and **Oyas Asimi**[†]

9.1 INTRODUCTION

The word "nutraceutical" was coined by Stephen De-Felice, MD, chairman and founder of the Foundation for Innovation in Medicine (FIM) in Cranford, New Jersey, derived from the words, "nourishment" and "pharmaceutical," in the year 1989. De-Felice described a nutraceutical as any portion of a food or a food providing health or medical benefits including treatment and/or prevention of a disorder. When functional food helps in treating and/or preventing disorder (s) and/or disease (s) except anaemia, then that functional food is termed a nutraceutical. The American Nutraceuticals Association (ANA) defines nutraceuticals as functional foods that are potentially disease-preventing and have health-promoting properties. It must be known that the word nutraceutical, generally utilised for selling, does not have any regulatory explanation. Hence, nutraceuticals are not the same as dietary supplements because of the following reasons:

1. Traditional foods can be used as nutraceuticals or as a solo item of a diet or a meal.
2. Apart from providing basic nutrition, dietetic constituents have a key positive role resulting in the designing of the concepts of nutraceuticals and functional foods. A functional food from one consumer may serve as a nutraceutical for another consumer.
3. Examples of nutraceuticals are citrus fruits (such as orange juices) or tropical fruits (like mango and pineapple) and fortified dairy products.

* Corresponding author E-mail: vijayreddy3@gmail.com;
† Corresponding author E-mail: oyasasim@gmail.com

DOI: 10.1201/9781003315100-9

The foodstuffs derived from food sources that endow both nutrition and medicinal benefits are known as nutraceuticals. The following are the other names of nutraceuticals (Laparra and Sanz, 2010):

- Practical foods
- Medicinal foods
- Designer foods
- Phytochemicals
- Nutritional supplements

Nutraceuticals contain a high concentration of bioactive compounds that have got physiological benefits and help in the prevention and treatment of disease. Nutraceuticals include dietary supplements, herbal products, genetically engineered foods and vitamins. Other nutraceuticals that are included in this group are beverages, pre- and probiotics, fortified cereals and processed foods.

9.1.1 History

"Let food be your medicine and medicine be your food," Hippocrates proclaimed 2500 years ago. The modern nutraceutical market began to develop in Japan during the 1980s, however, the concept of nutraceuticals was started from a survey in the UK, Germany and France. When discussions of fibre and fish oil were published in distinguished medical journals from that time, the nutraceutical revolution began, with a discussion on the actual or potential clinical benefits of calcium. In the US, the term "nutraceutical" was commonly used, but no dogmatic definition existed. Its meaning was adapted by the Health Ministry of Canada (Sarkar et al., 2021)

Primarily at the moment, there is no obvious globally customary definition of nutraceuticals, and if there was, it would vary from country to country depending on how they are categorised and synchronised. Hence there is no apparent global conventional definition of a nutraceutical. It is said that nutraceuticals can improve health, delay the ageing process, avert chronic disease, improve quality of life, or enhance the structure and functioning of the body. Nutraceuticals are also used in the anticipation and management of mental health issues and disorders. Fundamentally, nutraceuticals are foodstuffs that have a physiological advantage or provide fortification from chronic disease (Sarkar et al., 2021).

From the point of view of fisheries, fish nutraceuticals are used for improving health in different ways, such as delaying the process of ageing, preventing chronic and acute diseases, increasing the expectancy of life, and supporting the fundamental functions and structure of the body (Alshammari et al., 2019). The major reasons for promoting the growth of the market for fish nutraceuticals all over the world are population, urbanisation and awareness regarding the health of people that have either a stressed or sedentary lifestyle (Das et al., 2013). It has been newly reported that fish nutraceuticals provide positive strategies for managing human health exerting noteworthy beneficial impacts on health (Daneshkhan et al., 2020). An expansive variety of phytochemicals such as glucosinolates, terpenoids, limonoids, phytosterols, polyphenols, anthocyanidin, carotenoids, phytoestrogens, flavonoids and isoflavonoids have led to several therapeutic treatments for human health based on their antioxidant, anti-inflammatory, anti-allergic and antibacterial properties (Reddy et al., 2020). Such nutraceuticals not only prevent serious medical issues of the present time like cancer, diabetes, cholesterol, cardiovascular diseases, obesity, arthritis and osteoporosis but work in synergy with other curative strategies (Adnan et al., 2017). Fish nutraceuticals, which are cost-effective, can work on this basis, especially for people who receive a low income or who are economically vulnerable. In this way, fish or fish by-products can help in solving the crisis of starvation worldwide and other related problems by providing important macro and micronutrients, easily biodegradable proteins and high-density fats (Yao et al., 2006). All these important nutrients have several beneficial physiological functions in comparison to other proteins (Paitel, 2018).

In contemporary times, the seafood industry is recognised as being a creator of immense volumes of waste, comprising shrimp fins, bones, shrimp shells, skins, intestines and carcasses, along with the voluminous capacity of wastewater effluents. Seafood industry effluents comprise a huge quantity of lipids, proteins, amino acids, polyunsaturated fatty acids, minerals and carotenoids, which when disposed of in compost not only leads to an enormous waste of various nutrients but also leads to massive environmental contamination (Khalilii and Samples, 2018). Thus, the predicament of such seafood industry run-offs needs to be managed with a profit-making outlook, and the benefits of seafood waste need to be highlighted. Numerous functional compounds, such as enzymes like chitosan, chitinase, lipase, protease, hyaluronidase and phosphatase, and organic compounds such as bioactive peptides, collagen, gelatine and mineral-based nutraceuticals can be turned into value-added materials. The bioconversion of these compounds with a bio-refinery loom will allow environmentally friendly and cost-effective processing of the run-off generated from seafood, which could sustainably allow the manufacture of seafood (Adnan et al., 2017). The compounds generated could become dietary sources or nutraceuticals: foods with medicinal value. This would reduce the massive amounts of waste involved and improve the nutritional and medicinal quality of food while meeting the global demand for nutrients and nutraceuticals at a reasonable cost. Seafood is accredited for its uncultivated complex small molecules with potential health benefits in the field of nutraceutical research. The exceptional properties of marine compounds are a fundamental part of our diet and play an important therapeutic role in the treatment of various human diseases. Seafood, including finfish, shellfish, and marine algae, are rich in proteins, essential amino acids, omega-3 polyunsaturated fatty acids, carotenoids, vitamins and minerals. The by-products of fish-processing industries include shell, bones, intestines, fin and skin and contain important bioactive molecules such as fish oils, carotenoids, collagen, gelatin, peptides, chitin, chitosan, glycosaminoglycans and trace elements (Balami et al., 2019). There are various health benefits including cardioprotective, neuroprotective, antioxidant, anti-inflammatory, anticancer, anti-obesity, anticoagulant, antimicrobial and immunomodulatory activities. Thus, this chapter presents the currently available information/data on seafood bioactive components and their potential health benefits in human nutrition and health.

9.2 FISH-BASED BIOACTIVES AS POTENT NUTRACEUTICALS

Bioactive compounds are present in relatively small quantities in plants and are known to exhibit nutritional and healing characteristics. The bioactive compounds subsist as an ample variety of chemical structures, and the functions affecting the human body vary according to the health effect they induce. These include improved performance of targeted organs and/or bodily functions. Examples of these chemicals include phenolic and organosulphur compounds. Examples of phenolic compounds include flavonoids, phytoestrogens, hydroxytyrosols and lycopene. These compounds can be found in a variety of plant products such as cereal grains, fruits, vegetables, tea, nuts, legumes, herbs, spices, red wine and edible oils such as olive, flax and soy (Kris-Etherton et al., 2002). The plant structures from which these compounds are found vary from roots, stems (including tubers), leaves, flowers, barks, bulbs, fruits, seeds and clusters. The bioactive compounds are found as secretions such as resin, latex and sap.

Oceanic food products are bio-processed to improve the efficient characteristics of marine food ingredients, which allows them to be used as functional foods and nutraceuticals. Some of the functional products obtained from the bio-processing of marine food sources are discussed below. Table 9.1 and 9.2 indicates the principle fatty acid in some fish oils and **bioactive peptides from some fishery sources and their hydrolysates**

9.2.1 Bioactive Peptides

Proteins in oceanic foods contain sequences of bioactive peptides, which could have physiological effects on the body. Amino acid manacles are small and motionless

Table 9.1 Principle fatty acid in some fish oils

Fatty acid	Capelin	Atlantic mackerel	Atlantic sardine	Anchovies
Myristic (14.0)	7	18	8	9
Palmitic (16:0)	10	14	18	19
Palmitoleic (16:1)	10	7	10	9
Oleic (18:1)	14	13	13	13
Eicosanoid (20:1)	17	12	4	5
Cetoleic	14	15	3	2
EPA (20:5)	8	7	18	17
DHA (22:6)	6	8	9	9

Table 9.2 Bioactive peptides from some fishery sources and their hydrolysates (FPHs) (Abbas et al., 2009).

Shellfish	Gly-Pro-Leu	ACE inhibitory
Alaska pollock	Tyr-Asn-Lys-Leu	ACE inhibitory
Wakame	Gly-Asp-Leu-Gly-Lys-Thr-Thr-Thr-Val-Ser-Asn-T rp-Ser-Pro-Pro-Lys-Try-Lys-Asp-Thr-Pro	ACE inhibitory
Big eye tuna	Le-Phe-Val-Pro-Ala-Phe	ACE inhibitory
Shrimp (Protease)	Peptidea (molecular size,<10, 10 to 30, >30 kDa)	ACE inhibitory
Shrimp	Tyr-Asn-Lys-Leu	Anticancer
Wakame (Papain)	Cys, Leu, Glu, Asp, Phe, Tyr, Ile, Gly	-
American lobster	Gln-Tyr-Gly-Asn-Leu-Leu-Ser-Leu-Leu-Asn-Gly-T yr-Arg	Antimicrobial
Jumbo squid	Gelatine peptide	Antimicrobial
Blue mussel	Glu-Ala-Asp-Ile-Asp-Gly-Asp-Gly-Gln-Val-Asn-T yr-Glu-Glu-Phe-Val-Ala-Met-Met-Thr-Ser-Lys	Antioxidant

within the sequence of the parent protein but can be free during gastrointestinal digestion, food processing and fermentation.

The bioactive peptides present in our food have a key remedial role in relation to human diseases, which was discovered through recent research into nutraceuticals and functional foods. In addition, food scientists are focusing on the important task of identifying and producing food with enhanced bioactive activity, which could ultimately improve human health. Fish have always been considered important due to their rich nutritional values and by-product application in food industries, as well as in well-known foods with bioactive constituents. To uncover new bioactive molecules that could increase the potential and therapeutic benefits of these bioactive components, nutritionists, food scientists and other scientific communities have been working together. In spite of the incalculable benefits of fish and known fish bioactive molecules, its use by food or pharmaceutical industries is inadequate, and even research on fish-based nutraceuticals is not sufficient. Therefore, this chapter focuses on the current information/data available regarding fish bioactive components, their application as nutraceuticals for therapeutic purposes in the treatment of chronic diseases, and ethnic issues related to the consumption of fish or its by-products. Strong emphasis is given to the exploitation of fish waste and their by-products to accomplish the global demand for cheap dietary supplements specifically for underdeveloped/least-developed countries.

Fish is well-considered traditionally and is an inexpensive source of protein, helping a universal population as well as a source of nutraceutical significance (Paital, 2018). In addition, half of the biodiversity is comprised of fish and other maritime species, which are also significant sources of novel bioactive compounds that are relevant to the continuous improvement of human health (Chiesa et al., 2016). A variety of bioactive compounds are present in fish and fish by-products, such as proteins, lipids, vitamins and minerals, which are thought to be beneficial due to their therapeutic potential. During the utilisation of fish, the main therapeutic potential has been documented with the occurrence of long-chain omega-3 polyunsaturated fatty acids (PUFAs) at a high content. Various research has suggested that, as well as long-chain omega-3 PUFAs, other bioactive components from fish have been found to have a positive effect on human health (Khalili et al., 2018). One should consume fish on a regular basis (i.e., 1–2 servings weekly), which would provide approximately 200–500 mg of eicosapentaenoic acid (EPA) and docosahexaenoic acid (DHA), effortlessly obtained from fish that contain omega-3, for example salmon or trout. According to the World Health Organization (WHO, 2020) and the American Heart Association (AHA, 2020) this would protect against a range of coronary heart diseases and stroke. Fish contain considerable amounts of other vital nutrients with limitless health benefits as known by the nutritional status of fish and fishery products (Mohanty et al., 2016). According to reports, diseases related to living in urban areas have become a severe issue to the human population. Lifestyle and diet are major factors that are related to stress and strain in modern life, influencing lifestyle-related diseases, including middle-aged and older people as reported by various epidemiological data and clinical trials. In urbanised countries, there have been increases in chronic diseases such as obesity, diabetes, hypertension and hyperlipidaemia, affecting both wellness and well-being due to the imbalance in lifestyle and diet patterns (Hosomi et al., 2012). For that reason, fish is considered a solution to lifestyle-related diseases. The main purpose of this analysis is to summarise the nutritional importance of fish and fish-based nutraceuticals and functional foods and their therapeutic potential, as well as understand the risks of fish ingestion in our dietary behaviour.

Fish are an important and vital food in the diet of humans, due to their beneficial dietary value. Biochemical analysis of most fish or fish-related foods has been considered by researchers. Negligible differences in dietetic composition continue to exist based on fish age, sex, feeding habits, season, temperature, adaptation and hatching patterns (Pal et al., 2018). Further research has analysed the composition of fish and has suggested that fish have available protein of good quality with whole vital amino acids and fundamental sources of dietetic vitamins and minerals, including zinc (marine fish), iron, calcium, phosphorus, iodine, vitamin A, D, E, several B vitamins (B3, B6, and B12), important amounts of polyunsaturated fatty acids and a variety of other micronutrients (Suleria, 2015, Singh et al., 2016, Marques et al., 2019, Roos et al., 2007). Almost all small fish are eaten whole (with bone, scales and fins), as they are a rich resource of profusely bioavailable calcium, e.g. Amblypharyngodonmola, which has been reported to be a rich source of vitamin A in the form of dihydro-retinol and retinol (Roos et al., 2007). The bioavailability of zinc and iron from the other foods in a meal is enhanced by fish (Singh et al., 2016). The chief chemical composition of fish muscle is protein, water and lipids, which collectively hold around 98% of the total weight.

The various ways by which fish nutraceuticals are used to improve health are delaying the ageing process; delaying acute and chronic diseases, extending life expectancy and sustaining the basic structure and functions of the human body

(Nasri et al., 2014). The foremost cause of increased growth in the nutraceutical market worldwide is the urbanisation of the population and health awareness among people with either a sedentary lifestyle or a stressed lifestyle (Das et al., 2012). Recent reports suggested that nutraceuticals provide positive ways to manage health, with remarkable beneficial effects (Bagchi, 2014). A wide range of phytochemicals such as terpenoids, glucosinolatesphytosterols, limonoids, anthocyanidin, polyphenols, phytoestrogens, carotenoids, isoflavonoids and flavonoids has shown numerous therapeutic effects on human well-being, such as anti-inflammatory, antioxidants, antibacterial and anti-allergic properties (Gupta et al., 2014, Karwande and Borade, 2015, Mandadi et al., 2020, Adnan, 2017, Siddiqui et al., 2020). Nutraceuticals used alone or in conjunction with other therapies not only assist in maintaining health but also help in promoting quality of life and fighting serious medical illnesses of the current era, such as diabetes, cancer, cardiovascular diseases, cholesterol, arthritis, obesity and osteoporosis (Das et al., 2013, Adnan, 2017). Inexpensive nutraceuticals have always been in high demand, predominantly among economically vulnerable or reduced-income groups. Fish and fish by-products could therefore solve the global malnutrition problem and related disorders by providing vital micro and macronutrients and high-density fats, as well as being easily digestible proteins (Adnan, 2017).

The word “bioactive” is derived from the Greek and Latin words for “life” and “dynamic,” full of energy or involved in an activity (Adnan et al., 2017, Kundam, 2018). Fish, which is an inimitable fusion of bioactive compounds like long-chain PUFAs (EPA and DHA), omega-3 PUFAs, peptides, protein hydrolysates, amino acids, minerals, vitamins, gelatine, collagen, fish oil, fish bone and fat-soluble vitamins, is an important source of nutraceuticals (Kundam, 2018). Owing to the noteworthy therapeutic value of fish and fishery by-products, plentiful nutraceuticals are in use for preventative purposes in various fields of medicine. They can be used to reduce complications or to treat various diseases, such as cardiovascular diseases, cancer, viral infections, rickets, dermatologic problems and hypertension, especially during pregnancy, and parasitic infections (Siddiqui et al., 2015). In addition, fish-derived nutraceuticals also have anticoagulant, anti-inflammatory and antioxidant features (Sen et al., 2016). Generally, a large amount of fish by-products are either predisposed or used for low-value products (Muzaifa et al., 2012). It has been suggested that around 25% of fish waste coming from fishery industries is thrown away, which causes a significant impact on the environment as well as a potential loss of important bioactives present in fish waste (Anais et al., 2013). Fish waste is defined as any fish or fish species with very low or no commercial importance, or fish that are caught in minimal amounts that do not have a high market demand. Furthermore, around half of the fish body parts, such as head, fins, skin and viscera, are designated as “waste” in fishery industries. According to previous reports, each year the fishery industry generates more than 20 million tonnes of fish waste; this is a result of catching non-target fish, in addition to fish industry processing waste as well as by-products. These discards correspond to a gigantic amount of the total world fish production. In the European Union alone, these fish waste products total 5.2 million tonnes every year (Caruso, 2016). However, three major techniques (chemical or physical, enzymatic method, and microbial method) are engaged to recuperate the bioactive mechanism from these waste products. Amongst the three techniques, the enzymatic method is thought to be one of the best techniques for all types of fish waste (fish head, skin, viscera, and scales) (Wang et al., 2019). As a result, it is important to distinguish the value of the fish waste that can be utilised to develop novel nutraceuticals and functional foods. Consequently, an improvement in the research on fish nutraceuticals is required. Valued potent nutraceuticals can accomplish the world demand for dietary supplements, especially for underdeveloped/least-developed countries.

Omega fatty acid is a chain of carbon atoms with an acid group (COOH) at one end and an omega or methyl group (CH3) at the other end. They are categorised based on a variety of distinctiveness, i.e. length, weight, the presence of double bonds and the range of hydrogen atoms in double bonds (Abbas et al., 2009). The three types of fatty acids are (1) saturated fatty acids (SFAs), (2) monounsaturated fatty acids (MUFAs) and (3) polyunsaturated fatty acids (PUFAs). The saturated fatty acids and the mono-saturated fatty acids are synthesised endogenously, but the polyunsaturated fatty acids cannot be synthesised by humans from the other components by any known biochemical pathways. As a result, it must be obtained from the diet (Kaur et al., 2014). Polyunsaturated fatty acids are also known as "essential fatty acids" (EFAs), which are sourced externally through the diet (Mahan, 2000). They assist in the balance of healthy cell membranes, the correct expansion and functioning of the brain and nervous system, and the construction of hormone-like substances, eicosanoid, thromboxane, leukotriene and prostaglandin. EFAs are responsible for modifiable blood pressure, blood viscosity, vasoconstriction, and immune and inflammatory responses (Cholewski et al., 2018). Their PUFAs are classified as omega-3 (n-3) and omega-6 (n-6) based on the location of the last double bond relative to the terminal methyl end of the molecule (Patterson et al., 2012). Of these, n-3 and n-6 PUFAs play the most important biological roles, and the quantitative balance between n-3 and n-6 PUFAs is believed to be a crucial factor in many disease states including cardiovascular diseases (Reimers and Ljung, 2012). The major omega-3 fatty acids are α-linolenic acid (ALA), EPA and DHA. ALA is the precursor of EPA and DHA. EPA and DHA are found mainly in fatty fishes such as mackerel, salmon, herring, trout, bluefin tuna and in fish oils. Omega-3 fatty acids found in fatty fish are known to be essential in the growth of children and the prevention of coronary heart diseases. DHA is important for optimal brain and neurodevelopment in children, while EPA helps in improving cardiovascular health overall. Fatty acids also play an important role in membrane-mediated processes such as osmoregulation, nutrient assimilation and transport (Balami et al., 2019). Fishes are the most important source of fatty acids; fatty fish, such as sardines and mackerel, and some salmon species, contain high levels of EPA and DHA. The most important thing to know about fish is that fish themselves cannot synthesise these fatty acids; they obtain them from the food they consume (algae and plankton) (Tasbozan et al., 2017). It has been observed that therapeutically important omega-3 PUFAs prevent childhood asthma as well as attention-deficit/hyperactivity disorder in children. Moreover, in adults, it has been well-documented in the prevention of cardiovascular diseases, hypertension and idiopathic oligoasthenoteratozoospermia. Additionally, in geriatrics, it helps in preventing dementia, age-related macular degeneration, Alzheimer's disease and mood disorders (Kaarnivantan et al., 2011). The EPA and DHA fatty acids found in fish also exhibit anti-inflammatory properties through their impact on prostaglandin synthesis (Aucocin et al., 2017). For pregnant women, mothers who are breastfeeding and women of childbearing age, fish intake is highly recommended and important because it supplies DHA, a specific omega-3 fatty acid that is beneficial for the brain development of infants (Torpy et al., 2006). The benefits of nutraceuticals include anticoagulant properties, helping to thin the blood and acting as an antidepressant (Liao et al., 2019). The health benefits of omega-3 fatty acids are well-known to scientific, clinical and industry experts, with research examining the effects on most people and for numerous health conditions.

Fish is an important source of quality animal proteins (particularly the essential amino acids lysine and methionine). It has been reported by scientists that fish protein has a greater satiety effect than other sources of animal proteins like chicken and beef. Approximately 60% of people from developing countries depend upon fish, which amounts to 30% of their animal protein supplies (FAO, 2020). Furthermore, proteins obtained from fish are tremendously digestible and serve as a rich source of peptides and essential amino acids, which are found to be inadequate in terrestrial meat proteins. Proteins found in fish have high biological values because of the presence of essential amino acids in high-quality proportions. The important biomolecules that control key metabolic pathways and serve as precursors for the synthesis of biologically important substances are proteins and amino acids. Proteins and amino acids are indispensable for the appropriate growth and development of the body, which includes the maintenance and repair of worn-out tissues (Pal et al., 2018). Most prominently, proteins and amino acids play an important role in preventing protein-calorie malnutrition. They can act as a significant defence against bacterial and viral infections and help safeguard electrolyte and water balance in the human system. The regulatory factors are balanced by the proteins derived from fish (Balami et al., 2019). For example, a sardine protein diet lowered insulin resistance, leptin and tumour necrosis factor (TNF)-α, improved hyperglycemia and decreased adipose tissue oxidative stress in rats with induced metabolic syndrome. However, due to their undesirable fishy odour and flavour, they are mostly used in animal nutrition (Kundam, 2018). Amino acids are requisite biomolecules that serve as building blocks of proteins as well as intermediates in various metabolic pathways. Bioactive compounds are mainly obtained from a protein diet, and the quality of dietary protein is assessed from the essential to non-essential amino acid ratio. Deficiency or differences in amino acid ratio could affect body metabolism and homeostasis. We can avoid such deficiencies or metabolic disorders by ensuring adequate amounts of amino acids in the human diet.

Vitamins that are essential for human health are present in fish; however, the amount of vitamins may vary from species to species. Vitamins A, D and B are widely available in fish (Balami et al., 2019). Among the vitamins, vitamin A helps for normal growth, builds cells, prevents the problem of poor eyesight and helps in the treatment of many eye diseases. Vitamin D in fish is found in the form of vitamin D3 (cholecalciferol), and deficiency of vitamin D can lead to rickets, osteomalacia and low bone mineral density (BMD), thereby leading to osteoporosis. Vitamin D deficiency exacerbates osteopenia, osteoporosis and fractures in adults. In addition to bone-related issues, deficiency of vitamin D has been connected with diabetes, increased aggressiveness of certain cancers and increased occurrence of autoimmune diseases as well as cardiovascular diseases (Khalili et al., 2018). The profusion of important minerals and trace elements in fish is due to their capacity to obtain inorganic atoms from salt and fresh water. The essential minerals found in seafood are higher than that in terrestrial foods (Kundam, 2018). Minerals commonly found in fish flesh are sodium, potassium, calcium, magnesium, phosphorus, sulphur, iron, manganese, zinc, copper and iodine. Essential elements such as Ca, Mg, Na, K, Fe, Zn, Cu, Mn, Se, F and iodine participate in several biochemical reactions: calcium, magnesium and phosphorus are crucial in the formation of bones and teeth; sodium and potassium work together in the transmission of nerve impulses and maintaining electrolyte balance; zinc is mostly found as a cofactor in enzyme reactions; and iron forms part of the haemoglobin molecule which transports oxygen around the body.

9.3 ROLE IN HEALTHCARE

9.3.1 Benefits

Over the past few years, nutraceuticals have become much acknowledged. They are used as substitute or supplemental treatments along with pharmaceuticals and help to avert and treat a wide range of diseases. Nutraceuticals have received substantial attention because of their potential nutritional values, safety, affordability and multiple therapeutic effects and are often seen as a possible option to conventional treatments. Nutraceuticals can play an important role in the body's various biological processes, which help prevent various diseases and improve overall health and well-being (Yao et al., 2006).

9.3.2 Adverse Effects

Compared to pharmaceuticals, nutraceuticals are potentially safe and can recover health naturally. On the other hand, certain nutraceuticals can have side effects such as allergic reactions caused by interactions with other nutraceuticals or therapeutic drugs (Karwande et al., 2015). A few nutraceuticals can also cause toxicity. In addition, many people tend not to disclose nutraceutical use to their physicians, which can result in adverse drug-supplement reactions.

Nutraceuticals have a powerful effect on your body, and that's why it's important to consume only the recommended amounts of nutraceuticals. It is also important to talk to your health professional about your symptoms and nutraceutical use.

9.3.3 Prevention and Treatment of Chronic Diseases

Nutraceuticals play an important role in preventing the onset of chronic diseases and reducing the complications involved (Hosomi et al., 2012). Evidence suggests they are used in the prevention and treatment of cardiovascular diseases, cancer, diabetes, obesity and inflammatory-based diseases.

9.3.3.1 *Improve Eye Health*

Patients with eye disorders and conditions like age-related muscular degeneration, glaucoma and visual disorders can benefit from certain nutraceuticals (Gupta et al., 2014).

Nutraceuticals like lutein, zeaxanthin, vitamin C and vitamin E reduce the risk of cataracts. Essential fatty acids like omega-3 are important for visual development and retinal function.

9.3.3.2 *Improve Immune Function*

Immune booster nutraceuticals like green tea, blueberries, amino acids and vitamin D are found to be useful in improving and strengthening immune health and thus help to prevent disease (Chiesa et al., 2016).

Herbs and plant extracts like echinacea and astragalus help in the development and regeneration of stem cells. The anti-inflammatory effects of garlic can enhance the functioning of the immune system.

9.3.3.3 *Improve Gut Health*

Many nutraceuticals like prebiotics, probiotics, flavonoids, aloe vera, resveratrol and omega-3 and omega-6 acids can restore and stimulate the growth of good bacteria and reduce the effect of harmful bacteria in the gut (Gupta et al., 2014).

Nutraceuticals are shown to influence inflammation pathways and play an important role in reinforcing the body's natural gut defence mechanisms.

9.3.3.4 *Support Reproduction Health*

Many nutraceuticals play an important role in male and female reproductive health. Nutraceuticals help with male infertility and dysfunction, and prevent damage to sperm. Certain nutraceuticals can influence hormones and are also used to improve fertility, preconception, and pregnancy outcomes in women, and provide support during menstruation (Chiesa et al., 2016).

9.3.3.5 *Treat Inflammation*

Nutraceuticals are showing promising results in the prevention and treatment of inflammation due to the presence of phytochemicals. They can reduce oxidative stress in chronic inflammatory diseases like asthma, osteoarthritis, rheumatoid arthritis, cancer and autoimmune diseases (Hosomi et al., 2012).

9.3.3.6 Enhanced Sports Medicine

Nutraceuticals are useful in sports medicine, as they help athletes improve their performance and reduce oxidative stress and mitochondrial dysfunction, which occurs because of heavy exercise.

In particular, mitochondria-targeted nutraceuticals (MTNs), such as resveratrol and quercetin, have been shown to have antioxidant effects at a molecular level, which improves mitochondrial bioenergetics and leads to better exercise performance and recovery (Chiesa et al., 2016).

9.3.3.7 Prolong Life

Nutraceuticals provide preventative care to the body, which increases life expectancy, as it reduces the risk of many diseases and helps the body recover quickly (Hosomi et al., 2012).

Certain nutraceuticals like those present in citrus fruits and soybeans help renew healthier cells and improve gene expression. Some nutraceuticals can slow the ageing process and thus promote longevity.

9.3.3.8 Improve Mental Health

Research shows that nutrients and nutraceuticals play an important role in the prevention, management and treatment of mental health disorders and psychological functioning. Certain nutrients and nutraceutical compounds including omega-3, vitamins B and D, magnesium, iron, zinc, curcumin, lycopene and B-carotene are showing promising results in improving mood, stress, anxiety, depression and cognitive functioning.

Nutraceuticals with antioxidant properties help prevent neurodegenerative diseases, including Parkinson's and Alzheimer's disease (Chiesa et al., 2016).

9.3.3.9 Other Uses of Nutraceuticals

Here's a look at other benefits of nutraceuticals. They

- Improve overall health
- Increase energy
- Improve sleep
- Regulate bodily functions
- Enhance skincare
- Improve mental clarity
- Reduce cravings

9.4 CONCLUDING REMARKS AND FUTURE DIRECTION

The greatest scientific need in nutraceuticals pertains to the standardisation of compounds and/or products to carefully develop and execute studies/trials to provide the basis for health claims for nutraceuticals; to develop testing methods for ensuring and verifying the consistency of the dosage of ingredients included in the company's products; and to develop new products either by combining existing ingredients used in nutritional supplements or identifying new ingredients that can be used in nutritional supplements. The interaction of nutraceuticals with food and drugs is another area that should be taken into consideration. The effect of different processing methods on the biological availability and effectiveness of nutraceuticals remains to be determined. As with drugs, there should be strict regulatory controls for nutraceuticals.

To ensure food security and nutritional quality for a growing world population in the face of climate change, increasing aquaculture production and competition for natural resources, countries must be accountable for what they consume rather than what they produce. The State of World Fisheries and Aquaculture 2018 highlights the critical importance of fisheries and aquaculture for food and nutrition. Global fish production is about 167.2 million tonnes; 146.3 million tonnes are used for human consumption, and the remaining is used for non-food purposes or discarded as waste material. Consumption of fish per capita amplified in developing regions from 5.2 kg in 1961 to 18.8 kg in 2013, while in less-developed countries with food deficits, it increased from 3.5 kg to 7.6 kg in the same period. Therefore, fish accounts for about 17% of the intake of animal protein by the world population. Worth about $250 billion in 2014, the global market for nutraceuticals is comprised of functional foods and dietary supplements. Rapid increases in consumer demand for nutraceuticals will reach approximately $722.49 billion by 2027 according to a current report by Grand View Research, Inc. The demand for fish oil as a nutraceutical is very high, as it contains essential long-chain omega-3 fatty acids. Moreover, products like collagen, bioactive peptides, protein hydrolysate, carotenoids and glycosaminoglycans are obtained from nutraceuticals. African countries have the smallest consumption rate of fish or fish products among the least-developed countries. Nutraceutical-producing global leaders and the scientific community are urged to find the easiest possible way to use the discards of fish and its by-products for developing low-cost nutraceuticals that can fulfil almost all the nutritional necessities of humans, which could then be supplied to less-developed nations in Africa with accurate and prearranged awareness programmes by leading organisations like the World Health Organization (WHO).

Plant bioactive compounds are found in leaves, stems, bark, roots, flowers, seeds and fruits. When harvested for commercial processing, they are consumed when freshly cut and as extracts in juices, syrups, teas, wines and oils. They can also be processed and marketed as tablets, capsules, creams, gels, powders and granules. To increase the shelf life of these compounds, they can be packaged in small quantities in bottles, jars, cans, pouches, blister packs, cups and trays, or in bags, drums and pails for bulk storage. These packages can be made from metal, glass, plastic and paper products. There are factors that cause these plant-based compounds to deteriorate. It is therefore important to increase the shelf life of these products, the method of processing, storage conditions, maturity of the harvested plant, cleanliness and packaging selection, which should all be optimised. Because most commercial operations that process plant bioactive compounds may be subjected to federal regulations, knowledge of good manufacturing practices, labelling regulations and other relevant laws, including state and municipal ordinances, is essential. Depending on the package and product choice, an appropriate closure must be selected. These should be convenient to use, enhance the marketing of the product and should not compromise the safety of the product.

REFERENCES

Abbas, K.A., Saeed, M.A.E., Bakar, J. 2009. Fatty acids in fish and beef and their nutritional values: A review. *J. Food Agric. Environ.*, 7, 37–42.

Adnan, M., Patel, M., Hadi, S. 2017. Functional and health promoting inherent attributes of Enterococcus hirae F2 as a novel probiotic isolated from the digestive tract of the freshwater fish Catlacatla. *PeerJ*, 5, e3085.

Alshammari, E., Patel, M., Sachidanandan, M., Kumar, P., Adnan, M. 2019. Potential evaluation and health fostering intrinsic traits of novel probiotic strain *Enterococcus durans* F3 isolated from the gut of fresh water fish Catlacatla. *Food Sci. Anim. Resour.*, 39(5), 844–861. doi: 10.5851/kosfa.2019.e57.

American Heart Association (AHA). 2020. Eating fish twice a week reduces heart stroke risk. https://www.heart.org/en/news/2018/05/25/eating-fish-twice-a-weekreduces-heart-stroke-risk (accessed on 17 March).

Anais, P., Raul, P.G., Jean-Pascal, B. 2013. By-products from fish processing: Focus on French industry. In *Utilization of fish waste*. CRC Press. Chapter 1, pp. 1–25.

Aucoin, M., Cooley, K., Knee, C., Fritz, H., Balneaves, L.G., Breau, R., Fergusson, D., Skidmore, B., Wong, R., Seely, D. 2017. Fish-derived omega-3 fatty acids and prostate cancer: A systematic review. *Integr. Cancer Ther.*, 16(1), 32–62.

Balami, S., Sharma, A., Karn, R. 2019. Significance of nutritional value of fish for human health. *Malays. J. Halal Res. J. (MJHR)*, 2(2).

Caruso, G. 2016. Fishery wastes and by-products: A resource to be valorised. *J. Fish. Sci.*, 10(12–15), 265.

Chiesa, G., Busnelli, M., Manzini, S., Parolini, C. 2016. Nutraceuticals and bioactive components from fish for dyslipidemia and cardiovascular risk reduction. *Mar. Drugs*, 14(6), 113. doi: 10.3390/md14060113.

Cholewski, M., Tomczykowa, M., Tomczyk, M. 2018. A comprehensive review of chemistry, sources and bioavailability of omega-3 fatty acids. *Nutrients*, 10(11), 1662.

Daneshkhan, A., Siegel, A. P., Agarwal, M. 2020. Volatile organic compounds: Potential biomarkers for improved diagnosis and monitoring of diabetic wounds. In: Bagchi, D., Das, A., editors. *Roy tissue repair, and regeneration in diabetes*. Academic Press, pp. 491–512. doi: 10.1016/B978-0-12-816413-6.00023-X19.

Das, J.K., Salam, R.A., Kumar, R., Bhutta, Z.A. 2012. Micronutrient fortification of food and its impact on woman and child health: A systematic review. *Syst. Rev.*, 2, 67. doi: 10.1186/2046-4053-2-6718.

FAO. 2020. *The state of world fisheries and aquaculture: Sustainability in action*. Food and Agriculture Organization of the United Nations.

Gupta, C., Prakash, D. 2014. Phytonutrients as therapeutic agents. *J. Complement. Integr. Med.*, 11(3), 151–169.

Hosomi, R., Yoshida, M., Fukunaga, K. 2012. Seafood consumption and components for health. *Glob. J. Health Sci.*, 4(3), 72–86.

Kaarnivanthan, K., Salminen, A., Haapasalo, A., Soininen, H., Hiltunen, M. 2011. Age-related macular degeneration (AMD): Alzheimer's disease in the eye? *J. Alzheimers Dis.*, 24(4), 615–631.

Karwande, V., Borade, R. 2015. *Phytochemicals of nutraceutical importance.* Scitus Academics LLC.

Kaur, N., Chugh, V., Gupta, A.K. 2014. Essential fatty acids as functional components of foods- a review. *J. Food Sci. Technol.*, 51(10), 2289–2303.

Khalili Tilami, S., Sampels, S. 2018. Nutritional value of fish: Lipids, proteins, vitamins, and minerals. *Rev. Fish. Sci. Aquac*, 26, 243–253.

Kris-Etherton, P.M., Hecker, K.D., Bonanome, A. S., Coval, m., Binkoski, A.E., Hilpert, K.F., Griel, A.E., Etherton, T.D. 2002. Bioactive compounds in foods: Their role in the prevention of cardiovascular disease and cancer. *Am. J. Med.*, 113, 71S–88S.

Kundam, D.N. 2018. Israel OkpunyiAcham, and Abraham TartengerGirgih. *Asian Food Sci. J.*, 4, 1–14.

Laparra, J.M., Sanz, Y. 2010. Interactions of gut microbiota with functional food components and nutraceuticals. *Pharmacol. Res.*, 61(3), 219–225. doi: 10. 1016/j.phrs.2009.11.001.

Liao, Y., Xie, B., Zhang, H., He, Q., Guo, L., Subramaniapillai, M., Fan, B., Lu, C., McLntyer, R.S. 2019. Efficacy of omega-3 PUFAs in depression: A meta-analysis. *Transl. Psychiatry*, 9(1), 190.

Mahan, L.K., Escott-Stump, S. 2000. *Krause's food, nutrition, and diet therapy*, 10th ed. W.B. Saunders.

Mandadi, R., Mohd, A., Mousa, A., Mohd, S., Mitesh, P. 2020. Evaluation of anticancer, antibacterial and antioxidant properties of a medicinally treasured fern tectaria coadunata with its phytoconstituents analysis by HR-LCMS. *Anti Cancer Agents Med. Chem.*, 20, 1–12.

Marques, I., Botelho, G., Guiné, R. 2019. Comparative study on nutritional composition of fish available in Portugal. *Nutr. Food Sci.*, 49(5), 925–941.

Mohanty, B.P., Ganguly, S., Mahanty, A., Sankar, T.V., Anandan, R., Chakraborty, K., Paul, B.N., Sarma, D., Syama Dayal, J., Venkateshwarlu, G. 2016. DHA and EPA content and fatty acid profile of 39 food fishes from India. *BioMed Res. Int.*, 4027437. doi: 10.1155/2014/269797.

Muzaifa, M., Safriani, N., Zakaria, F. 2012. Production of protein hydrolysates from fish by-product prepared by enzymatic hydrolysis. *Int. J. Bioflux Soc.*, 5, 36–39.

Nasri, H., Baradaran, A., Shirzad, H., Rafieian-Kopaei, M. 2014. New concepts in nutraceuticals as alternative for pharmaceuticals. *Int. J. Prev. Med.*, 5(12), 1487–1499.

Paital, B. 2018. Nutraceutical values of fish demand their ecological genetic studies: A short review. *J. Basic Appl. Zool.*, 79(1), 16. doi: 10.1186/s41936-018-0030-x.

Pal, J., Shukla, B.N., Maurya, A.K., Verma, H.O., Pandey, G., Amitha, A. 2018. A review on role of fish in human nutrition with special emphasis to essential fatty acid. *Int. J. Fish. Aquat. Stud.*, 6, 427–430.

Patterson, E., Wall, R., Fitzgerald, G.F., Ross, R.P., Stanton, C. 2012. Health implications of high dietary omega-6 polyunsaturated fatty acids. *J. Nutr. Metab.*, 539426. doi: 10.1155/2012/539426

Reddy, M.N., Adnan, M., Alreshidi, M.M., Saeed, M., Patel, M. 2020. Evaluation of anticancer, antibacterial and antioxidant properties of a medicinally treasured fern Tectaria coadunata with its phytoconstituents analysis by HR-LCMS. *Anti Cancer Agents Med. Chem.*, 20(15), 1845–1856. doi: 10.2174/1871520620666200318101938 20.

Reimers, A., Ljung, H. 2019. The emerging role of omega-3 fatty acids as a therapeutic option in neuropsychiatric disorders. *Ther. Adv. Psychopharmacol.*, 9, 2045125319858901.

Roos, N.C., Chamnan, C., Loeung, D., Jakobsen, J., Thilsted, S.H. 2007. Freshwater fish as a dietary source of vitamin A in Cambodia. *Food Chem.*, 103(4), 1104–1111.

Sen, A.R., Datta, S., Mahapatra, B.K., Sardar, P. 2016. Bioactive compounds from fishery resources—A boon for human health. Presented at International Conference on Aquatic Resources and Sustainable Management, Kolkata, India, 17 February.

Siddiqui, A.J., Bhardwaj, J., Goyal, M., Prakash, K., Soni, A., Tiwari, V., Puri, S.K. 2015. Assessment of real-time method to detect liver parasite burden under different experimental conditions in mice infected with Plasmodium yoeliisporozoites. *Microb. Pathog.*, 89, 35–42.

Sarkar, T., Salauddin, M., Sheikh, H.I., Pati, S., Chakraborty, R. 2021. Effect of drying on vitamin, carotene, organic acid, mineral composition and microstructural properties of mango (Mangiferaindica). *J. Food Process. Preserv.*, 46, e16237. doi: 10.1111/jfpp.1623713.

Singh, D.K., Ranjan, A. 2016. Comparative study on macro and micro nutrient profiling of selected freshwater, brackish water and marine water food fishes available in Kerala, India. *Food Nutr. Sci.*, 1, 1–7.

Suleria, H.A., Osborne, S., Masci, P., Gobe, G. 2015. Marine-based nutraceuticals: An innovative trend in the food and supplement industries. *Mar. Drugs*, 13(10), 6336–6351.

Tasbozan, O., Gökçe, M.A. 2017. Fatty acids in fish. In *Fatty acids*. IntechOpen.

Torpy, J.M., Lynm, C., Glass, R.M. 2006. Eating fish: Health benefits and risks. *JAMA*, 296(15), 1926.

Wang, C.-H., Doan, C.T., Nguyen, V.B., Nguyen, A.D., Wang, S.-L. 2019. Reclamation of fishery processing waste: A mini-review. *Molecules*, 24(12), 2234.

World Health Organization (WHO). 2020. Population nutrient intake goals for preventing diet related chronic diseases. https://www.who.int/nutrition/topics/5_population _nutrient/en/index13.html (accessed on 17 March).

Yao, J.-J., Zhao, Y.-L., Wang, Q., Zhou, Z.-L., Hu, X.-C., Duan, X.-W., An, C.-G. 2006. Biochemical compositions and digestive enzyme activities during the embryonic development of prawn, Macrobrachiumrosenbergii. *Aquaculture*, 253(1–4), 573–582.

CHAPTER 10

Probiotic Foods

Divya Aggarwal, Upasana Yadav*, Pallavi Singh, Bindvi Arora and **Srishti Khurana**

10.1 INTRODUCTION

Probiotics are live microbial food supplements that benefit the host animal by enhancing the microbial balance in its intestine (Fuller, 1989). Probiotics are not a new notion; in fact, they have been taken by humans for thousands of years in the form of fermented foods. Their health benefits have long been recognised, with Hippocrates and other early scientists reporting that fermented milk may treat some disorders of the digestive system (Senok, Ismaeel & Botta, 2005; Kopp-Hoolihan, 2001). The use of probiotics in the treatment of physical illnesses is even described in the Bible. The notion of probiotics as we know it now was initially presented by Russian scientist Elie Metchnikoff in 1907 (Mackowiak, 2013). He observed that consumption of large amounts of fermented milk products containing lactobacilli prolonged life and gave the first scientific explanation for the beneficial effects of lactic acid bacteria present in fermented milk. Today, it is well known that taking these probiotics regularly helps to maintain healthy gut flora and avoids gastrointestinal problems.

Although bacteria and other microorganisms are sometimes thought of as "destructive germs," many of them are really beneficial to humans.

Some bacteria help with digestion, diseases and nutrient manufacturing. Many of the bacteria found in probiotic supplements are similar to or identical to those present in our bodies naturally. Probiotics are part of the microbiome, which is a bigger picture of microorganisms and human bodies. A microbiome may be thought of as a varied collection of creatures, similar to a forest, that collaborate to keep our bodies healthy. This community is made up of organisms called microbes. We have trillions of microbes in our body. These microbes are a combination of:

* Corresponding author E-mail: upasana.ndri@gmail.com

DOI: 10.1201/9781003315100-10

- Bacteria
- Fungi (including yeasts)
- Viruses
- Protozoa

Every person's microbiome is different. Microbial cells are unique to each individual, including twins. A microbe must possess certain qualities to be classified as a probiotic. These include the ability to:

- Be isolated from a human
- Survive in our intestine after ingestion (being eaten)
- Have a proven benefit to you
- Be safely consumed

10.2 PROBIOTICS HEALTH BENEFITS AND THERAPEUTIC POTENTIAL

Probiotics may have therapeutic implications in the prevention and treatment of disorders of the gastrointestinal, respiratory, and urogenital tracts, according to various studies (Gardiner et al., 2002). Probiotics may provide a variety of health advantages, including improving the body's ability to resist pathogen invasion and sustaining the host's well-being by maintaining the balance and composition of gut flora. Improvement of the immune system, reduction of lactose intolerance, (Gilliland, 1990) reduction of serum cholesterol level and blood pressure, maintenance of normal intestinal gut microflora, defence against gastrointestinal pathogens (Lourens-Hattingh & Viljoen, 2001), anti-carcinogenic activity (Gilliland, 1990; Ouwehand, Kirjavainen, Shortt & Salminen, 1999) improved nutrient utilisation and improvement of nutritional value of food are some of the identified benefits. Therefore, probiotics have been used extensively in therapeutic applications, such as the prevention of urogenital diseases candidal vaginitis (Lourens-Hattingh & Viljoen, 2001) relief from constipation, prevention of infantile diarrhoea, reduction of antibody-induced diarrhoea, control of inflammatory bowel diseases and irritable bowel syndrome, reduction of hypercholesterolemia, prevention of colon and bladder cancer and prevention of osteoporosis. The growth and action of the probiotic during the production of cultured foods, as well as the growth and action of specific probiotic species in the intestinal tract after consumption of foods containing them, may all contribute to these potential benefits, though not all of them have yet been conclusively proven (Ranadheera et al., 2017; Lourens-Hattingh & Viljoen, 2001).

To be categorised as a probiotic, a food product must meet the following criteria:

- Must contain > 10^7 colony-forming units (cfu) /mL of a living organism
- Organisms are members of the lactic acid bacteria (LAB) family
- The organisms are resistant to gastric acidity and bile salts
- No negative nutritional effects on the human body

10.3 PROBIOTIC MICROORGANISMS IN FOOD

Probiotics are now made up of a variety of microorganisms. However, bacteria belonging to the genera *Lactobacillus* and *Bifidobacterium*, the first and biggest group of microorganisms to be classified as probiotics, are the most often utilised (Mombelli & Gismondo, 2000). These bacteria are found only in the human gastrointestinal tract (Bielecka, Biedrzycka & Majkowska, 2002). In contrast to other gut bacteria, they are known to have no negative consequences. Species of *Lactobacilli* include *L. acidophilus*, *L. rhamnosus*, *L. casei*, *L. delbrueckii ssp. bulgaricus*, *L. johnsonii*, *L. reuteri*, *L. brevis*, *L. cellobiosus*, *L. curvatus*, *L. fermentum*, *L. gasseri* and *L. plantarum* (Krasaekoopt et al., 2003). The most recognised *Bifidobacteria* species used are *Bifidobacterium breve* and *B. animalis subsp.* (Masco, Huys, De Brandt, Temmerman & Swings, 2005). Other lactic acid bacteria (LAB) from genera including *Streptococcus*, *Lactococcus*, *Enterococcus*, *Leuconostoc*, *Propionibacterium* and *Pediococcus* are now included as probiotics (Krasaekoopt et al., 2003; Boylston et al., 2004).

Other non-related microbes used include bacteria such as non-pathogenic *E. coli* and *Clostridium butyricum*, yeasts (*Saccharomyces cereviciae*, *Saccharomyces boulardii*), filamentous fungi (*Aspergillus oryzae*) and some spore-forming *bacilli*. (Mombelli & Gismondo, 2000; Fioramonti, Theodorou & Bueno, 2003).

According to FAO/WHO, probiotics are defined as "live microorganisms which when administered in adequate amounts confer a health benefit on the host." The demand for these foods is constantly rising due to the increased awareness about their health and nutritional benefits. Consumers are now becoming concerned about their food habits and adopting a healthy lifestyle, and they have started incorporating these foods into daily diets to meet their body's nutritional requirements. Probiotics are generally available in different forms, including capsules, powder sachets and tablets. Their consumption is more encouraged via functional foods such as traditional dairy products (buttermilk, curd, yoghurt, cream, ice cream, shrikhand and cheese) as well as non-dairy products including fruits and fruit juices, cereals, meat and meat products (Roy & Kumar, 2018; Chugh & Kamal-Eldin, 2020). Various microorganisms express their probiotic potential by establishing a symbiotic relationship with the human host. Some of the probiotic microorganisms commonly employed in the formulation of functional foods are discussed here.

Among all the available probiotic strains, *Lactobacillus* and *Bifidobacterium* spp. are the most widely studied species. They are regarded as the autochthonous microbiota in the human intestinal tract. They exert major functional effects, such as the production of bacteriocins (anti-microbial peptides), assimilation of dietary fibres, regulation of fat storage and regulation of intestinal microflora by alleviating the colonisation of pathogens (Ku et al., 2016).

The genus *Lactobacillus* comprises several Gram-positive, rod-shaped, facultatively anaerobic or microaerophilic bacteria. They belong to the major group of lactic acid bacteria (LAB) that includes other bacterial species like *Enterococcus*, *Oenococcous*, *Lactococcus*, *Leuconostoc*, *Pediococcus* and *Streptococcus*, which can convert hexose sugars to lactic acid, thus creating an acidic environment that can consequently inhibit the growth of harmful bacterial species (Fijan, 2014). Patel et al. (2019) developed a carotenoid-enriched probiotic yoghurt using fresh biomass of *Spirullina platensis*, which exhibited an accelerated increase in the *L. acidophilus* counts during fermentation and perpetuated its viability upon storage.

Bifidobacterium spp. is yet another Gram-positive, predominantly rod-shaped, anaerobic, non-motile and non-sporulating organism. It usually provides many beneficial effects such as anti-carcinogenic activity, the improvement of lactose digestibility, reduction of serum cholesterol levels, enhanced calcium absorption and synthesis of B vitamins. *B. adolescentis*, *B. bifidum*, *B. breve*, *B. catenulatum*, *B. pseudocatenulatum*, *B. longum subsp. infantis (B. infantis)*, *B. longum subsp. longum (B. longum)*, *B. angulatum and B. dentium* are regarded as the most plentiful species found in the human gastrointestinal tract. Specifically, the bacterial species including *B. catenulatum*, *B. longum and B. adolescentis* share a significant part of the adult intestinal flora, while *B. breve*, *B. infantis and B. longum* are known to predominate in the intestinal tracts of infants (Prasanna et al., 2014). The incorporation of *Bifidobacteria* in cheeses provides an appropriate medium to support their long-term survival. The matrix of the cheese and its high-fat content endows protection to the acid-sensitive probiotic bacteria during passage through the gastrointestinal tract (Boylston et al., 2004). Wang et al. (2009) developed a probiotic noni juice using lactic acid bacteria (*L. casei* and *L. plantarum*) and *Bifidobacteria* (*B. longum*). The juice fermented with *B. longum* expressed high antioxidant activity, which was comparatively insignificant to that of lactic acid bacteria, suggesting *B. longum* and *L. plantarum* as optimal probiotics for fermenting noni juice.

Conventionally, from an economic perspective, yeasts have also been widely used as a starting culture for the development of functional food and beverages. *Saccharomyces cerevisiae var. boulardii* is a non-pathogenic yeast, which is a regular probiotic and marketed as a dietary supplement. It is usually used as a therapeutic agent in the treatment of various gastrointestinal disorders, such as diarrhoea, colitis and irritable bowel syndrome. Moreover, like other probiotic strains, it helps in the production of short-chain fatty acids, degradation of pathogenic toxins and boosting the immune system. It has a broad spectrum of antioxidant, anti-carcinogenic, antibacterial and antiviral activities (Arslan et al., 2015; Lazo-Vélez et al., 2018; Ansari et al., 2021). It is commercially used in various food applications, such as in the preparation of probiotic yoghurts (Kalkan et al., 2018), probiotic cornflakes (Singu et al., 2020), lentil and adzuki bean sprouts (Swieca et al., 2019). The craft beer produced with this probiotic yeast resulted in higher antioxidant activity, more acidification and higher yeast viability.

10.4 PROBIOTIC FOOD PRODUCTS

Probiotics can be consumed as food components or as non-food preparations. Others refer to probiotic-containing foods as functional foods. This term refers to foods that have nutritional or non-nutrient components that influence specific bodily functions, resulting in a favourable health effect. Functional foods, therefore, have physiological or psychological effects in addition to their nutritional worth. Probiotics are typically found in dairy products such as milk powders, yoghurt, soft, semi-hard, and hard cheeses and ice cream. These items provide an ideal environment for probiotic viability and development. Other foods are increasingly being used as probiotic carriers. This is largely because certain customers are allergic to milk products. Non-dairy products such as malt-based beverages and fruit juices, meat sausages, capsules and freeze-dried preparations are among these alternatives.

10.4.1 Dairy Probiotic Products

For thousands of years, fermented and non-fermented dairy foods have played an essential part in human diets throughout the world. Since Mechinikov's observations in the early 1900s, there has been a growing interest in the advantages of certain microbes, such as probiotic gut flora and LAB, and their impact on human general health, bodily functioning and life expectancy. Pasteurised milk, ice cream, fermented milks, cheeses and infant milk powder are just a few examples of probiotic dairy products that are now created and eaten across the world (Kailasapathy, 2006; Cruz et al., 2009; Awaisheh, 2011). Milk is a great way to transport or create living, active cultured dairy products. Milk's buffering capability aids probiotic bacterial survival in the gastrointestinal system (Kailasapathy, 2006). For several reasons, fermented dairy products are excellent for fostering the probiotic bacteria's favourable health reputation. Because fermented foods contain live microorganisms (starter cultures), consumers already have a favourable perception of them as being healthy. Additionally, probiotic bacteria utilised as beginning organisms combine the favourable perceptions of fermentation and probiotic cultures (Lourens-Hattingh & Viljoen, 2001).

10.4.1.1 Fermented Milk and Yoghurt (Bio-Yoghurt) Probiotic Products

Different types of food products have been proposed as carriers for probiotic bacteria, allowing customers to eat vast volumes of viable probiotic cells for optimum probiotic bacteria viability and therapeutic benefits. As a fermented milk product, yoghurt is one of the most common food carriers for probiotic delivery (Lourens-Hattingh & Viljoen, 2001). Yoghurt has long been recognised as a product having a variety of positive benefits for customers, and it is also crucial to note that most people believe yoghurt to be "healthy," especially when probiotic bacteria are included. Commercial yoghurt typically has a pH between 3.7 and 4.3 (Awaisheh, 2011). Since recent discoveries in several areas of bioscience support the idea that, beyond nutrition, diet may modulate various functions in the body, new yoghurt products, known as "bio-yoghurt," have been created by adding live probiotic strains in addition to the standard cultures, *S. thermophilus* and *L. bulgaricus*, to yoghurt (Cruz et al., 2009). The probiotic strains used in the formulation of bio-yoghurt products, which are mostly *Lactobacillus* and *Bifidobacteria* species, include. *L. acidophilus*, *L. casei*, *L. gasseri*, *L. rhamnosus*, *B. animalis*, *B. infantis* and *B. longum*. In light of this, bio-yoghurt is yoghurt made with living probiotic cultures, which may result in the stated health benefits of probiotics. Around the world, several textures of yoghurt and products similar to yoghurt are produced, including natural-set yoghurt, stirred yoghurt and drink yoghurt. The amounts of non-fat solids in these products vary substantially, with 16–18%, 13–14% and 11–12% being the respective percentages (Kailasapathy, 2006).

10.4.1.2 Ice Cream and Frozen Probiotic Products

Ice cream is a frozen dairy product made up of a variety of ingredients such as milk, flavours, sweeteners, stabilisers and emulsifiers. Ice cream's makeup, which includes milk proteins, fat, lactose and other substances, makes it an excellent carrier for probiotic microorganisms. Furthermore, the comparatively high pH values of ice cream (5.5 to 6.5) contribute to improved probiotic bacteria survival during storage (Douglas, Goff & W. Hartel, 2018). Several research studies have shown that ice cream may be used to provide probiotic microorganisms. New types of ice cream products have entered the market during the past several decades; these products were created by adding probiotic cultures to existing ice cream goods. As a result of the addition of probiotic cultures, ice cream gained value and came to be seen as a functional product in addition to being a nutrient-dense food due to the inclusion of dairy-based ingredients, vitamins and minerals (Goff, 2008). Ice cream and frozen dairy desserts have shown significant promise for use as carriers of probiotic microorganisms due to their composition/structure, production processes and storage conditions. Probiotic bacteria are more viable due to the ice cream freezing storage temperature and little danger of temperature damage during storage (Cruz et al., 2009).

10.4.1.3 Cheese Probiotic Products

Cheese is a general term for a variety of fermented and non-fermented milk-based dairy products that come in a wide range of tastes, textures and shapes and are made and enjoyed all over the world (Awaisheh, 2011). According to moisture content, all cheeses, whether rennet or acid set, can be classed as soft, semi-soft (semi-hard), hard or extremely hard (Mahony, McAuliffe, Cotter & Fitzgerald, 2019). As evidenced by the low probiotic bacteria viability in commercial yoghurts, these cultured products may not be the best food carriers for probiotic bacteria to humans due to cultural considerations and the technologies involved in the production of fermented milks and yoghurt. These include relatively short fermentation times, low pH values, oxygen residues and antagonistic activity of yoghurt starter cultures against probiotic bacteria. As a dietary delivery method for probiotics in this situation, cheese offers a useful option. Probiotic bacteria have a better chance of surviving throughout the gastric transit when cheese's high protein content offers them decent buffering protection against the gastrointestinal tract's high acidity state. Furthermore, probiotic microorganisms in the stomach may benefit from the thick matrix and relatively high-fat content of cheese (Patel et al., 2008)).

10.4.2 Non-Dairy Probiotic Products

Dairy products, as previously stated, are the primary dietary transporters for probiotic bacteria in humans. Because of the limitations of existing products, such as allergies, high lactose and cholesterol content, and the necessity for cold storage, new probiotic product lines based on non-dairy substrates have been developed (Ranadheera et al., 2017). Additionally, the trend in non-dairy probiotic product development has been sparked by the rise in consumer vegetarianism throughout industrialised nations, which has led to an increase in demand for vegetarian probiotic products (Prado, Parada, Pandey & Soccol, 2008). Fruits and vegetables, juices, non-dairy drinks, cereal-based goods, chocolate-based products and meat are examples of non-dairy probiotic items available on the market (Granato et al., 2010; Stadnik & Dolatowski, 2013; Vasudha & Mishra, 2013).

10.4.3 Probiotic Foods from Fruits and Vegetables

Different types of fruits and vegetables can have probiotics added to them in the form of juices, purees, pulp, drinks and dried fruit. Fruits and vegetables are regarded as nutritious foods because they are high in carbs, dietary fibre, vitamins, polyphenols and certain minerals (Sutton, 2007). Fruits and vegetables contain micronutrients like calcium, retinol, vitamin E, folate, nicotine acid, riboflavin, pantothenic acid and biotin for genetic damage restoration. Combining these micronutrients with the probiotic strain in probiotic juices will be a future goal and a great challenge, as it may provide more health benefits. Certain berries, including blueberries and raspberries, have antibacterial and pathogenic defences against diseases that are spread through food (Vasudha & Mishra, 2013).

10.4.4 Cereal-Based Probiotic Products

Cereals such as wheat, maize, oats and barley have dietary fibres and a variety of other physiological features that aid in the nourishment of the stomach through the use of prebiotic carbohydrates (Capozzi, Russo, Dueñas, López & Spano, 2012). This is despite the fact that cereal grains have a lower nutritional quality than milk and meat due to their lower protein content, deficiency of certain essential amino acids (lysine), low starch availability, anti-nutrients substances (phytic acid, tannins and polyphenols), and coarse nature. Cereal grains are still considered one of the most important food sources of protein, carbohydrates, vitamins, minerals and fibres for large segments of the population worldwide. Prebiotics, probiotics and dietary fibres all have an impact on fermented cereal goods, whereas probiotics are frequently linked to fermented dairy-based products. Many beneficial bioactive substances and fibres can be employed as probiotic carriers in the bran and germ. Oats and barley are grains that help lower blood cholesterol levels and maintain a healthy heart because of their high levels of glucan and hypo-cholesterolemic impact (Lamsal & Faubion, 2009).

10.4.5 Soy-Based Probiotics

Soy is from the soy plant. Soy can be processed into soy protein, soy milk or soy fibres. They are used as a substitute for milk for lactose-intolerant people. Soy is suitable for the growth *of Lactobacillus* and *Bifidobacterium* and is a functional food, being used for the treatment of diseases like menopausal disorder, cancer, and atherosclerosis. Consumption of soy oligosaccharide may cause bloating or cramping. This can be prevented by fermenting soy with probiotics like *Lactobacillus* and *Bifidobacterium* (Granato et al., 2010). In order to not lose the viability of the probiotic strain, the soy product should be freeze-dried, stored at a temperature of about 4°C and packed in a Tetra Pak, as freeze-drying is preferred over spray-drying or other techniques.

10.4.6 Meat Probiotic Products

Meat is a highly nutritious food with a high level of nutrient absorption, and people appreciate the taste, flavour and texture of meat. As a result of the content and structure of meat, it is considered to be an effective carrier of probiotics (Stadnik & Dolatowski, 2014). Furthermore, meat was discovered to protect LAB against the fatal effects of bile. Dry fermented sausages without heat are one of the most researched and processed probiotic meat products. In addition to having a high nutritional value, this type of meat product has properties that make it an ideal food matrix for delivering probiotics to humans (Seo et al., 2009). For example, because it is a fermented product, adding probiotic bacteria won't change its sensory qualities. Additionally, because it hasn't been heated, the probiotic bacteria's viability won't be diminished. For the probiotic to ensure positive effects on the host, it is crucial to use probiotic bacteria from meat-dominating strains, which supports the requirement for greater numbers of viable cells at the time of intake.

10.4.7 Chocolate Probiotic Products

Due to its great taste and flavour, high nutritional energy, quick metabolism and good digestion, chocolate is one of the most popular products all over the globe. The presence of cocoa butter, milk and milk-based ingredients, and sugar in its formula helps provide proper protein, carbohydrate, lipid, mineral and vitamin absorption. One of the most significant trends in chocolate production today is driven by consumer demand for chocolate that is functional or health-promoting, i.e., chocolate that not only doesn't negatively impact consumer health but also treats or prevents diseases like heart disease, osteoporosis, cancer and diabetes (Egan, Lakenx, Donovan & Woolson, 2010). Given that it has a substantial amount of polyphenolic antioxidants and flavonoid components, chocolate is a functional food. These advantageous components in chocolate are thought to be responsible for cocoa's positive health effects. To create chocolate with probiotics, it is important to have a thorough awareness of the probiotic strains that were chosen, the procedure used to make chocolate, and the many aspects that are crucial for microbial survival (Patel et al., 2008). Since probiotic bacteria must endure food processing, be maintained in supplement shelf life, transit through the stomach's highly acidic environment and interact with enzymes and bile salts in the small intestine, their viability at the point of consumption is a crucial factor in determining a product's efficacy (Coman et al., 2012).

10.5 EFFECT OF PROCESSING AND STORAGE OF PROBIOTIC FOODS

Commercially accessible viable probiotic strains include fermented foods, freeze-dried supplements and medicinal treatments. The most basic need for probiotics is that they contain enough bacteria to last until the expiration date. Probiotics must include specific strains and maintain a particular number of live cells to give health benefits to the host. Products containing diverse probiotic organisms should include at least a million of each of them per gram to get the intended good effects (Laranjo, Potes & Elias, 2019). A daily dose of 10^9 to 10^{10} cfu viable cells has been found to have beneficial impacts on the host's health and is regarded as the minimum dose to exhibit positive effects on the host's health. This could be accomplished by consuming 100 g of a product containing between 10^6 and 10^7 viable cells per gram on a regular basis. The viability of probiotic microorganisms in yoghurt has been discovered to be poor. Maintaining the viability of probiotic bacteria for them to be employed in functional meals is a major commercial and technological challenge. During the manufacturing, storage or shipping of the finished product, as well as during intestinal transit, many active cultures perish. As a result, the vast majority die before the consumer receives any health advantages. A major issue of shelf instability has been observed with dried cultures. Because of the unfavourable effects of low temperatures and the production of crystals on bacterial organisms, refrigerated items have a short shelf life.

The number of live bacteria steadily decreases during refrigerated storage. Market surveys revealed significantly decreased numbers even before the product's expiry date. Shelf life for probiotics is thus unpredictable, therefore the industry has had trouble substantiating label claims and difficulty backing up label claims (Lahtinen et al., 2005). After surviving food processing, probiotics are exposed to the circumstances in the stomach and small intestine before reaching their intended site in the colon. The bacteria may perish on their journey from the upper intestine to the colon, preventing them from colonising it. They must therefore then survive gastric acidity and bile salts that they encounter during their passage through the GIT. The strain and species-specific tolerance to low pH (pH values range from 1.3 to 3.0) in gastric juice and bile salts present in the small intestine determine their survival in the GIT. Probiotic bacteria can only function when they are given the right environment and are shielded from stresses (such as intense temperatures, high pressure and shear forces) that they may face during their production at the factory or in the GIT.

Temperature, moisture, light and oxygen all have an effect on viability during storage and should be taken into account. Another important factor in probiotic bacteria survival in dairy foods is oxygen toxicity. The viability of anaerobic bacteria is affected by excessive levels of oxygen in the product.

Probiotic cultures are often marketed as frozen or dried forms. However, frozen starters possess certain limitations such as the requirement of subzero storage temperatures, which may exert bacterial cellular injuries due to the subsequent freezing and thawing operations. This ultimately utilises high energy consumption and high transportation costs. Hence, dehydration methods are more favoured, out of which spray-drying and freeze-drying are extensively applied (Cassani et al., 2020).

10.5.1 Spray-Drying

This technique is widely used to convert liquid products (solutions, emulsions or suspensions) into low-moisture powders with high stability throughout the shelf life. The obtained powders possess desirable bulk characteristics and remarkable functional properties including excellent reconstitution behaviour (Schutyser et al., 2019). This technology is based on the key principle of atomisation wherein bacterial suspensions are pumped through the atomiser and disintegrated into a large number of fine and uniform droplets. In the drying chamber, the exposure of these droplets to a stream of hot air promotes heat transfer due to convection and initiates the drying process. Water starts evaporating from inside of the droplets, resulting in the separation of dried product from humid air. Nevertheless, spray-drying is an inexpensive and highly reproducible technique that produces powders of low water activity, thereby safeguarding the microbial activity and facilitating its efficient transportation and commercialisation due to reduced dimensions (in terms of weight and volume) (Dyvelkov & Sloth, 2014; Barbosa & Teixeira, 2017). Due to the shorter drying time, this process can also be used for treating various heat-sensitive or thermolabile materials. These advantages make spray-drying an effective technology to be employed as the most common encapsulation process in preserving bioactive compounds and producing probiotics on a larger scale (Guiné, 2018). Table 10.1 below shows spray-drying parameters in the production of probiotic foods.

The survival of probiotic microorganisms is mainly influenced by the drying medium and by certain drying parameters including drying temperatures.

Drying medium: During the microencapsulation process, several protective agents are added to the drying medium to protect the probiotic cultures and increase their cell viability. These agents include carbohydrate-based systems i.e., polysaccharides (agar, carrageenan, sodium alginate, chitosan, gelatine, gum arabic, starch etc.), oligosaccharides (corn syrup, maltodextrin, sucrose, resistant starch etc.), proteins (gluten, egg albumin, casein, whey protein etc.) and lipids (fats, oils, diglycerides, paraffin, waxes etc) (Vandeker & Ortega, 2017). Skim milk is the most common carrier in exerting the protective effect on different microbial species. Among polysaccharides, maltodextrin is commonly preferred as a carrier agent due to its inexpensive cost, good solubility, non-toxicity and easy availability. The use of maltodextrin at a concentration of 25% in spray-drying of a non-dairy-based probiotic sohiong fruit enhanced the solid content of the feed and thereby increased its yield. This assisted in elevating the glass transition temperature of the food components and resolving the problem of powder stickiness

(Vivek et al., 2021). Due to its low emulsifying capacity, it is recommended to use in combination with other encapsulating agents such as gum arabic. It is known to prevent the complete dehydration of cell components and improve the stability of bacterial cells upon drying and storage. The proteins present in GA constitute a protective layer over the bacterial cell wall while the fibres serve as a partial replacement of sites of water molecules, thus preventing the disruption of the cell membrane and increasing the encapsulation efficiency (Arepally & Goswami, 2019).

Currently, fermented foods inoculated with starter cultures, especially *Lactobacillus* sp., bring immense health benefits and promote food safety. They are known to utilise several types of carbohydrates to produce various antagonistic substances like lactic acid against pathogenic bacteria. Tantratian and Pradeamchai (2020) studied the selection of five different protective agents that could provide appreciable protection for *Lactobacillus plantarum* FT 35 (LPFT 35) obtained from Pla Som, a sour fermented fish product. The prepared suspensions were subjected to spray-drying at a feeding rate of 20 ml $min^{-1,}$ inlet temperature of 85°C and an outlet temperature of 85 ± 5°C. Due to the highest glass transition temperature, the soluble starch exhibited the best cell protection followed by maltodextrin, lactose, sucrose and glucose. Furthermore, it enhanced the cell viability with the least number of cell injuries in the encapsulated powder, thus indicating soluble starch as a suitable protective agent and a preferred carbon source for bacterial cultures.

Drying parameters: In spray-drying, the application of high temperatures results in the exposure of bacterial cells to heat and oxidative and osmotic stresses. As a consequence, these harsh conditions result in their thermal inactivation. Moreover, it damages the cytoplasmic membrane, cell wall, DNA and RNA of cells, resulting in the reduction of their metabolic activity (Vandeker & Ortega, 2017). These effects are attributed due to the different drying process parameters including feed rate, inlet and outlet temperature, and drying time. The outlet air temperature is considered to be the major drying parameter affecting the survival of probiotic cultures. Zhang et al. (2016) investigated the effect of outlet temperature on the survival of *Lactobacillus salivarius* NRRL B-3051 during spray-drying. However, higher outlet temperatures resulted in lower values of water activity that favours the survival of microbial cells. Conversely, it led to a significant reduction in the number of viable cells. On lowering the outlet temperatures from 90–92°C to 84–86°C, 84–86°C to 80–82°C, and from 80–82°C to 74–76°C, a notable increase in the cell viable counts was observed.

As discussed before, atomisation is the key step in the spray-drying process wherein the bacterial suspensions are exposed to many shear forces, which generally do not cause a remarkable cell inactivation. However, Ghandi et al. (2012) examined the effect of shear rates and oxygen injury on the survival of *Lactococcus lactis subsp. cremoris* ASCC930119. The quantitative analysis demonstrated that the lower shear rates could enhance bacterial survival by minimising bacterial death. In addition to this, the process of atomisation also governs the initial droplet size distribution, which can affect the cell viability by altering the droplet's temperature, moisture trajectories and oxygen exposure by varying the surface:volume ratio. The occurrence of bacterial death due to oxygen injury could be alleviated by creating an anaerobic environment using oxygen scavengers such as ascorbic acid and thereby improving the efficacy of spray-drying. The drying rate is yet another parameter in determining the cell viability of probiotics. In a comparative study, spray-drying expressed the highest residual viability due to the higher drying rates and shorter drying time of about 1.3 s than that of single droplet drying (30 mins) and freeze-drying (24 hrs). The results revealed that the higher drying rates could instantly stabilise the bacterial cells and effectively prevent the dehydration inactivation of *Lb. plantarum* (Perdana et al., 2013).

10.5.2 Freeze-Drying

Freeze-drying, also known as lyophilisation, is predominantly used for the encapsulation of several bioactive compounds including probiotics that exert functional effects when ingested. This processing technique comprises three stages: freezing, primary drying and secondary drying. Firstly, the biological materials are frozen under atmospheric pressure, leading to the formation of ice crystals. The morphology of these crystals determines the rate of sublimation both in primary and secondary drying. In the case of large ice crystals, conduction resistance prevents rapid freezing, and hence, in order to speed up the process, it is recommended to use supercooling. The greater the degree of supercooling, the increased the rate of ice nucleation and the faster the freezing process. One of the ways to achieve this state of supercooling is by adding cryoprotectants. During the primary drying phase, the pressure in the drying chamber is significantly lower than the vapour pressure of ice, facilitating the process of sublimation to take place. Free water crystallises in this

Table 10.1 Spray-drying parameters in the production of probiotic foods

Probiotic foods	Microorganism	Inlet temperature (°C)	Feed rate (ml/ min)	Reference
Probiotic yoghurt	*L. paracasei* NFBC 338, *S. thermophilus* and *L. delbrueckii subsp. bulgaricus*	170±2°C	n.d.	(Kearney et al., 2009)
Fermented soy milk	*Lactobacillus acidophilus*	135°C	20 mL/ min	(Telang & Thorat, 2010)
Raspberry juice	*Lactobacillus acidophilus* NRRL B-4495 and *Lactobacillus rhamnosus* NRRL B-442	100°C	40 mL/ min	(Anekella & Orsat, 2013)
Cashew apple juice	*L. casei* NRRL B-442	120°C	5 mL/min	(Pereira et al., 2014)
Sweet whey	*Lactobacillus casei* BL23 and *Propionibacterium freudenreichii* ITG P20	140 ± 5°C	n.d.	(S. Huang et al., 2017)
Sohiong fruit powder	*L. plantarum* subsp. *plantarum* MCC 2974	120°C	3.35 ml/ min	(Vivek et al., 2021)

phase while the remaining bound water is removed in the secondary phase of drying (Nowak & Jakubczyk, 2020; Rezvankhah et al., 2020).

Reddy et al. (2009) evaluated the effect of three different cryoprotectants (lactose, maltodextrin, skimmed milk) on the viability and functional properties of freeze-dried *Pediococcus acidilactici*, *Lactobacillus plantarum* and *Lactobacillus salivarius*. The findings demonstrated that freeze-dried LAB with maltodextrin exhibited a high degree of viability and desired probiotic properties, thereby making maltodextrin a suitable cryoprotective agent for the production of shelf-stable probiotic foods. Unipectine RS 150 has also been found to be a superior agent in improving the enumerated viable counts of *L. acidophilus* and *Bifidobacteria* by 1 log in freeze-dried yoghurts. These agents can inhibit the formation of intracellular or extracellular ice crystals by binding to water, thus enhancing probiotic survival (Capela et al., 2006). The survival rate of sorbitol-enriched *L. plantarum* microcapsules was reported to be 91.2% in freeze-dried yoghurt powder (Jouki et al., 2021). Henceforth, freeze-drying presents a unique technique for developing these synbiotic powders with high solubility and similar characteristics to that of fresh commodities. The introduction of freeze-dried microencapsulated *L. acidophilus* LA-2 in soy protein bars was noted to effectively reduce the content of α-galactoside via enzymatic hydrolysis. The high cell viability of this bacterium was remarkably maintained throughout storage (Chen & Mustapha, 2012). The major limitation associated with freeze-drying is the high operational cost and higher energy consumption. In order to tackle these shortcomings, several hybrid methods of drying have been considered. For instance, freeze-drying coupled with microwave drying elicited better results in the development of probiotic-enriched apple snacks with appreciable bacterial stability and viability along with maintaining their colour, texture and sensorial attributes (Li et al., 2018).

10.5.3 Spray-Freeze Drying (SFD)

This is a relatively new technique for preparing biopharmaceutical powders. As the name suggests, it exploits the advantages of both spray-drying and freeze-drying. It is a tri-step process of drying that involves the atomisation of droplets, followed by their solidification when contacted by cold fluid, and sublimation at conditions of low pressure and temperature. This technique allows the formation of cavities on the outer surfaces of powders, enabling easier rehydration in various applications (Dutta et al., 2018). A probiotic strain, *Lactobacillus plantarum* (MTCC 5422), isolated from fermented cereals/legumes was subjected to spray-freeze drying. Compared to conventional freeze-drying, the microcapsules obtained by SFD took approximately 12 h shorter processing time and exhibited excellent flowability with lower hygroscopicity. This investigation study revealed that microencapsulation by SFD proves to be an effective process of obtaining microcapsules with good encapsulation efficiency (87.92% to 94.86%) and better retention of cell viability (Rajam & Anandharamakrishnan, 2015).

In a research study, *Lactobacillus casei* (IFO 15883) was inoculated into MRS broth with 1% mannose and 0.1% CaCO3. The air pressure was maintained at 20 kPa and the use of 100% buffered peptone water as the protective agent was selected as the optimised parameter. Under such optimised conditions, the survival rate of the final obtained SFD powder was reported to be 97.7%, which dictates that it did not affect the viability of probiotic powder significantly (Her et al., 2015).

10.5.4 High-Pressure Processing (HPP)

This is a non-thermal processing technology that utilises the application of ultra-high pressures (100-600 MPa) to achieve pasteurisation. This technique can be used to increase the diversity of probiotic foods. In comparison to thermal processing methods, it not only eradicates pathogenic microorganisms but also preserves the sensorial and nutritional values of ingredients, thereby helping in achieving both food safety and flavour retention. It has been observed that consumers also tend to prefer high-pressure processed products over others. The processing of juçara and Ubá mango juice mixture using high isostatic pressure with the addition of a probiotic strain *L. rhamnosus* GG showed positive results in terms of probiotic protection and its increased viability (Moreira et al., 2017; Huang et al., 2020). High hydrostatic pressures are also reported to increase the cell hydrophobicity of probiotic cultures, as in the case of probiotic-enriched mandarin juice, cell hydrophobicity of *L. salivarius* spp. *Salivarius* CECT 4063 expressed high values for samples homogenised in MRS broth at 100 MPa (Betoret et al., 2017).

10.6 TECHNOLOGICAL CHALLENGES FOR FUTURE PROBIOTIC FOODS

Probiotic cultures must be able to persist for a longer period and be resistant to acidity in the upper digestive system to colonise the epithelium in large numbers, which is a challenge for manufacturers. As a result, probiotic cultures should be produced in such a way that these sensitive bacteria are protected from detrimental interactions. Given the potential health benefits of probiotics, it's no wonder that research into their viability is gaining traction. In their active form, probiotics do not have a long shelf life (Ross, Desmond, Fitzgerald & Stanton, 2005). In most circumstances, refrigeration is essential to extend the shelf life of probiotic cultures, as high temperatures might kill them. Even when stored at moderate temperatures, most probiotics have a short shelf life. And there is a poor recovery of viable bacteria in products that claim to contain probiotic bacteria. Therefore, the preservation or storage of these probiotic microorganisms presents a challenge because they are affected by exposure to temperature, oxygen, and light.

However, industrial-scale production of probiotics and prebiotics faces several challenges, including (i) the use of novel techniques and cost-effective sources for prebiotic production, (ii) low-cost probiotic production, and (iii) improving probiotic viability after storage, during the manufacturing process of functional foods and during stomach transit.

The following are the major issues that need to be addressed:

10.6.1 Prebiotic Manufacturing Strategies and Sources that are Both Novel and Cost-Effective

The majority of prebiotic oligosaccharides are produced by enzymatic treatment of inexpensive basic materials such as sucrose, lactose and plant derivatives. Several variables influence the amount and kind of oligosaccharides generated, including the enzyme source, the concentration and nature of the substrate and the reaction circumstances. Nonetheless, the present methods for obtaining oligosaccharides have relatively poor yields, increasing the expense of production (Del-Val & Otero, 2003; Williams & Stevensen, 1990). The yield of oligosaccharides can be enhanced by lowering the water content in the reaction medium, according to Panesar et al. (2006). Del-Val and Otero (2003) investigated the enzymatic production of galacto-oligosaccharides from lactose in aqueous/polyethylene glycol medium in this respect. They discovered that lowering the water level enhanced both overall galacto-oligosaccharide yield and selective synthesis of 6-galactosyl lactose. Germán, Fabián and Carlos (2021) also studied the production of galacto-oligosaccharides from lactose in an aqueous/acetone medium employing a hyper-thermophilic -glycosidase. They discovered that water activity (a_w) had a significant impact on oligosaccharide yields, with the highest yield attained at an a_w of 0.57. As a result, enzymatic prebiotic biosynthesis in organic media is a promising study area for increasing prebiotic yields above standard aqueous media synthesis. When the synthesis medium comprises mostly of an organic solvent, the reviewed literature already concentrates on such research to determine appropriate aqueous/organic volume ratios or aw/substrate concentration ratios. Non-dairy oligosaccharide manufacturing is another promising alternative for low-cost prebiotic manufacture on a large scale. Cereals have been studied in recent years to see if they may be used as prebiotic sources. According to, Charalampopoulos et al. (2002), cereal grains contain at least two forms of naturally occurring oligosaccharides: galactosyl derivatives of sucrose, stachyose and raffinose, and fructosyl derivatives of sucrose. Cereal grains, on the other hand, are high in dietary fibre, which is made up of a variety of complex polysaccharides that are not fully digested in the small intestine and pass through to the colon. As a result, dietary fibre is a possible source of prebiotics in cereals. Dietary fibre concentration in cereals like rice, rye and maize can range from 150 to 190 g/ kg dry matter. Dietary fibre is mostly made up of glucans, xylans, xylo-oligosaccharides and arabinoxylans. Many *Bifidobacterium species* and *Lactobacillus brevis* were able to grow at high yields utilising xylo-oligosaccharides, according to Crittenden et al. (2002). They also discovered that *B. longum* strains could thrive with only arabinoxylan as a carbon source. Using autohydrolysis procedures with maize cobs as the raw substrate, Moura et al. (2008) recently generated oligosaccharides with degrees of polymerisation of 3–4 and 5–6. Furthermore, the acquired oligosaccharides encouraged *Bifidobacterium* and *Lactobacillus* species' growth to the same amount as commercial xylo-oligosaccharides. Many cereal grains, such as corn, wheat, rice and oat, have a significant amount of starch (785, 650, 820 and 636 g/ kg dry matter, respectively) (Snow, 1981). Despite the fact that resistant starch occurs naturally in cereal grains, it is typically destroyed during modern food processing. Partial acid hydrolysis, hydrothermal treatment, heating, retrogradation, extrusion cooking, chemical modification and repolymerisation are some of the industrial procedures for producing resistant starch (Charalampopoulos et al., 2002). Studies in animals and human volunteers have demonstrated that a portion of total starch is not digested in the small intestine and goes through to the colon, similar to the case with dietary fibre. Resistant starch is the name given to this fraction, which has been described as a prebiotic. Many *Bifidobacterium* strains, including *B. adolescentis*, *B. bifidum*, *B. breve*, *B. infantis*, *B. lactis* and *B. longum*, were able to hydrolyse resistant starch, according to Crittenden et al. (2002). Only *B. lactis* (strain Lafti B94™) was able to retain viability in circumstances comparable to those found after transit through the gastrointestinal tract. Finally, a synbiotic yoghurt based on resistant starch and *B. lactis* was presented. Topping et al. (2003), on the other hand, found that evidence for resistant starch as a prebiotic is still limited, owing to the fact that much of the experimental work has been done in animals, with the existing human studies being of very short duration. However, the authors acknowledged that there is a lot of potential and that further study is needed to examine resistant starch's potential as a prebiotic in humans.

10.6.2 Production of Probiotics at a Low Cost

To meet the market's growing need for probiotics, the low-cost development of concentrated cultures of probiotics is a major problem (Mattila-Sandholm et al., 2002). Because most vitamins and amino acids are not biosynthetically produced by probiotic microorganisms, a culture medium for probiotic production must be supplied with both amino acids and vitamins (Yang et al., 2016; Kontula, Suihko, Von Wright & Mattila-Sandholm, 1999). The use of various agro-industrial residual effluents as probiotic growth media might be a starting point for lowering production costs on an industrial scale. Fruit and vegetable processing effluents often include significant levels of proteins, carbs, lipids and vitamins in this regard. Mono-, di- and oligosaccharides can also be discovered in effluents following grain processing (Federici, Fava, Kalogerakis & Mantzavinos, 2009). The wastewater from industrial corn processing is a good example of this type of effluent. The wastewater from the production process of commercial fresh masa and dried masa flour (also known as nejayote) has a significant load of maize solids, according to Topping et al. (2003). About half of these solids are suspended, and they comprise 64, 20 and 1.4% non-starch polysaccharides, starch and protein, respectively. Proteins, carbohydrates, vitamins and phytochemicals high in phenolics and carotenoids make up the remaining 50%. According to other research, nejayote includes polysaccharides of arabinose, xylose, glucose, galactose and D-glucuronic acid, as well as about 23% crude fibre (whose dietary fibre fraction may stimulate the growth of probiotics) (Crittenden et al., 2002; Charalampopoulos et al., 2002). As a result, agro-industrial effluents such as cereal processing wastewater might be a rich source of prebiotic chemicals. Efforts must be undertaken to evaluate the viability of employing these effluents in the formulation of culture medium, either partially or entirely.

10.6.3 Improving Probiotic Viability

The survivability of probiotic microorganisms, which is required for them to reach and colonise the human large intestine, is critical to their quality (Lahtinen et al., 2005; Mazzantini et al., 2021). Probiotics must stay alive during three crucial stages: storage, synthesis of the functional meal, and transit through the stomach and small intestine. As a result, probiotic viability is a critical problem from both an economic and technological standpoint. In this context, Figueroa-González, Quijano, Ramírez and Cruz-Guerrero (2011) discovered that freezing a probiotic at 40°C reduced its viability from 1.8 1015 to 1.6 1010 CFU/ mL, whereas freeze-drying pre-treatment followed by storage at 4°C reduced vitality from 8.91014 to 2.4109 CFU/ mL. Similarly, Lahtinen et al. (2005) suggested that probiotics might become dormant during storage. The authors discovered that after 25 days of storage at 4° C in an oat-in-water solution, *Bifidobacterium* strains can remain entirely inactive. These findings show that storing probiotics might dramatically reduce their vitality. Probiotic viability has been suggested to be improved by freeze-drying in food matrices during storage and transit through the stomach and small intestine. For example, Celik and O'Sullivan (2013) found that freeze-drying a *Bifidobacterium* strain in a matrix of low-fat milk or fruit juice dramatically increased its viability. They discovered that during storage, the viability in the juice matrix was higher than in low-fat milk and that after storage, tolerance to acid and bile was better in cells with the low-fat milk matrix. The scientists concluded that viability should not be the only characteristic to consider when selecting a food matrix as a cryoprotectant for a probiotic but also the capacity to withstand unfavourable circumstances after storage. As a result, research into low-cost food matrices as cryoprotectants to boost probiotic viability after storage is a worthwhile subject to pursue. Microencapsulation, on the other hand, is a potential technology that has recently been researched to increase the survivability of probiotic microorganisms. Microencapsulation is a technique for encapsulating solids, liquids or gases in small, sealed capsules that may release their contents at regulated rates under controlled conditions (Anal & Singh, 2007). Microencapsulation has been shown to protect probiotics during storage and the production process of a specific functional food, with increased viability against heat stress and acidic environments (Anal & Singh, 2007; Kostyuchenok et al., 1982). Spray-drying, spray-congealing, fluidised bed coating/air suspension, extrusion, coacervation/phase separation and electrostatic methods have all been utilised to preserve probiotic microorganisms thus far. Anal and Singh (2007) provide a full explanation of various microencapsulation methods. The coating substance in which the probiotic is encased is a common material used in all microencapsulation procedures. Water-soluble/insoluble polymers, waxes, fatty acids and lipids are all common coating materials. The coating composition might include prebiotics like fructo-oligosaccharides or resistant starch. In the case of *B. bifidum* microencapsulation, Kreuter (1996) added fructo-oligosaccharides in the coating formulation. The author discovered that a formulation including prebiotics, gellan gum and peptides dramatically increased probiotic viability when compared to microcapsules coated just with alginate. As a result, novel coating materials and microencapsulation methods are being investigated as an important research focus. Available research appears to focus on the development of low-cost, plentiful coating materials as well as the optimisation of the coating process (e.g. optimum prebiotic percentages).

10.7 GROWTH PATTERNS, EMERGING OPPORTUNITIES, AND FUTURE TRENDS IN PROBIOTIC FOODS

In essence, probiotic and prebiotic-based functional foods are important biotechnological subjects with tremendous innovation potential. Here, three main challenges to the industrial production of probiotic and prebiotic-based functional foods are identified.

1. The improvement in prebiotic synthesis production techniques
2. Lower production costs for probiotics
3. The preservation of probiotic viability during storage, the functional food's production process and transit through the upper gastrointestinal system. A multidisciplinary strategy is required to solve such obstacles

The following are some of the possibilities considered:

- Prebiotics are produced in two-liquid-phase systems by adding an organic solvent, which improves prebiotic production. With particular degrees of polymerisation, such multiphase systems may also boost the production of prebiotics.
- Oligosaccharide production is from non-dairy, low-cost raw materials, with cereals appearing to be the most appealing choice since they include naturally occurring oligosaccharides such as galactosyl sucrose derivatives, stachyose and raffinose, as well as fructosyl sucrose derivatives. Moreover, cereal grains are also high in dietary fibre and starch, the latter of which can be manipulated to produce resistant starch.
- Corn-processing effluents, with their high concentrations of non-starch polysaccharides, starch, protein, sugars, vitamins, phytochemicals rich in phenolics and carotenoids, and polysaccharides of arabinose, xylose, glucose, galactose and D-glucuronic acid, appear to be a very interesting option for probiotic production from agro-industrial effluents. As a result, this type of effluent might be utilised to make low-cost culture medium in whole or in part.
- Probiotic viability can be increased by adopting low-cost food matrices or microencapsulation methods. As a result, research on matrices such as milk or fruit juice may boost probiotic viability by acting as cryoprotectants throughout the freeze-drying process.
- Numerous environmental variables must be overcome in order to obtain functioning probiotic microorganisms. Microencapsulation of probiotic bacteria might be a more effective way to address these issues.

  Microencapsulation of probiotic bacteria could be a better solution to deal with those problems (Kostyuchenok et al., 1982). The encapsulation approach maintains cell viability during processing, storage and ingestion, as well as ensuring effective activity in the gut. Encapsulation is a technique for immobilising cells where the sensitive bioactive compound is separated from the adverse environment by closing the cells within the encapsulating matrix or membrane. The encapsulation approach reduces cell injury, enhances cell resistance in growth conditions and improves cell viability throughout the storage period. Encapsulation is a developed and successful method for protecting probiotic microorganisms. Encapsulation techniques are commonly used in a variety of industries for a variety of purposes, including preserving food active compounds, controlling the oxidative process, masking unpleasant flavours and odours, and acting as a barrier between sensitive bioactive compounds and the harsh environment (Anal & Singh, 2007). Spray-drying, emulsions, coacervation, electrospinning/electro spraying, impinging aerosol method and extrusions are all examples of encapsulation techniques. In encapsulation, different polymers are utilised as excipients. Starch and cellulose are two examples. During storage, encapsulated probiotics lose cell viability. When the relative humidity is high, the water content rises, triggering oxidative processes and causing cell damage. Prebiotics (e.g., inulin and high amylase maize starch powder) are incorporated into encapsulation to reduce water activity and promote stability throughout storage time and conditions. Lactic acid bacteria (*Lactobacillus acidophilus*) were encapsulated with and without polymers. The results demonstrate that alginate/starch is effective. Various authors discovered that starch-encapsulated lactic acid bacteria protect sensitive strains, preserve the number of viable bacteria and maintain cell viability quality during processing, storage and in the digestive system.

10.8 CONCLUDING REMARKS AND FUTURE DIRECTIONS

In recent trends, consumers give importance to healthier foods with reduced chemicals that work as preventive medicine. This thereby increases the need for probiotic, prebiotic and synbiotic food products. Probiotic organisms benefit the host by reducing harmful bacteria in the stomach and provide a variety of health advantages whereas. Prebiotics are dietary fibres beneficial for probiotic bacteria. The term "synbiotic" refers to a product that contains both prebiotic and probiotic bacteria. When probiotics are subjected to diverse processing conditions, storage and digestion, they lose their bioavailability. This is a major technological challenge. The search for methods that can protect and retain the viability of probiotic cultures is continuously ongoing. The encapsulation strategy is a superior answer to this problem in this case. The encapsulation approach acts as a protective covering, preventing the active chemical from being exposed directly. Bioavailability is increased with encapsulated probiotic meals. Encapsulated synbiotic meals have been proven in several studies to have higher functional activity and better sensory qualities.

Future probiotic bacteria research will focus on finding new and more particular strains that are beneficial to the host (age groups, healthy populations, disease-specific). The future scientific and technological research trends will be to:

- Investigate the mechanisms of action of probiotics in the GI-tract and develop diagnostic tools and biomarkers for their assessment

- Examine the impacts of probiotics on GI diseases, GI infections and allergies
- Develop technology that will assure the stability and viability of probiotic products (e.g., process and material development for microencapsulation)
- Develop technology for non-dairy, novel or artificial probiotic applications
- Evaluate the function of probiotics in healthy consumer groups and address consumer aspects

REFERENCES

Anal, A. K. and Singh, H. 2007. Recent advances in microencapsulation of probiotics for industrial applications and targeted delivery. *Trends in Food Science and Technology 18*(5): 240–251. https://doi.org/10.1016/j.tifs.2007.01.004.

Anekella, K. and Orsat, V. 2013. Optimization of microencapsulation of probiotics in raspberry juice by spray drying. *LWT - Food Science and Technology 50*(1): 17–24. https://doi.org/10.1016/j.lwt.2012.08.003.

Ansari, F., Alian Samakkhah, S., Bahadori, A., Jafari, S. M., Ziaee, M., Khodayari, M. T. and Pourjafar, H. 2021. Health-promoting properties of Saccharomyces cerevisiae var. boulardii as a probiotic; characteristics, isolation, and applications in dairy products. *Critical Reviews in Food Science and Nutrition 0*(0): 1–29. https://doi.org/10.1080/10408398.2021.1949577.

Arepally, D. and Goswami, T. K. 2019. Effect of inlet air temperature and gum arabic concentration on encapsulation of probiotics by spray drying. *LWT - Food Science and Technology 99*(September 2018): 583–593. https://doi.org/10.1016/j.lwt.2018.10.022.

Arslan, S., Erbas, M., Tontul, I. and Topuz, A. 2015. Microencapsulation of probiotic Saccharomyces cerevisiae var: Boulardii with different wall materials by spray drying. *LWT - Food Science and Technology 63*(1): 685–690. https://doi.org/10.1016/j.lwt.2015.03.034.

Awaisheh, S. S. 2011. Development of probiotic soft cheese manufactured using goat's milk with the addition of thyme. *Milchwissenschaft 66*(1): 51–54.

Barbosa, J. and Teixeira, P. 2017. Development of probiotic fruit juice powders by spray-drying: A review. *Food Reviews International 33*(4): 335–358. https://doi.org/10.1080/87559129.2016.1175016.

Betoret, E., Calabuig-Jiménez, L., Patrignani, F., Lanciotti, R. and Dalla Rosa, M. 2017. Effect of high pressure processing and trehalose addition on functional properties of mandarin juice enriched with probiotic microorganisms. *LWT - Food Science and Technology 85*: 418–422. https://doi.org/10.1016/j.lwt.2016.10.036.

Bielecka, M., Biedrzycka, E. and Majkowska, A. 2002. Selection of probiotics and prebiotics for Synbiotics and confirmation of their in vivo effectiveness. *Food Research International 35*(2–3): 125–131. https://doi.org/10.1016/S0963-9969(01)00173-9.

Boylston, T. D., Vinderola, C. G., Ghoddusi, H. B. and Reinheimer, J. A. 2004. Incorporation of bifidobacteria into cheeses: Challenges and rewards. *International Dairy Journal 14*(5): 375–387. https://doi.org/10.1016/j.idairyj.2003.08.008.

Capela, P., Hay, T. K. C. and Shah, N. P. 2006. Effect of cryoprotectants, prebiotics and microencapsulation on survival of probiotic organisms in yoghurt and freeze-dried yoghurt. *Food Research International 39*(2): 203–211. https://doi.org/10.1016/j.foodres.2005.07.007.

Capozzi, V., Russo, P., Dueñas, M. T., López, P. and Spano, G. 2012. Lactic acid bacteria producing B-group vitamins: A great potential for functional cereals products. *Applied Microbiology and Biotechnology 96*(6): 1383–1394. https://doi.org/10.1007/s00253-012-4440-2.

Cassani, L., Gomez-Zavaglia, A. and Simal-Gandara, J. 2020. Technological strategies ensuring the safe arrival of beneficial microorganisms to the gut: From food processing and storage to their passage through the gastrointestinal tract. *Food Research International 129*: 108852. https://doi.org/10.1016/j.foodres.2019.108852.

Celik, O. F. and O'Sullivan, D. J. 2013. Factors influencing the stability of freeze-dried stress-resilient and stress-sensitive strains of bifidobacteria. *Journal of Dairy Science 96*(6): 3506–3516. https://doi.org/10.3168/jds.2012-6327.

Charalampopoulos, D., Wang, R., Pandiella, S. S. and Webb, C. 2002. Application of cereals and cereal components in functional foods: A review. *International Journal of Food Microbiology 79*(1–2): 131–141. Charalampopoulos_2002_Application-of-cereals-and-cereal-components-in-functional-foods-a-review.pdf.

Chen, M. and Mustapha, A. 2012. Survival of freeze-dried microcapsules of α-galactosidase producing probiotics in a soy bar matrix. *Food Microbiology 30*(1): 68–73. https://doi.org/10.1016/j.fm.2011.10.017.

Chugh, B. and Kamal-Eldin, A. 2020. Bioactive compounds produced by probiotics in food products. *Current Opinion in Food Science 32*: 76–82. https://doi.org/10.1016/j.cofs.2020.02.003.

Coman, M. M., Cecchini, C., Verdenelli, M. C., Silvi, S., Orpianesi, C. and Cresci, A. 2012. Functional foods as carriers for SYNBIO®, a probiotic bacteria combination. *International Journal of Food Microbiology 157*(3): 346–352. https://doi.org/10.1016/j.ijfoodmicro.2012.06.003.

Crittenden, R., Karppinen, S., Ojanen, S., Tenkanen, M., Fagerstrm, R., Mättö, J. and Poutanen, K. 2002. In vitro fermentation of cereal dietary fibre carbohydrates by probiotic and intestinal bacteria. *Journal of the Science of Food and Agriculture 82*(8): 781–789. https://doi.org/10.1002/jsfa.1095.

Cruz, A. G., Antunes, A. E. C., Sousa, A. L. O. P., Faria, J. A. F. and Saad, S. M. I. 2009. Ice-cream as a probiotic food carrier. *Food Research International 42*(9): 1233–1239. https://doi.org/10.1016/j.foodres.2009.03.020.

Del-Val, M. I. and Otero, C. 2003. Biphasic aqueous media containing polyethylene glycol for the enzymatic synthesis of oligosaccharides from lactose. *Enzyme and Microbial Technology 33*(1): 118–126. https://doi.org/10.1016/S0141-0229(03)00098-X.

Douglas Goff, H. and Hartel, W. R. 2018. *Ice cream*, 7th edition. In Angewandte Chemie International Edition *6*(11): 951–952.

Dutta, S., Moses, J. A. and Anandharamakrishnan, C. 2018. Modern frontiers and applications of spray-freeze-drying in design of food and biological supplements. *Journal of Food Process Engineering 41*(8): 1–8. https://doi.org/10.1111/jfpe.12881.

Dyvelkov, K. N. and Sloth, J. 2014. New advances in spray-drying processes. *Microencapsulation in the Food Industry* (Edition?) 57–63. https://doi.org/10.1016/b978-0-12-404568-2.00006-6.

Egan, B. M., Lakenx, M. A., Donovan, J. L. and Woolson, R. F. 2010. Does dark chocolate have a role in the prevention and management of hypertension?: Commentary on the evidence. *Hypertension 55*(6): 1289–1295. https://doi.org/10.1161/HYPERTENSIONAHA.110.151522.

Federici, F., Fava, F., Kalogerakis, N. and Mantzavinos, D. 2009. Valorisation of agro-industrial by-products, effluents and waste: Concept, opportunities and the case of olive mill waste waters. *Journal of Chemical Technology and Biotechnology 84*(6): 895–900. https://doi.org/10.1002/jctb.2165.

Figueroa-González, I., Quijano, G., Ramírez, G. and Cruz-Guerrero, A. 2011. Probiotics and prebiotics-perspectives and challenges. *Journal of the Science of Food and Agriculture 91*(8): 1341–1348. https://doi.org/10.1002/jsfa.4367.

Fijan, S. 2014. Microorganisms with claimed probiotic properties: An overview of recent literature. *International Journal of Environmental Research and Public Health 11*(5): 4745–4767. https://doi.org/10.3390/ijerph110504745.

Fioramonti, J., Theodorou, V. and Bueno, L. 2003. Probiotics: What are they? What are their effects on gut physiology? *Bailliere's Best Practice and Research in Clinical Gastroenterology 17*(5): 711–724. https://doi.org/10.1016/S1521-6918(03)00075-1.

Fuller, R. 1989. Probiotics in man and animals. *Journal of Applied Bacteriology*: *66*(5): 365–378.

Gardiner, G. E., Bouchier, P., O'Sullivan, E., Kelly, J., Kevin Collins, J., Fitzgerald, G., Paul Ross, R. and Stanton, C. 2002. A spray-dried culture for probiotic Cheddar cheese manufacture. *International Dairy Journal 12*(9): 749–756. https://doi.org/10.1016/S0958-6946(02)00072-9.

Germán, C., Fabián, R. and Carlos, S. J. 2021. Optimization of galactooligosaccharides reaction conditions through commercial galactosidases. *Food Processing & Nutritional Science 2*(1): 34–46.

Ghandi, A., Powell, I. B., Howes, T., Chen, X. D. and Adhikari, B. 2012. Effect of shear rate and oxygen stresses on the survival of Lactococcus lactis during the atomization and drying stages of spray drying: A laboratory and pilot scale study. *Journal of Food Engineering 113*(2): 194–200. https://doi.org/10.1016/j.jfoodeng.2012.06.005.

Gilliland, S. E. 1990. Health and nutritional benefits from lactic acid bacteria. *FEMS Microbiology Letters 87*(1–2): 175–188. https://doi.org/10.1016/0378-1097(90)90705-U.

Goff, H. D. 2008. 65 years of ice cream science. *International Dairy Journal 18*(7): 754–758. https://doi.org/10.1016/j.idairyj.2008.03.006.

Granato, D., Branco, G. F., Nazzaro, F., Cruz, A. G. and Faria, J. A. F. 2010. Functional foods and nondairy probiotic food development: Trends, concepts, and products. *Comprehensive Reviews in Food Science and Food Safety 9*(3): 292–302. https://doi.org/10.1111/j.1541-4337.2010.00110.x.

Guiné, R. P. F. 2018. The drying of foods and its effect on the physical-chemical, sensorial and nutritional properties. *ETP International Journal of Food Engineering 4*(2): 93–100. https://doi.org/10.18178/ijfe.4.2.93-100.

Her, J. Y., Kim, M. S. and Lee, K. G. 2015. Preparation of probiotic powder by the spray freeze-drying method. *Journal of Food Engineering 150*(October): 70–74. https://doi.org/10.1016/j.jfoodeng.2014.10.029.

Huang, H. W., Hsu, C. P. and Wang, C. Y. 2020. Healthy expectations of high hydrostatic pressure treatment in food processing industry. *Journal of Food and Drug Analysis 28*(1): 1–13. https://doi.org/10.1016/j.jfda.2019.10.002.

Huang, S., Méjean, S., Rabah, H., Dolivet, A., Le Loir, Y., Chen, X. D., Jan, G., Jeantet, R. and Schuck, P. 2017. Double use of concentrated sweet whey for growth and spray drying of probiotics: Towards maximal viability in pilot scale spray dryer. *Journal of Food Engineering 196*: 11–17. https://doi.org/10.1016/j.jfoodeng.2016.10.017.

Jouki, M., Khazaei, N., Rezaei, F. and Taghavian-Saeid, R. 2021. Production of synbiotic freeze-dried yoghurt powder using microencapsulation and cryopreservation of L. plantarum in alginate-skim milk microcapsules. *International Dairy Journal 122*: 105133. https://doi.org/10.1016/j.idairyj.2021.105133.

Kailasapathy, K. 2006. Survival of free and encapsulated probiotic bacteria and their effect on the sensory properties of yoghurt. *LWT - Food Science and Technology 39*(10): 1221–1227. https://doi.org/10.1016/j.lwt.2005.07.013.

Kalkan, S., Öztürk, D. and Selimoğlu, B. S. 2018. Determining some of the quality characteristics of probiotic yogurts manufactured by using microencapsulated Saccharomyces cerevisiae var. boulardii. *Turkish Journal of Veterinary and Animal Sciences 42*(6): 617–623. https://doi.org/10.3906/vet-1804-5.

Kearney, N., Meng, X. C., Stanton, C., Kelly, J., Fitzgerald, G. F. and Ross, R. P. 2009. Development of a spray dried probiotic yoghurt containing Lactobacillus paracasei NFBC 338. *International Dairy Journal 19*(11): 684–689. https://doi.org/10.1016/j.idairyj.2009.05.003.

Kontula, P., Suihko, M. L., Von Wright, A. and Mattila-Sandholm, T. 1999. The effect of lactose derivatives on intestinal lactic acid bacteria. *Journal of Dairy Science 82*(2): 249–256. https://doi.org/10.3168/jds.S0022-0302(99)75230-6.

Kopp-Hoolihan, L. 2001. Prophylactic and therapeutic uses of probiotics: A review. *Journal of the American Dietetic Association 101*(2): 229–241. https://doi.org/10.1016/S0002-8223(01)00060-8.

Kostyuchenok, B. M., Karlov, V. A. and Goloborodko, N. K. 1982. Cleansing of purulent wound with intermittent stream of antiseptics. *Khirurgiya 58*(8): 16–19.

Krasaekoopt, W., Bhandari, B. and Deeth, H. 2003. Evaluation of encapsulation techniques of probiotics for yoghurt. *International Dairy Journal 13*(1): 3–13. https://doi.org/10.1016/S0958-6946(02)00155-3.

Kreuter, J. 1996. Nanoparticles and microparticles for drug and vaccine delivery. *Journal of Anatomy 189*(Pt 3(II)): 503–505. https://doi.org/10.1002/bit.

Ku, S., Park, M. S., Ji, G. E. and You, H. J. 2016. Review on Bifidobacterium bifidum bgn4: Functionality and nutraceutical applications as a probiotic microorganism. *International Journal of Molecular Sciences 17*(9). https://doi.org/10.3390/ijms17091544.

Lahtinen, S. J., Gueimonde, M., Ouwehand, A. C., Reinikainen, J. P. and Salminen, S. J. 2005. Probiotic bacteria may become dormant during storage. *Applied and Environmental Microbiology 71*(3): 1662–1663. https://doi.org/10.1128/AEM.71.3.1662-1663.2005.

Lamsal, B. P. and Faubion, J. M. 2009. The beneficial use of cereal and cereal components in probiotic foods. *Food Reviews International 25*(2): 103–114. https://doi.org/10.1080/87559120802682573.

Laranjo, M., Potes, M. E. and Elias, M. 2019. Role of starter cultures on the safety of fermented meat products. *Frontiers in Microbiology 10*(April). https://doi.org/10.3389/fmicb.2019.00853.

Lazo-Vélez, M. A., Serna-Saldívar, S. O., Rosales-Medina, M. F., Tinoco-Alvear, M. and Briones-García, M. 2018. Application of Saccharomyces cerevisiae var. boulardii in food processing: A review. *Journal of Applied Microbiology 125*(4): 943–951. https://doi.org/10.1111/jam.14037.

Li, C. U. I., Li-Ying, N. I. U., Da-Jing, L. I., Chun-Quan, L. I. U., Ying-Ping, L. I. U., Chun-Ju, L. I. U. and Jiang-Feng, S. 2018. Effects of different drying methods on quality, bacterial viability and storage stability of probiotic enriched apple snacks. *Journal of Integrative Agriculture 17*(1): 247–255. https://doi.org/10.1016/S2095-3119(17)61742-8.

Lourens-Hattingh, A. and Viljoen, B. C. 2001. Yogurt as probiotic carrier food. *International Dairy Journal 11*(1–2): 1–17. https://doi.org/10.1016/S0958-6946(01)00036-X.

Mackowiak, P. A. 2013. Recycling Metchnikoff: Probiotics, the intestinal microbiome and the quest for long life. *Frontiers in Public Health 1*(V): 1–3. https://doi.org/10.3389/fpubh.2013.00052.

Mahony, J., McAuliffe, O., Cotter, P. D. and Fitzgerald, G. F. 2019. Starter cultures. In *Food microbiology: Fundamentals and frontiers*. https://doi.org/10.1128/9781555819972.ch30.

Masco, L., Huys, G., De Brandt, E., Temmerman, R. and Swings, J. 2005. Culture-dependent and culture-independent qualitative analysis of probiotic products claimed to contain bifidobacteria. *International Journal of Food Microbiology 102*(2): 221–230. https://doi.org/10.1016/j.ijfoodmicro.2004.11.018.

Mattila-Sandholm, T., Myllärinen, P., Crittenden, R., Mogensen, G., Fondén, R. and Saarela, M. 2002. Technological challenges for future probiotic foods. *International Dairy Journal 12*(2–3): 173–182. https://doi.org/10.1016/S0958-6946(01)00099-1.

Mazzantini, D., Celandroni, F., Calvigioni, M., Panattoni, A., Labella, R. and Ghelardi, E. 2021. Microbiological quality and resistance to an artificial gut environment of two probiotic formulations. *Foods 10*(11): 1–13. https://doi.org/10.3390/foods10112781.

Mombelli, B. and Gismondo, M. R. 2000. The use of probiotics in medical practice. *International Journal of Antimicrobial Agents 16*(4): 531–536. https://doi.org/10.1016/S0924-8579(00)00322-8.

Moreira, R. M., Martins, M. L., Leite Júnior, B. R. de C., Martins, E. M. F., Ramos, A. M., Cristianini, M., Campos, A. N. da R., Stringheta, P. C., Silva, V. R. O., Canuto, J. W., de Oliveira, D. C. and de Souza 64.Pereira, D. C. 2017. Development of a juçara and Ubá mango juice mixture with added Lactobacillus rhamnosus GG processed by high pressure. *LWT - Food Science and Technology 77*: 259–268. https://doi.org/10.1016/j.lwt.2016.11.049.

Moura, P., Cabanas, S., Lourenço, P., Gírio, F., Loureiro-Dias, M. C. and Esteves, M. P. 2008. In vitro fermentation of selected xylo-oligosaccharides by piglet intestinal microbiota. *LWT - Food Science and Technology 41*(10): 1952–1961. https://doi.org/10.1016/j.lwt.2007.11.007.

Nowak, D. and Jakubczyk, E. 2020. The freeze-drying of foods⇔the characteristic of the process course and the effect of its parameters on the physical properties of food materials. *Foods 9*(10). https://doi.org/10.3390/foods9101488.

Ouwehand, A. C., Kirjavainen, P. V., Shortt, C. and Salminen, S. 1999. Probiotics: Mechanisms and established effects. *International Dairy Journal 9*(1): 43–52. https://doi.org/10.1016/S0958-6946(99)00043-6.

Panesar, P. S., Panesar, R., Singh, R. S., Kennedy, J. F. and Kumar, H. 2006. Microbial production, immobilization and applications of β-D-galactosidase. *Journal of Chemical Technology and Biotechnology 81*(4): 530–543. https://doi.org/10.1002/jctb.1453.

Patel, P., Jethani, H., Radha, C., Vijayendra, S. V. N., Mudliar, S. N., Sarada, R. and Chauhan, V. S. 2019. Development of a carotenoid enriched probiotic yogurt from fresh biomass of Spirulina and its characterization. *Journal of Food Science and Technology 56*(8): 3721–3731. https://doi.org/10.1007/s13197-019-03844-0.

Patel, P., Parekh, T. and Subhash, R. 2008. Development of probiotic and synbiotic chocolate mousse: A functional food. *Biotechnology(Faisalabad) 7*(4): 769–774. https://doi.org/10.3923/biotech.2008.769.774.

Perdana, J., Bereschenko, L., Fox, M. B., Kuperus, J. H., Kleerebezem, M., Boom, R. M. and Schutyser, M. A. I. 2013. Dehydration and thermal inactivation of Lactobacillus plantarum WCFS1: Comparing single droplet drying to spray and freeze drying. *Food Research International 54*(2): 1351–1359. https://doi.org/10.1016/j.foodres.2013.09.043.

Pereira, A. L. F., Almeida, F. D. L., Lima, M. A., da Costa, J. M. C. and Rodrigues, S. 2014. Spray-drying of probiotic cashew apple juice. *Food and Bioprocess Technology 7*(9): 2492–2499. https://doi.org/10.1007/s11947-013-1236-z.

Prado, F. C., Parada, J. L., Pandey, A. and Soccol, C. R. 2008. Trends in non-dairy probiotic beverages. *Food Research International 41*(2): 111–123. https://doi.org/10.1016/j.foodres.2007.10.010.

Prasanna, P. H. P., Grandison, A. S. and Charalampopoulos, D. 2014. Bifidobacteria in milk products: An overview of physiological and biochemical properties, exopolysaccharide production, selection criteria of milk products and health benefits. *Food Research International 55*: 247–262. https://doi.org/10.1016/j.foodres.2013.11.013.

Rajam, R. and Anandharamakrishnan, C. 2015. Spray freeze drying method for microencapsulation of Lactobacillus plantarum. *Journal of Food Engineering 166*: 95–103. https://doi.org/10.1016/j.jfoodeng.2015.05.029.

Ranadheera, C. S., Vidanarachchi, J. K., Rocha, R. S., Cruz, A. G. and Ajlouni, S. 2017. Probiotic delivery through fermentation: Dairy vs. non-dairy beverages. *Fermentation 3*(4): 1–17. https://doi.org/10.3390/fermentation3040067.

Reddy, K. B. P. K., Awasthi, S. P., Madhu, A. N. and Prapulla, S. G. 2009. Role of cryoprotectants on the viability and functional properties of probiotic lactic acid bacteria during freeze drying. *Food Biotechnology 23*(3): 243–265. https://doi.org/10.1080/08905430903106811.

Rezvankhah, A., Emam-Djomeh, Z. and Askari, G. 2020. Encapsulation and delivery of bioactive compounds using spray and freeze-drying techniques: A review. *Drying Technology 38*(1–2): 235–258. https://doi.org/10.1080/07373937.2019.1653906.

Ross, R. P., Desmond, C., Fitzgerald, G. F. and Stanton, C. 2005. Overcoming the technological hurdles in the development of probiotic foods. *Journal of Applied Microbiology 98*(6): 1410–1417. https://doi.org/10.1111/j.1365-2672.2005.02654.x.

Roy, P. and Kumar, V. 2018. Functional food: Probiotic as health booster. *Journal of Food, Nutrition and Population Health* 2(2): 10–13. https://doi.org/10.21767/2577-0586.100042.

Schutyser, M. A. I., Both, E. M., Siemons, I., Vaessen, E. M. J. and Zhang, L. 2019. Gaining insight on spray drying behavior of foods via single droplet drying analyses. *Drying Technology 37*(5): 525–534. https://doi.org/10.1080/07373937.2018.1482908.

Senok, A. C., Ismaeel, A. Y. and Botta, G. A. 2005. Probiotics: Facts and myths. *Clinical Microbiology and Infection 11*(12): 958–966. https://doi.org/10.1111/j.1469-0691.2005.01228.x.

Seo, K. C., Kim, M. J., Kwon, M. J., Kim, H. J., Noh, J. S. and Song, Y. O. 2009. Antioxidative activities of soymilk fermented with Bacillus subtilis. *Food Science and Biotechnology 18*(5): 1298–1300.

Singu, B. D., Bhushette, P. R. and Annapure, U. S. 2020. Thermo-tolerant Saccharomyces cerevisiae var. boulardii coated cornflakes as a potential probiotic vehicle. *Food Bioscience 36*(June 2019): 100668. https://doi.org/10.1016/j.fbio.2020.100668.

Snow, P. 1981. Factors affecting in food1' 2 of hydrolysis of starch. *Time* (February): 2721–2727.

Stadnik, J. and Dolatowski, Z. J. 2013. Changes in selected parameters related to proteolysis during ageing of dry-cured pork loins inoculated with probiotics. *Food Chemistry 139*(1–4): 67–71. https://doi.org/10.1016/j.foodchem.2013.01.079.

Stadnik, J. and Dolatowski, Z. J. 2014. Effect of inoculation with probiotics and ageing time on selected functional properties and oxidation of proteins in dry-cured pork loins. *International Journal of Food Properties 17*(4): 866–876. https://doi.org/10.1080/10942912.2012.685679.

Sutton, K. H. 2007. Considerations for the successful development and launch of personalised nutrigenomic foods. *Mutation Research/Fundamental and Molecular Mechanisms of Mutagenesis 622*(1–2): 117–121. https://doi.org/10.1016/j.mrfmmm.2007.03.007.

Swieca, M., Kordowska-Wiater, M., Pytka, M., Gawlik-Dziki, U., Seczyk, L., Złotek, U. and Kapusta, I. 2019. Nutritional and pro-health quality of lentil and adzuki bean sprouts enriched with probiotic yeast 90.Saccharomyces cerevisiae var. boulardii. *LWT - Food Science and Technology 100*(October 2018): 220–226. https://doi.org/10.1016/j.lwt.2018.10.081.

Tantratian, S. and Pradeamchai, M. 2020. Select a protective agent for encapsulation of Lactobacillus plantarum. *LWT - Food Science and Technology 123*(January): 109075. https://doi.org/10.1016/j.lwt.2020.109075.

Telang, A. M. and Thorat, B. N. 2010. Optimization of process parameters for spray drying of fermented soy milk. *Drying Technology 28*(12): 1445–1456. https://doi.org/10.1080/07373937.2010.482694.

Topping, D. L., Fukushima, M. and Bird, A. R. 2003. Resistant starch as a prebiotic and synbiotic: State of the art. *Proceedings of the Nutrition Society 62*(1): 171–176. https://doi.org/10.1079/pns2002224.

Vandeker, V. and Ortega, R. 2017. Comparative viability of Spray dried lactobacilli affected by different protective agents and storage conditions recommended citation comparative viability of Spray dried lactobacilli affected by different protective agents and storage conditions. https://digitalcommons.lsu.edu/gradschool_theses.

Vasudha, S. and Mishra, H. N. 2013. Non dairy probiotic beverages. *International Food Research Journal 20*(1): 7–15.

Vivek, K., Mishra, S. and Pradhan, R. C. 2021. Optimization of spray drying conditions for developing nondairy based probiotic Sohiong fruit powder. *International Journal of Fruit Science 21*(1): 193–204. https://doi.org/10.1080/15538362.2020.1864567.

Wang, C. Y., Ng, C. C., Su, H., Tzeng, W. S. and Shyu, Y. T. 2009. Probiotic potential of noni juice fermented with lactic acid bacteria and bifidobacteria. *International Journal of Food Sciences and Nutrition 60*(suppl. 6): 98–106. https://doi.org/10.1080/09637480902755095.

Williams, P. C. and Stevensen, S. G. 1990. Near-infrared reflectance analysis: Food industry applications. *Trends in Food Science and Technology 1*(C): 44–48. https://doi.org/10.1016/0924-2244(90)90030-3.

Yang, L., Chen, Y., Li, Z., Shi, Y., Li, Z. and Zhao, X. 2016. Complete genome sequence of Lactobacillus acidophilus MN-BM-F01. *Genome Announcements 4*(1): 4–5. https://doi.org/10.1128/genomeA.01699-15.

Zhang, Y., Lin, J. and Zhong, Q. 2016. Effects of media, heat adaptation, and outlet temperature on the survival of Lactobacillus salivarius NRRL B-30514 after spray drying and subsequent storage. *LWT - Food Science and Technology 74*: 441–447. https://doi.org/10.1016/j.lwt.2016.08.008.

CHAPTER 11

Spices as a Functional Food for Health

Makdud Islam*, **Praveen Saini** and **Kamlesh Prasad**

11.1 INTRODUCTION

Since ancient times, spices have been used in various traditional forms of medicine such as Ayurveda, Unani and Sino-Tibetan systems. The Vedic literature (in 2500 BC) was a reliable source of information that was suggested in the development of Ayurveda by mentioning the usage of various plant components as medicines. The use of spices was a very common exercise around the world including in ancient Greece, Rome and the European colonisation of America and Africa. Spices are an essential component of the human diet and the demand growth rate for spices is around 3.19%, which is slightly above the population growth rate of the world (Rajanbabu and Ganesan 2015). Natural herbs, especially spices, were consumed as food or utilised as remedies to maintain appropriate health, sanitisation and hygiene, as well as extend life anticipation (Wani et al. 2022a). India is considered "The origin and Land of spices" due to its huge availability and variability. The developing countries like India, 80% of people rely on traditional medicine from spices and herbs. Spices are aromatic compounds obtained from various parts of plants such as roots, stems, entire grasses, leaves, flowers and fruits in a fresh and dried form. The most commonly used spices are black pepper, cardamom, turmeric, ginger, ajowan, annatto, nutmeg, fennel, anise, saffron, chilli, garlic, fenugreek, onion, rosemary, clove, celery seeds, sage and thyme (Wani et al. 2022b; Perussello 2020). In addition to improving digestion and the organoleptic qualities of food, they also add flavour and cut down on the need for salt and fatty condiments. They also have antioxidants to prevent physiologic and metabolic changes in the human body. Spices are widely known for their naturally occurring bioactives that have been generally recognised as safe (GRAS). These bioactives include phenolic compounds, alkaloids, glycosides, sulfur-containing compounds, vital fats, oleoresins, carotenoids and vitamins (Sharif, Ejaz and Pasha 2018).

* Corresponding author E-mail: makubkv96@gmail.com

DOI: 10.1201/9781003315100-11

Recently, consumers, researchers and the food business communities have shown a greater interest, as it preserves organic health, and the importance that diet plays in the prevention and treatment of many ailments has been widely recognised.

The classical concept of "adequate nutrition" has been changed to "optimum nutrition" where besides the food providing all nutrients (carbohydrates, proteins, fats, vitamins and minerals), it also promotes good health and reduces the risk of diseases. Hence, terminology like functional foods, medicinal foods, superfoods and therapeutic foods have become common (Pasupuleti et al. 2017). The phrase "functional food" is used in a variety of contexts, including technology advancements, food marketing, and food regulatory norms (Palou, Serra and Pico 2003). According to a policy document published by the Academy of Nutrition and Dietetics (AND), functional foods are whole foods, as well as fortified, enriched or enhanced foods, that have the possibility to improve health when consumed as part of a diverse diet regularly at effective levels based on significant standards of evidence (Crowe and Francis 2013). According to the International Food Information Council (IFIC) "Functional foods are foods or dietary components that may provide a health benefit beyond basic nutrition, such as reducing or minimising the risk of certain diseases and other health issues" (Sikand et al. 2015). A functional food's beneficial effects may include the preservation of health or wellness, as well as a reduction in the chance of contracting a certain ailment. Modifying one or more ingredients, or omitting them altogether, can result in functional food. Establishing food as functional means the bioactive components of these foods must be identified to better describe their potential health benefits, and aromatic herbs and spices fit this criterion in a number of ways. This chapter mainly focuses on the functional properties of different spices and their application as functional food ingredients for healthcare.

11.2 NUTRACEUTICALS PROPERTIES OF SPICES

Spices have a unique flavour that derives from volatile oils, fixed oils and a small quantity of resins. The flavour is sometimes enhanced with a mixture of alcohols, phenols, esters, organic acids, alkaloids and sulfur. Apart from that, spices contain protein (4.5–31.5%), fat content (0.6–2.6%), minerals (2.3–16.7%), tannins and vitamins, minerals which are all common components of plant products (Viuda et al. 2007). Some spices are also good sources of dietary fibres (both soluble and insoluble), including chilli (43.3%), black pepper (27.8%), coriander (36.2%), cumin (23.0%), fennel (28.7%) and fenugreek (33.5%), etc. Natural colour extracted from spices also plays a vital role in building a good economy for the country. Due to the less consumption of spices, their nutrition cannot be as effective as the other ingredients (Wani et al. 2022b).

11.2.1 General

Spices have a significant effect on human health and disease. Spices, in addition to playing an essential role as a flavour enhancer, also show promise in the treatment of diseases such as cancer, arthritis, diabetes, rheumatic, metabolic and immunological problems. The primary health issues facing humanity today are diabetes, cardiovascular disease, arthritis and cancer, hence the antiproliferative, anti-diabetic, anti-hypercholesterolemic and anti-inflammatory property of spices are of paramount relevance. Spices or their active ingredients may be utilised as potential therapeutic or preventative measures for several medical conditions (Wani et al. 2022b). Most of the spices that are accessible in India have been thoroughly investigated, and their active components and therapeutic properties are discussed in Table 11.1.

Spices are susceptible to microbial or insect assault leading to deterioration. Different conventional (steam distillation, solvent extraction) and novel extraction (microwave, supercritical, ultrasonic extraction) methods are used for the isolation of all the active ingredients in spices and preserved as spice oils. Oleoresins (essential oil and resin) are used as flavouring agents to change or improve the flavour of curries, baked goods, pickles, processed meat, drinks, liquors and other foods. It can be extracted by pulverised spices with different solvents such as acetone, isopropanol, methanol and hexane. Spices that have been processed into finished goods are simple to use, contaminant-free, have a longer shelf life and are portable. Different functional properties of spices have been documented by different researchers. Some spices have antioxidant effects, while others are used as carminatives. Some spices are employed as preservatives, whereas others, such as clove and mustard, guard against food spoiling because of their antibacterial qualities. Important physiological and therapeutic characteristics are also found in several spices. In addition to having antimicrobial qualities, spices also contain anti-inflammatory, analgesic, anti-obesity, anti-pollution and anti-mutagenic effects (Wani et al. 2022b). The biological activities and therapeutic properties of different spices have been summarised in Figure 11.1. These qualities of spices support their traditional employment in a variety of folkloric and traditional treatments in diverse countries and societies.

11.2.2 Antioxidant/Anti-Inflammatory Activity

Nowadays, researchers are more focused on certain biomaterials that have the capability to prevent carcinogenic, anti-inflammatory, antioxidant and anti-atherosclerotic effects in biological systems in both animals and humans, both in vitro and in vivo. Researchers have suggested that spices are high in antioxidants and effective regulators of tissue damage and inflammation induced by elevated blood sugar and circulating lipids. As spices are low in calories and relatively affordable, it is considered a good source of antioxidants and other potential bioactive substances in the diet (A.K De 2004). Spices like chilli, turmeric, curcumin, garlic, clove and onion show antioxidant and anti-inflammatory activity. Turmeric is one of the most exploited Indian traditional spices used for wound healing, rheumatoid arthritis, gastrointestinal problems, deworming, rhinitis and as a cosmetic. According to Srinvasan (2005), the anti-inflammatory properties of spices are a result of their active components, which prevent the synthesis of metabolites like prostaglandin E2 (PgE2), leukocytes or liposomal enzymes by macrophages such collagenase and elastase.

Table 11.1 Principal phenolic, flavonoids compounds and their therapeutic property of different spices

Scientific name	Traditional name	Major phenolic and flavonoids	Therapeutic property	Reference
Rosmarinus officinali	Rosemary	Rosmarinic acid, carnosic acid, total phenols	Antimicrobial, anti-inflammatory, antioxidant, anti-apoptotic, anti-tumorigenic, antinociceptive and neuroprotective properties	(Rodríguez-rojo et al. 2012)
Thymus vulgaris	Thyme	p-Coumaric acid,4-O-glucoside kaempferol, ferulic acid, phenols, thymol, carvacrol	Antioxidant, antifungal, antimicrobial, expectorant, spasmolytic, mucolytic and antitussive	(Munekata et al. 2020)
Curcuma longa	Turmeric	Curcuminoids, saponin, alkaloid, sterol, tannin, flavonoid, phytic acid, phenol	Antioxidant, anti-inflammatory, anti-arthritic, anti-mutagenic/ cancer preventive, antimicrobial, blood cholesterol-lowering	(Sharma et al. 2019)
Cinnamomum verum	Cinnamon	Cinnamaldehyde, eugenol, vitamin C	Antioxidant, antimicrobial, anti-inflammatory, anticancer, cholesterol-lowering, immunomodulatory and cardiovascular	(Willis et al. 2019)
Zingiber officinal	Ginger	Phenolic acids, gingerols, paradols, shogaols	Antioxidant, antimicrobial, anti-diabetic, anti-inflammatory, neuro-protective, cardiovascular, gastrointestinal, anticancer, antihypertensive	(Idris, Yasin and Usman 2019)
Allium sativum	Garlic	Allicin, phenolic, polysaccharides compounds	Antioxidant, antimicrobial, anticancer, for ear infection, antidyspeptic, antiflatulent, lowering of blood cholesterol, duodenal ulcers, as a rubefacient in skin diseases	(Toledano Medina et al. 2019)
Coriandrum sativum	Coriander	Antidyspeptic, flavourant, geraniol, linalool, monoterpenes, borneol, citronellol, camphor, coriandrin, coriandrons A–E, flavonoids	Antimicrobial, antioxidant, hypoglycaemia, hypolipidemic, anxiolytic, anti-inflammatory, analgesic anti-inflammatory	(Laribi at al. 2015)

(*Continued*)

Table 11.1 (Continued)

Scientific name	Traditional name	Major phenolic and flavonoids	Therapeutic property	Reference
Trigonella foenum graecum	Fenugreek	Steroidal sapogenins, saponins, trigonellin, galactomannan, flavonoids (quercetin, rutin, vitexin, isovitexin)	Anti-diabetic, anticarcinogenic, antioxidant, hypocholesterolemic and immunological activities	(Wani and Kumar 2018)
Allium cepa	Onion	Onionin, isoalliin, cycloalliin, methiin, rutin, quercetin	Anti-obesity, antimicrobial activity, anticancer activity, neuroprotection, digestive system protection	(Zhao et al. 2021)
Cuminum cymminum	Cumins	Cuminaldehyde, p-1, 3-mentha-dienal, gallic, cinnamic, rosmarinic, coumaric and vanillic acids, flavonoids	Digestive stimulant action, hypoglycaemic potential, hypolipidemia, cancer and diabetes	(Mnif and Aifa 2015)
Piper nigrum	Papper	Capsaicinoids, phenolic compounds, carotenoids and vitamins C and E	Antipyretic, rubefacient, analgesic, anti-obesity, cardioprotective, pharmacological, neurological and dietic properties	(Mendes and Gonçalves 2020)
Capsicum annuum	Red pepper	Capsaicin, Vitamin C, dihydrocapsaicin, carotenoids	Anti-inflammatory, pain reliever (rheumatism/ neuralgia), useful in indigestion, rubefacient	(Shaha, Rahman and Asrul 2013)
Syzygium aromaticum	Clove	Eugenol, eugenyl acetate, α-humulene, 2-heptanone, β-caryophyllene	Antioxidant, antimicrobial, anti-inflammatory, anti-mutagenic, anti-allergic and anticancer	(Cortés-Rojas et al. 2014)

In animal studies, turmeric and curcumin were found to boost detoxification enzymes, limit DNA damage, promote DNA repair, reduce mutations and also have antioxidative properties. The usefulness of curcumin extracts in osteoarthritic patients has been studied in numerous clinical trials. After consuming curcumin, patients with osteoarthritis reported improved physical function (pain) and quality of life. During treatment, they also reported less use of analgesics and fewer adverse effects. Curcumin has been shown in vitro to prevent chondrocyte death, inhibit the release of proteoglycans and metalloproteases, and limit the production of cyclooxygenase, prostaglandin E-2 and inflammatory cytokines in chondrocytes.

Ginger has traditionally been used as a pain reliever. It is capable of curing all sorts of pain. A headache can be relieved by applying ginger ointment created by rubbing dry ginger with a little water on a grinding stone and applying it to the forehead. A few drops of ginger juice can help relieve an earache. Cassumurins A, B and C, which have both antioxidant and anti-inflammatory properties, were isolated from the rhizomes of a tropical ginger Zingibercassumunar using assay-guided isolation.

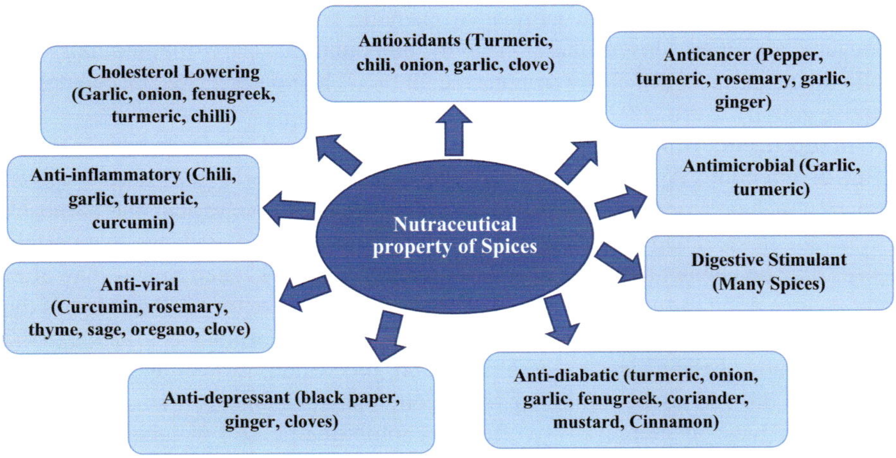

Figure 11.1 Nutraceutical and therapeutic properties of different spices

Additionally, capsaicin has been recommended for the first treatment of neuralgia brought on by herpes infection.

11.2.3 Antiviral Activity

The antiviral action of spices and herbs is another useful characteristic. There are mainly two primary components of rosemary extract, carnosol and carnosic acid, which have been shown inhibitory effects against the human immunodeficiency virus (HIV) (Aruoma et al. 1996). The flavonoid apigenin, found in rosemary, thyme, sage and oregano, suppressed HIV-1 in latent infection models by a mechanism that most likely included suppression of viral transcription (Critchfield et al. 1996).

The antiviral properties of curcumin have been examined extensively. Curcumin was shown to suppress the activity of histone acetyltransferase in an experiment that prevented the type 1 herpes simplex virus (HSV-1) from being promoted genetically (Kutluay et al. 2008). Essential oils, which are the main ingredients in spices, have also been shown to have antiviral properties. For instance, eugenol, the primary component of clove bud (*Eugenia caryophyllus*) essential oil, was shown to have antiviral activity against types 1 and 2 herpes simplex viruses (HSV-1 and HSV-2) for both in vitro and in a mouse model (Benencia and Courreges 2000). Several herbs and spices, including hyssop (*Hyssopus officinalis*), thyme (*Thymus vulgaris L.*), anis (*Pimpinella anisum*), ginger (*Zingiber officinale*), camomile (*Chamaemelum nobile*) and sandalwood (*Santalum album*) have essential oils that have antiviral properties against HSV-2 (Koch et al. 2008).

11.2.4 Anticancer Activity

Recently, researchers have focused on discovering naturally occurring substances that are capable of inhibiting, retarding and inverting multi-stage carcinogenesis.

The spices have a substantial number of phenolic compounds that shows anticarcinogenic activities. Most of these phenolic compounds have anti-inflammatory and antioxidant activities which are responsible for their chemoprotective and chemopreventive activity (Surh 2002).

In this regard, pepper, turmeric and its main component curcumin have received a lot of attention. Apart from those, other spices and herbs that have received attention include rosemary and its constituents including ursolic acid and carnosol. Numerous in vitro and in vivo studies on rodents have found evidence that certain spices may be chemopreventive against the early stages of cancer. Spices may offer cancer prevention through a variety of methods. Rosemary extract (30 g/mL) has the ability to protect DNA against oxidative damage (O_2) (Slamenov'a et al. 2002). One or more steps of the cancer process (initiation, promotion, growth and metastasis) have been proven to be inhibited by specific phytochemicals found in spices. Spices may have chemopreventive effects by inhibiting phase I metabolic enzymes (engaged in procarcinogen activation) and inducing phase II metabolic enzymes (involved in carcinogen deactivation). Inflammation and oxidative stress, both of which increase the risk of developing and progressing cancer, may be prevented by spices. Anti-mutagenic spices may also be anticarcinogenic, and antioxidant-rich spices can prevent mutation since mutagenesis directly affects the development of cancer. Clove flavouring eugenol and mustard seed dithiolthione content have an anti-mutagenic action by shielding cells from DNA damage. Turmeric and curcumin also reduced the ability of tobacco-based dentifrice to nitrosate methylurea in vitro, as well as the ability of cigarette and beedi smoke condensates to cause mutations. According to studies on smokers, taking curcumin for 30 days at a dose of 1.5 g per day significantly decreased the number of mutagens that were excreted in the urine (Polasa et al. 2004). There is evidence that dithiolthiones can protect the liver from damage brought on by certain chemicals and the fungus toxin aflatoxin.

11.2.5 Anti-Diabetic Activity

Diabetes is one of the most common metabolic diseases on the planet, and its prevalence is on the rise. According to WHO report, it is the third most prevalent disease, after cardiovascular and oenological disease. Proper dieting is one of the strategies to regulate diabetes mellitus, and here spices also can play a significant role to regulate diabetes. Some spices have shown several beneficial effects including anti-diabetic activity. Among those, fenugreek seeds, garlic and onion, and their sulfur compounds, as well as turmeric and its yellow component, curcumin, have been reported to improve the glycemic status and glucose tolerance in diabetic patients and type-2 diabetes individuals. Apart from that cumin seeds, coriander, curry leaves, ginger and mustard have shown positive effects on hypoglycaemia (Srinivasan 2005).

Cinnamon (*Cinnamomum zeylanicum*) is by far one of the most popular spices for lowering blood sugar levels (Muhammad and Dewettinck 2017). In vitro tests of cinnamon showed a raising impact on insulin because of high concentrations of chromium, and it is known as a well-known insulin potentiator (Khan et al. 1990). Another researcher also investigates the intake of 1–6 g of cinnamon in the diet and found that it reduces the blood serum glucose levels after 40 days (Qin, Panickar and Anderson 2010). Cinnamon extracts activate glucogen synthase, increase glucose absorption, activate insulin receptor kinase and inhibit insulin receptor dephosphorylation, resulting in maximum phosphorylation of the insulin receptor. All of these effects would lead to increased insulin sensitivity. Other research showed that turmeric (*Curcuma longa*) has a potential against diabetic activity. Turmeric's inclusion of sesquiterpenoids and curcuminoids has demonstrated a synergistic benefit against diabetes. Daily consumption of curcumin decreased the amount of insulin needed to maintain normoglycemia as well as fasting blood sugar levels (Srinivasan 2005). In alloxan-induced diabetic rats, curcumin exhibited a greater hypoglycaemic effect than turmeric. Curcumin's anti-diabetic effects are attributable in part to a decrease in hepatic glucose synthesis caused by AMP kinase activation and inhibition of hepatic glucose-6-phosphatase and phosphoenolpyruvate carboxykinase activity (Fujiwara et al. 2008). Fenugreek also has anti-diabetic properties. In a human experiment including 60 non-insulin-dependent diabetes mellitus patients, the hypoglycaemic impact was anticipated over the course of 24 weeks, fenugreek whole seed powder was distributed in two doses of 25 g/day throughout the diet, and glucose tolerance and fasting blood glucose levels were both improved. The amount of sugar excreted was dramatically reduced, as were insulin levels (Sharma et al. 2019). The effect of oral fenugreek extracts on diabetic mice found a decrease in serum glucose levels. The anti-diabetic impact was comparable to that of the anti-diabetic medication glibenclamide (Eidi, Eidi, and Sokhteh 2007). Coriander has long been used as traditional cholesterol and diabetes therapy. In a study, the effect of coriander seed on lipid metabolism was examined for rats fed with a high-fat diet and increased cholesterol levels. There was a significant increase in beta-hydroxy, beta-methyl glutaryl CoA reductase and plasma lecithin cholesterol acyl transferase activity noted in the experimental group. When compared to the control group, the amount of low-density lipoprotein (LDL) and very low-density lipoprotein (VLDL) cholesterol in the experimental group decreased while the level of high-density lipoprotein (HDL) cholesterol increased.

11.2.6 Traditional Nutraceutical Relevance

Since ancient times, spices have been used as the basic remedy for different common diseases including joint pain, diarrhoea, constipation, tooth decay, heart disease, asthma, tumours, cough, vomiting and allergies. Black paper is used in traditional medicine to treat snake bites and epilepsy (Szallasi 2005). Black pepper (or maybe long pepper) is also used for a variety of ailments, including toothaches, gangrene, earaches, heart disease, hernias, hoarseness, insect bites, sleeplessness, joint discomfort, liver issues and lung diseases. Another important spice is cardamon which is also used in various traditional applications such as tooth infection, asthma, colds, kidney disease, heart disease, constipation and digestive disorders. There are some household spices such as ginger, garlic, fenugreek and coriander that can be used to rectify conditions such as cough, fevers, headache, diabetes and respiratory disorders. Peppermint plants are frequently employed as a herbal remedy to treat nausea, vomiting, fever, bronchitis, sinusitis and lack of appetite (Wani et al. 2022;

Kunnumakkara et al. 2009). In folk medicine, rosemary is used to cure a variety of illnesses, including dermatitis, respiratory disorders, diabetes mellitus, poor circulation, migraines, epilepsy and inflammatory diseases as well as to promote hair growth. Turmeric has been used traditionally for centuries to treat a wide range of conditions, including rheumatism, body aches, skin issues (including wounds, burns and acne), intestinal worms, diarrhoea, intermittent fevers, hepatic diseases, urinary discharges, dyspepsia, inflammations, constipation, leukoderma, amenorrhea, dental diseases and digestive disorders like dyspepsia and acidity, indigestion flatulence (Aggarwal et al. 2007; Thangapazham, Sharma and Maheshwari 2007).

11.2.7 Other Nutraceutical Properties

Apart from the traditional nutraceutical property of spices, in the situations of hypercholesterolemia and hypertriglyceridemia, it was shown that dietary spices with hypolipidemic properties, such as curcumin, capsaicin and garlic, helped preserve the structural integrity and liquidity of erythrocytes. In hypercholesterolemic and hypertriglyceridemic circumstances, they restore the antioxidant status of erythrocytes and reverse the altered cholesterol-to-phosphorus ratio in erythrocytes. Through the formation of compact and sticky aggregates, platelets in blood play a vital function in haemostasis following thrombosis and blood vessel injury. Inhibitors of platelet aggregation play a preventative function in thrombotic diseases. In vitro studies using the spice components cumin aldehyde, eugenol and zingerone show that pre-treatment with garlic or onion prevents platelet aggregation. Additionally, coriander and cumin water extracts were shown to significantly suppress platelet aggregation.

11.3 RECENT ADVANCEMENTS IN THE DEVELOPMENT OF FUNCTIONAL SPICES

Phytochemical-rich functional spices have increased their market share over the past few years due to both their health advantages and business potential. The most recent comprehensive Antioxidant Food Database based on the evaluation of 3,100 foods, beverages, spices and herbs reveals that spices and herbs both account for the majority of the antioxidant-rich foods and beverages in the average person's diet. The application of spices as functional food ingredients in different food products is summarised in Table 11.2. In every segment of the food industry including dairy, bakery, beverage and active food packaging, spices have a substantial role in the production of functional food and also active packaging materials. Some researchers have developed functional chocolate with a mixture of some common spices like ginger, cinnamon, aniseed and lemon peels, which show good antimicrobial and antioxidant activity (Albak and Tekin 2014). The combining of herbs or spices in dairy products (ice cream, ghee, butter, cheeses yoghurt) also provides high nutritional and medicinal value without compromising the taste, colour, flavour, aroma or appearance (El-Sayed and Youssef 2019).

Table 11.2 Spices as a functional ingredient for different food products

Spices	Developed food products	Effect on the product quality	Reference
Rosemary extract	Raw and precooked pork sausages	Retard the thiobarbituric acid-reactive substances (TBARS) value and loss of red colour in frozen sausage.	(Sebranek et al. 2005)
Thyme	Ice cream	It shows a natural flavouring and antioxidant activity.	(Ramadan et al. 2013)
Fenugreek seed flour	Beef burger,	It shows antioxidant and antimicrobial properties.	(Hegazy 2011)
Clove powder	Chicken patties	It shows antimicrobial properties.	(Kumar and Tanwar 2011)
Cardamom, black pepper	Paneer	Increasing in phenolic compound compared to the control sample.	(Badola et al. 2018)
Turmeric	Lassi	It shows higher antioxidant activity with good sensory properties.	(Maji et al. 2018)
Coriander extracts	Ghee	It shows higher antioxidant activity compared to oleoresin.	(Patel et al. 2013)
Ginger extract	Lassi	It shows higher antioxidant activity compared to lassi incorporate with pudina, carrot extracts.	(Maji et al. 2018); (Dhumal et al. 2018)
Ginger extract	Ice cream	It shows antibacterial and antioxidant activity.	(Chamchan et al. 2017)

11.4 CONCLUDING REMARKS AND FUTURE DIRECTIONS

Nowadays, consumers have a better understanding of how nutrition impacts people from a healthcare view of point. To provide comprehensive medical care, they are therefore investigating how medical therapy and diet might be linked. In the past few decades, scientific research has shown that spices can be a popular food additive and offer a number of health benefits, including the ability to stimulate digestion, prevent cancer, anti-atherogenic and cardioprotective effects, and antilithogenic properties. Spices' antioxidant and hypolipidemic qualities have significant nutritional significance. Each spice has a unique potential health benefit, and there is a chance that they might work together synergistically to improve both flavour and wellness. Although herbs and spices have a wide range of possible health benefits, further research is required to determine exactly how exposure to their constituent parts causes a particular biochemical reaction. It will only be feasible to establish intervention techniques to maximise the advantages of herbs and spices while avoiding unfavourable side effects if such knowledge is accessible.

REFERENCES

Aggarwal, B. B., Sundaram, C., Malani, N., & Ichikawa, H. (2007). Curcumin: The Indian solid gold. In *The Molecular Targets and Therapeutic Uses of Curcumin in Health and Disease*, 1–75. Springer Science & Business Media.

Albak, F., & Tekin, A. R. (2014). Development of functional chocolate with spices and lemon peel powder by using response surface method: Development of functional chocolate. *Akademik Gıda*, 12(2), 19–25.

Aruoma, O. I., Spencer, J. P. E., Rossi, R., Aeschbach, R., Khan, A., Mahmood, N., ... & Halliwell, B. (1996). An evaluation of the antioxidant and antiviral action of extracts of rosemary and Provencal herbs. *Food and Chemical Toxicology*, *34*(5), 449–456. https://doi.org/10.1016/0278-6915(96)00004-X

Badola, R., Danish, M., Kumar, S., Fahad, M., Kanade, P. P., Upadhayay, S., Badola, R., & Rautela, I. (2018). Effect of incorporation of black pepper and cardamom on quality characteristics of paneer. *International Journal of Applied Science and Engineering*, 6(2), 121–127. https://doi.org/10.30954/2322-0465.2.2018.4

Chamchan, R., Sinchaipanit, P., Disnil, S., Jittinandana, S., Nitithamyong, A., & On-nom, N. (2017). Formulation of reduced sugar herbal ice cream using lemongrass or ginger extract. *British Food Journal*. https://doi.org/10.1108/BFJ-10-2016-0502

Crowe, K. M., & Francis, C. (2013). Position of the academy of nutrition and dietetics: Functional foods. *Journal of the Academy of Nutrition and Dietetics*, 113(8), 1096–1103. https://doi.org/10.1016/j.jand.2013.06.002

Cortés-Rojas, D. F., de Souza, C. R. F., & Oliveira, W. P. (2014). Clove (Syzygium aromaticum): A precious spice. *Asian Pacific Journal of Tropical Biomedicine*, *4*, 90–96.

Critchfield, J.W., Butera, S.T., and Folks, T.M. (1996). Inhibition of HIV acti- vation in latently infected cells by flavonoid compounds. *Res. Human, Retro- viruses*, *12*, 39–486.

de Sá Mendes, N., & de Andrade Gonçalves, É. C. B. (2020). The role of bioactive components found in peppers. *Trends in Food Science and Technology*, 99, 229–243. https://doi.org/10.1016/j.tifs.2020.02.032

De, A. K. (2004). *Spices: Traditional Uses and Medicinal Properties*. Asian Books.

Dhumal, V. S., Padghan, P. V., Shinde, S. P., & Maske, T. A. (2018). Effect of pudina extract on physico chemical properties of Lassi with optimized the level of pudina leaves. *Journal of Pharmacognosy and Phytochemistry*, 7(1), 2763–2766.

El-Sayed, S. M., & Youssef, A. M. (2019). Potential application of herbs and spices and their effects in functional dairy products. *Heliyon*, 5(6), e01989. https://doi.org/10.1016/j.heliyon.2019.e01989

Eidi, A., Eidi, M., & Sokhteh, M. (2007). Effect of fenugreek (Trigonella foenum-graecum L) seeds on serum parameters in normal and streptozotocin-induced diabetic rats. *Nutrition Research*, 27(11), 728–733. https://doi.org/10.1016/j.nutres.2007.09.006

Fujiwara, H., Hosokawa, M., Zhou, X., Fujimoto, S., Fukuda, K., Toyoda, K., Nishi, Y., Fujita, Y., Yamada, K., Yamada, Y., Seino, Y., & Inagaki, N. (2008). Curcumin inhibits glucose production in isolated mice hepatocytes. *Diabetes Research and Clinical Practice*, 80(2), 185–191. https://doi.org/10.1016/j.diabres.2007.12.004

Hegazy, A. I. *World Journal of Agricultural Sciences*, 20117, 391.

Idris, N. A., Yasin, H. M., & Usman, A. (2019). Voltammetric and spectroscopic determination of polyphenols and antioxidants in ginger (Zingiber officinale Roscoe). *Heliyon*, 5(5), e01717. https://doi.org/10.1016/j.heliyon.2019.e01717

Khan, A., Bryden, N. A., Polansky, M. M., & Anderson, R. A. (1990). Insulin potentiating factor and chromium content of selected foods and spices. *Biological Trace Element Research*, 24(2), 183–188. https://doi.org/10.1007/BF02917206

Koch, C., Reichling, J., Schneele, J., & Schnitzler, P. (2008). Inhibitory effect of essential oils against herpes simplex virus type 2. *Phytomedicine*, *15*(1-2), 71–78. https://doi.org/10.1016/j.phymed.2007.09.003

Kumar, D., & Tanwar, V. K. (2011). Utilization of clove powder as phytopreservative for chicken nuggets preparation. *Journal of Stored Products and Postharvest Research*, 2, 11–14.

Kunnumakkara, A. B., Koca, C., Dey, S., Gehlot, P., Yodkeeree, S., Danda, D., & Aggarwal, B. B. (2009). Traditional uses of spices: An overview. In *Molecular Targets and Therapeutic Uses of Spices: Modern Uses for Ancient Medicine*, 1–24.

Kutluay, S. B., Doroghazi, J., Roemer, M. E., & Triezenberg, S. J. (2008). Curcumin inhibits herpes simplex virus immediate-early gene expression by a mechanism independent of p300/CBP histone acetyltransferase activity. *Virology*, 373(2), 239–247. https://doi.org/10.1016/j.virol.2007.11.028

Laribi, B., Kouki, K., M'Hamdi, M., & Bettaieb, T. (2015). Coriander (Coriandrum sativum L.) and its bioactive constituents. *Fitoterapia*, 103, 9–26. https://doi.org/10.1016/j.fitote.2015.03.012

Maji, S., Ray, P. R., Ghatak, P. K., & Chakraborty, C. (2018). Total phenolic content (TPC) and quality of herbal lassi fortified with turmeric (Curcuma longa) extract. *Asian Journal of Dairy and Food Research*, 37, 273–277. https://doi.org/10.18805/ajdfr.DR-1391

Mnif, S., & Aifa, S. (2015). Cumin (Cuminum cyminum L.) from traditional uses to potential biomedical applications. *Chemistry and Biodiversity*, 12(5), 733–742. https://doi.org/10.1002/cbdv.201400305

Muhammad, D. R. A., & Dewettinck, K. (2017). Cinnamon and its derivatives as potential ingredient in functional food—A review. *International Journal of Food Properties*, 20(Suppl. 2), 2237–2263. https://doi.org/10.1080/10942912.2017.1369102

Munekata, P. E., Alcántara, C., Žugčić, T., Abdelkebir, R., Collado, M. C., García-Pérez, J. V., ... & Lorenzo, J. M. (2020). Impact of ultrasound-assisted extraction and solvent composition on bioactive compounds and in vitro biological activities of thyme and rosemary. *Food Research International*, *134*, 109242. https://doi.org/10.1016/j.foodres.2020.109242

Palou, A., Serra, F., & Pico, C. (2003). General aspects on the assessment of functional foods in the European Union. *European Journal of Clinical Nutrition*, 57(1), S12–S17. https://doi.org/10.1038/sj.ejcn.1601822

Pasupuleti, V. R., Sammugam, L., Ramesh, N., & Gan, S. H. (2017). Honey, propolis, and royal jelly: A comprehensive review of their biological actions and health benefits. *Oxidative Medicine and Cellular Longevity*, 2017. https://doi.org/10.1155/2017/1259510

Patel, S., Shende, S., Arora, S., & Singh, A. K. (2013). An assessment of the antioxidant potential of coriander extracts in ghee when stored at high temperature and during deep fat frying. *International Journal of Dairy Technology*, 66(2), 207–213. https://doi.org/10.1111/1471-0307.12023

Perussello, C. A. (2020). Non-thermal processing of herbs and spices. In *Herbs, Spices and Medicinal Plants: Processing, Health Benefits, and Safety*, 23–44. https://doi.org/10.1002/9781119036685.ch2

Polasa, K., Naidu, A. N., Ravindranath, I., & Krishnaswamy, K. (2004). Inhibition of B (a) P induced strand breaks in presence of curcumin. *Mutation Research/Genetic Toxicology and Environmental Mutagenesis*, 557(2), 203–213. https://doi.org/10.1016/j.mrgentox.2003.10.016

Qin, B., Panickar, K. S., & Anderson, R. A. (2010). Cinnamon: Potential role in the prevention of insulin resistance, metabolic syndrome, and type 2 diabetes. *Journal of Diabetes Science and Technology*, 4(3), 685–693. https://doi.org/10.1177/193229681000400324

Rajanbabu, R., & Ganesan, S. (2015). Growth and instability in production of Indian spices. *International Journal of Management and Social Development*, 2, 42–50.

Ramadan, K. M., Ashoush, I. S., & El-Batawy, O. I. (2013). Comparative evaluation of three essential oils as functional antioxidants and natural flavoring agents in ice cream. *World Applied Sciences Journal*, 23(2), 159–166. https://doi.org/10.5829/idosi.wasj.2013.23.02.13065

Rodríguez-Rojo, S., Visentin, A., Maestri, D., & Cocero, M. J. (2012). Assisted extraction of rosemary antioxidants with green solvents. *Journal of Food Engineering, 109*(1), 98–103. https://doi.org/10.1016/j.jfoodeng.2011.09.029

Sebranek, J. G., Sewalt, V. J. H., Robbins, K., & Houser, T. A. (2005). Comparison of a natural rosemary extract and BHA/BHT for relative antioxidant effectiveness in pork sausage. *Meat Science*, 69(2), 289–296. https://doi.org/10.1016/j.meatsci.2004.07.010

Shaha, R. K., Rahman, S., & Asrul, A. (2013). Bioactive compounds in chilli peppers (Capsicum annuum L.) at various ripening (green, yellow and red) stages. *Annals of Biological Research*, 4(8), 27–34.

Sharif, M. K., Ejaz, R., & Pasha, I. (2018). Nutritional and therapeutic potential of spices. In *Therapeutic, Probiotic, and Unconventional Foods*. Academic Press, 181–199. https://doi.org/10.1016/B978-0-12-814625-5.00011-X

Sharma, S., Ghataury, S. K., Sarathe, A., Dubey, G., & Parkhe, G. (2019). Curcuma angustifolia Roxb, (Zingiberaceae): Ethnobotany, phytochemistry and pharmacology: A review. *Journal of Pharmacognosy and Phytochemistry*, 8, 1535–1540.

Sikand, G., Kris-Etherton, P., & Boulos, N. M. (2015). Impact of functional foods on prevention of cardiovascular disease and diabetes. *Current Cardiology Reports*, 17(6), 1–16. https://doi.org/10.1007/s11886-015-0593-9

Slameňová, D., Kubošková, K., Horváthová, E., & Robichová, S. (2002). Rosemary-stimulated reduction of DNA strand breaks and FPG-sensitive sites in mammalian cells treated with H2O2 or visible light-excited Methylene Blue. *Cancer letters*, *177*(2), 145–153. https://doi.org/10.1016/S0304-3835(01)00784-4

Srinivasan, K. (2005). Plant foods in the management of diabetes mellitus: Spices as beneficial antidiabetic food adjuncts. *International Journal of Food Sciences and Nutrition*, 56(6), 399–414.

Surh, Y. J. (2002). Anti-tumor promoting potential of selected spice ingredients with antioxidative and anti-inflammatory activities: a short review. *Food and Chemical Toxicology*, *40*(8), 1091–1097.

Szallasi, A. (2005). Piperine: Researchers discover new flavor in an ancient spice. *Trends in Pharmacological Sciences*, 26(9), 437–439. https://doi.org/10.1016/j.tips.2005.07.001

Thangapazham, R. L., Sharma, A., & Maheshwari, R. K. (2007). Beneficial role of curcumin in skin diseases. In *The Molecular Targets and Therapeutic Uses of Curcumin in Health and Disease*, 343–357. https://doi.org/10.1007/978-0-387-46401-5_15

Toledano Medina, M. Á., Merinas-Amo, T., Fernández-Bedmar, Z., Font, R., del Río-Celestino, M., Pérez-Aparicio, J., Moreno-Rojas, R., Alonso-Moraga, Á., & Moreno-Rojas, R. (2019). Physicochemical characterization and biological activities of black and white garlic: In vivo and in vitro assays. *Foods*, 8(6), 220.

Viuda-Martos, M., Ruíz-Navajas, Y., Fernández-López, J., & Pérez-Álvarez, J. A. (2007). Chemical composition of the essential oils obtained from some spices widely used in Mediterranean region. *Acta Chimica Slovenica*, 54(4), 921. https://doi.org/10.3390/foods8060220

Wani, S. A., & Kumar, P. (2018). Fenugreek: A review on its nutraceutical properties and utilization in various food products. *Journal of the Saudi Society of Agricultural Sciences*, 17(2), 97–106. https://doi.org/10.1016/j.jssas.2016.01.007

Wani, S. A., Naik, H. R., Wagay, J., Ganie, N. A., Mulla, M. Z., & Dar, B. N. (2022a). Mentha: A review on its bioactive compounds and potential health benefits. *Quality Assurance and Safety of Crops and Foods*, 14(4), 154–168.

Wani, S. A., Singh, A., & Kumar, P. (Eds.). (2022b). *Spice Bioactive Compounds: Properties, Applications, and Health Benefits* (1st ed.). CRC Press. https://doi.org/10.1201/9781003215387

Willis, S., Sunkara, R., Hester, F., Shackelford, L., Walker, L. T., & Verghese, M. (2019). Chemopreventive and anti-inflammatory potential of select herbal teas and cinnamon in an in-vitro cell model. *Food and Nutrition Sciences*, 10(9), 1142–1156.

Zhao, X. X., Lin, F. J., Li, H., Li, H. B., Wu, D. T., Geng, F., Ma, W., Wang, Y., Miao, B. H., & Gan, R. Y. (2021). Recent advances in bioactive compounds, health functions, and safety concerns of onion (Allium cepa L.). *Frontiers in Nutrition*. https://doi.org/10.3389/fnut.2021.669805

CHAPTER 12

Medicinal Herbs as a Functional Food for Health

Gholamreza Abdi*, **Mukul Machhindra Barwant** and **Abu Saeid**

12.1 INTRODUCTION

Over the past two centuries, functional foods have made enormous contributions to traditional medicine. Functional foods are now seen as an alternative form of medication among the general public as a result of economic expansion and a strong desire among Malaysian consumers to maintain a healthy lifestyle. Despite Malaysia seeing an increase in the consumption of functional foods, surprisingly little is known about the variables that influence customer purchase intentions for naturally functional foods. A study assessed Malaysian consumers' intent to buy natural functional foods (Rezai et al. 2017). It is always a component of a healthy diet, and to be clear, it never appears in the form of a capsule or pill but rather as food. A functional food can be either natural, unaltered or processed food with a specialised design to achieve the intended result. In this chapter, the suitability of herbs and spices as functional components for the creation of such functional foods will be examined. In every case, it is necessary to determine whether a spice has a measurable positive impact on the consumer's health or well-being or whether it helps lower the likelihood that a certain sickness would manifest itself (Buckenhueskes 2009). Herbs can be used in many different ways. In the broadest sense, herbs are plants that are frequently cultivated and harvested for their special therapeutic and fragrant properties. A herb is a plant or portion of a plant utilised for its distinctive flavour, scent or medical characteristics. The parts of many plants that are grown for their aromatic, spicy or other desirable properties are called spices and herbs. Rhizomes, stigmas, fruits, seeds, flower buds, bulbs, barks and leaves make up spices and herbs. They are frequently separated into the categories of herbs, spices and spice seeds. Herbal

* Corresponding author E-mail: abdi@pgu.ac.ir

DOI: 10.1201/9781003315100-12

medicines are typically consumed as dietary supplements The idea of an herb garden was heavily influenced by Islamic, Christian and Egyptian traditions. In the majority of the world, herbs are primarily farmed as field crops or on a limited scale as a catch crop with vegetables. An old-age question about the distinction between a spice and a herb was adequately addressed in Rodale's Illustrated Encyclopedia of Herbs. Herbs are typically fresh or dried leaves that are more frequently grown in temperate climates, while spices are typically seeds, roots, fruits, flowers or bark that are native to tropical climates (Bishnoi 2016). Since the middle of the 1990s, a "new" product category known as "functional foods" has helped to enrich the food market. Such goods must at least meet the following criteria, according to the definition provided by the "Concerted Action on Functional Food Science in Europe Group" (FUFOSE group) in 1999: a functional meal must have a demonstrable beneficial influence on the consumer's health or well-being, or it must help lower the likelihood that a particular illness may manifest. It is always a component of a healthy diet, and to be clear, it never appears in the form of a capsule or pill but rather as food. A useful food may be natural, naturally changed, processed, or both (Buckenhueskes 2009).

Market segment growth for functional food is 10% annually, which is faster than the worldwide market segment growth of 10% for conventional food, which increases by 2–3% per year. Dietary factors have a part in the acceptance of functional foods, such as knowledge level, cultural norms and the development of technology in the food industry anticipation, sensory attributes and cost (Frewer et al. 2003, Wulan and Ihalauw 2020). Herbal remedies are some of our oldest treatments, and the rise in their popularity in recent years shows that people are interested in finding alternatives to traditional treatment. For a variety of purposes, including culinary, aromatic, fibrous and medicinal ones, herbs are planted and harvested. Some herbs have the ability to alter how our bodies work, which has transformed medicine and made those who grow, prepare and use them wealthy. Herbs are becoming more valuable economically since they are used in a variety of items, including food, medicine, perfumes and cosmetics. For many years, herbal knowledge has been passed down from one generation to the next (Bown 1995). Ayurveda, the oldest medical system in the world, was created in India more than 3,000 years ago. It is the most conventional and established type of medical care. Nootropics are herbs and drugs that help with memory, cognition, mood and other elements of mental function. Some plants that have nootropic effects include Shankhpushpi, ginkgo (*Ginkgo biloba*), gotu kola (*Centella asiatica*), Ashwagandha (*Convolvulus arvensis/Withania somnifera*) and Brahmi (*Bacopa monnieri*). Water hyssop, also known as *Bacopa monnieri (BM)*, *Herpestis monniera* or *Jalanimba* in India, has been used by Ayurveda, an all-encompassing medicinal system that has its roots in India, for millennia.

The most important uses of spices and herbs in the past have been as aphrodisiacs, in the creation of holy oils and lubricants, and in the practice of medicine. They were employed in rituals, incantations and acts of devotion by priests and shamans alike. Shamans utilised them as charms to fend off bad spirits. To refresh and smell the home, aromatic plants were used. In Cathay (northern China), Assyria, Sumer, Greece, Egypt and Rome, herbals from antiquity have been discovered. Herbals are manuals for identifying plants and making curative combinations. Hippocrates,

Galen and Pedanius Dioscorides were among the individuals who used them. Pliny the Elder commended the effectiveness and curative properties of spices and herbs in the treatment of practically every ailment known at the time in his *Natural History*, which was published in the first century of the Christian period (Singh and Dhawan 1997 and Devendra et al. 2018). Cooking, alcoholic drinks, medicine, cosmetics, fragrances, colouring and garden plants all benefit greatly from the use of herbs and spices. Herbs and spices give food its flavour, pungency and colour. In addition, herbs and spices have dietary, therapeutic and antimicrobial properties (Bishnoi 2016). Spices not only enhance flavour and taste but also play a significant function in food preservation by postponing food degradation. Additionally, antioxidants aid the body's defences against arthritis, asthma, certain types of cancer (epithelial) and cardiovascular disease. Flavonoids and other phenolic chemicals may assist to prevent bowel cancer and cardiovascular disease (black pepper, oregano, thyme and marjoram). Gingerol, a component of ginger, is also an intestine stimulant and a medication bioactivity promoter (Peter 2004).

Products that mimic traditional foods yet have physiological advantages are called functional foods. Conventional, enriched or dietary supplements and fortified, enhanced foods are some of its examples. Nutraceuticals, on the other hand, are products made from food that are used medicinally as pills, capsules or liquids and again show physiological benefits. Herbs that are offered in a capsule, powder, soft gel, gel cap or another form that isn't advertised as a typical food are regarded as dietary supplements and are subject to different regulations than foods or food additives. Despite this, the rising interest in herbal goods may have peaked due to worries about one's health and well-being. For the preparation of poultices, teas, tinctures, broths and other remedies, traditional herbalists used plant roots, leaves, bark, flowers and seeds. Today, capsules and other delivery systems for exceedingly pure extracts and powders are tablets. The basis of contemporary herbal medicine (or phototherapy) is standardisation and established correlations between constituents and efficacy. There are numerous functional foods and nutraceuticals that are becoming more widely available on the market, but this presents a difficulty for the makers of functional foods because these goods must address the problem of sensory perception (Shahidi 2012). Nutraceuticals are foods or components of foods that provide health advantages, such as the ability to treat or prevent disease. Because they avoid side effects and contain naturally occurring dietary supplements, nutraceuticals offer an advantage over pharmaceuticals. Nutraceuticals are divided into three main groups based on their chemical makeup and natural sources: nutrients, herbs, dietary supplements, and fibre, among others. Dietary supplements and natural/herbal goods saw the industry's fastest growth rates (19.5% annually) (11.6% per year). In order to effectively combat the diabetes epidemic, preventative strategies must consider the complexity of disease development, increasing access to and knowledge of natural methods of prevention and management for those who are affected, and appropriately intervening while also taking into account the costs the public health sector is currently facing (Ballali and Lanciai 2012). When plant material, such as the complete plant, roots, bark, leaves, flowers, berries or seeds, is used for medicinal purposes, it is known as "herbal medicine," "traditional herbal medicine" or "phytomedicine" (WHO 2005). Many traditional and historic herbal medicines have produced significant modern therapeutic agents, from which the main bioactive components have been identified and characterised. These include drugs like aspirin, whose precursor salicin is derived from willow (Salix species) and was used as an analgesic for the first time over 2,000 years ago (Smith et al. 2014). Herbal medicine is gaining popularity in new and growing industries and can be added to healthy diets and preventative medications. A growing number of functional food, nutraceutical and natural health products contain them as well. Plants, herbs and ethnobotanicals have been utilised for health promotion and disease treatment since the dawn of humans and are being used today in many parts of the world. Some people prefer using herbal remedies as a form of medicine. Others utilise herbs as a complementary therapy to traditional medications. Herbal medicine is a vital component of traditional medicine, which is the only system of healthcare that is accessible or inexpensive in many underdeveloped nations (Chauhan et al. 2013; Wachtel-Galor and Benzie 2011).

12.2 NUTRACEUTICAL PROPERTIES OF HERBS

Functional foods and dietary supplements generally contain nutraceutical components, which are made up of naturally occurring compounds with therapeutic characteristics that can help maintain health or prevent disease. Herbal products can also be included in this category (Sendker and Sheridan 2017). Nutraceuticals, in their simplest form, are food products (such as fortified food or dietary supplement) that offer health advantages (Chauhan et al. 2013). It has been reported that Stephen DeFelice coined the term "nutraceutical" in 1989; it is a combination of the terms "nutrition" and "pharmaceutical." A food or food component with linked medicinal or health advantages is included in the definition. These may go beyond the simple nutritional worth of foods and include their medicinal advantages (Brower 2005) The human diet contains a variety of plant-derived carotenoids, but carotene, carotene, lutein, zeaxanthin and lycopene appear to provide the greatest health benefits. Plant antioxidants are among the most potent food ingredients. Plant matter is the principal source of these chemicals (Kris-Etherton et al. 2002, Choudhary and Tandon 2009). Theoretically, nutritional therapy and nutraceuticals help to attain this goal by aiding in the body's detoxification processes, preventing vitamin and mineral shortages, and re-establishing normal digestive and dietary patterns. Phytonutrients are essentially planted nutrients with specific biological functions that benefit human health (Zhao 2007).

Coriander leaves have a nutritional potential that has been widely reported, and people typically cook with its fresh leaves. Many Indian cuisines have a garnish of chopped coriander leaves. (Chadha 1976). The leaves enhance assimilation and digestion, especially in people with chronic illnesses or poor digestion (Suneetha and Krishnakantha 2005). Dieresis is also heightened by the leaves. The coriander leaf has also demonstrated anti-inflammatory and antiplatelet properties. Table 12.1 lists the nutritional composition of 100 g of the leaves (Mahendra and Bisht 2011).

Table 12.1 Nutritional potential of some common herbs used as spices (Suneetha and Krishnakantha, 2005)

Sr no	Content	Black pepper drupe per 100 g	Black mustard seed per 100 g	Coriander leaves per 100 g
1	Water	12.46	5.27	92.21
2	Energy	251	508	23
3	Protein	10.39	26.08	2.13
4	Total lipid (fat)	3.26	36.24	0.52
5	Carbohydrate	63.95	28.09	3.67
6	Fibre, total dietary	25.3	12.2	2.8
7	Sugars, total	0.64	6.79	0.87
8	**Minerals**			
9	Calcium	443	266	67
10	Iron	9.71	9.21	1.77
11	Magnesium	171	370	26
12	Phosphorus	158	828	48
13	Potassium	1329	738	521
14	Sodium	20	13	46
15	Zinc	1.19	6.08	0.5
16	**Vitamins**			
17	Vitamin C, total ascorbic acid	-	7.1	27
18	Thiamine	0.108	0.805	0.067
19	Riboflavin	0.18	0.261	0.162
20	Niacin	1.143	4.733	1.114
21	Pantothenic acid	-	-	0.57
22	Vitamin B-6	0.291	0.397	0.149
23	Folate (DFE)	17	162	62
24	Vitamin A (RAE)	27	2	337
25	Vitamin A (IU)	547	31	6748
26	Vitamin E (alpha-tocopherol)	1.04	5.07	2.5
27	Vitamin K (phylloquinone)	163.7	5.4	310
28	**Lipids**			
29	Fatty acids, total saturated	1.392	1.989	0.014
30	Fatty acids, total monounsaturated	0.739	22.518	0.275
31	Fatty acids, total polyunsaturated	0.998	10.088	0.04

Beginning in the fifth century, numerous sources advised using black pepper to cure eye issues, frequently by dabbing the eye directly with pepper-based salves or poultices. There is currently no medical proof, however, that any of these treatments are helpful. Numerous conditions, including sleeplessness, oral abscesses, sunburn and toothaches, can be cured with black pepper. Due to its antioxidant, anti-obesity, gastroprotective, anti-cancer and antibacterial capabilities, it is a crucial healthy food. Black pepper is known to be an effective natural remedy for a variety of conditions, including strep throat, coma, migraine headaches, arthritis, nausea, fever and poor digestion (Vallette et al. 2016)

It has additionally been employed as an insecticide in some types of conventional medicine. Although this research is preliminary and the methods by which piperine acts in the human body are yet unknown, piperine is being investigated for a number of potential physiological effects (Srinivasan 2007). Along with other substances, piperine improves the absorption of selenium, vitamin B12, carotene and curcumin. It has been demonstrated that piperine has bioavailability-improving properties with a variety of structurally and therapeutically varied medicines. Between 4.6 and 9.7% of black pepper has piperine. Black pepper's pungency is caused by the alkaloid piperine and its isomer chavicine, as well as piperidines and amides (Khajuria et al. 2002). Rats exposed to black mustard showed a hypoglycaemic impact, and the mucilage in the plant helped the rats' postprandial glucose and insulin levels. The nutritional value of black mustard seed per 100 g has been reported and is shown in Table 12.1 (Srinivasan 2005). It also has been reported that extracts and concentrations are made from herbs or other botanical items. Common plants have the potential for healing due to their nutritive potential. Nutrients are substances with well-established nutritional roles, including vitamins, minerals, amino acids and fatty acids. Herbal extracts or concentrates are made from plants or herbs. Substances obtained

Table 12.2 Herbs and their therapeutic properties (Chauhan et al. 2013 and Chintale et al., 2013)

Sr no	Name of herb	Therapeutic property
1	*Aloe Vera*	Capillary dilation, anti-inflammatory, emollient and wound-healing properties.
2	*Ephedra sinica Stapf.*	Reduces bronchial oedema and is a bronchodilator and vasoconstrictor.
3	*Allium sativum L*	Anti-inflammatory, anti-thrombotic, antifungal and antibacterial
4	*Glycyrrhiza glabra L.*	Expectorant, secretolytic and peptic ulcer therapy
5	*Zingiber officinale Rosc.*	Cholagogue, carminative, antiemetic and positive inotropic
6	Ginkgo biloba	Memory booster, PAF antagonist and antioxidant
7	Echinacea purpurea	Anti-inflammatory, immunomodulator, antiviral
8	Panax ginseng	Qualities that stimulate the immune, neurological and adaptogenic systems
9	Hypericum perforatum	HIV and hepatitis C prevention, antidepressant
10	Curcuma Longa	Antiseptic, anti-arthritic, anti-cancer and anti-inflammatory
11	Allium cepa Linn	Antibiotic, hypoglycaemic action and antiatherosclerosis
12	Cassia angustifolia	Purgative
13	Ferula assafoetida L.	Stimulating, digestive and expectorant
14	Aegle marmelos Corr.	Diarrhoea and dysentery treatment, digestive and appetiser
15	Centella asiatica	Anti-anxiety, spasmolytic and nerve tonic

from other sources (such as pyruvate, chondroitin sulphate and steroid hormone precursors) are used for specific purposes, such as meal replacements, weight-loss aids and sports nutrition. A list of herbs and their therapeutic property are mentioned in Table 12.2 (Wani et al. 2022; Chauhan et al. 2013 and Chintale et al. 2013).

12.3 NUTRACEUTICAL HEALTH BENEFITS

Nutraceutical health benefits: (Chauhan et al. 2013)

- Have negligible negative impacts
- Can intensify the favourable impact on health
- Can contain naturally occurring dietary supplements, therefore no undesirable side effects
- Can enhance our diet's nutritional worth and better our overall health
- Can be easily accessible and reasonably priced
- Can be used to boost wellness, slow down ageing, stop chronic diseases from occurring and lengthen life expectancy
- Helpful in supporting the body's structure or functions
- Due to their potential for having nutritional, safe and therapeutic impacts, nutraceuticals have recently attracted a lot of attention
- Boost energy, reduce anxiety and increase general mental clarity
- Enhance the quantity and quality of your sleep
- Help to avoid chronic illnesses
- Decrease drug cravings
- Help prolong life, and support and control body functions

12.4 ANTIOXIDANT POTENTIAL

Oxidant-antioxidant equilibrium is crucial for the health of the human body because it preserves the proteins, nucleic acids and integrity of cell membranes (Knight 2000). Free radicals, such as reactive oxygen and nitrogen species, are controlled in healthy individuals. The idea of oxidative stress, on the other hand, postulates that contact with harmful physiochemical, environmental and pathogenic factors disturbs the body's normal equilibrium. Unchecked free radical damage can be done to vital extracellular or cellular components by antioxidants. The negative effects of antioxidant supplementation (vitamin E and selenium) have called into question the effectiveness of antioxidants in preventing illness (Knight 2000).

Unbalanced levels of oxidants and antioxidants have the ability to harm cells and weaken immunity as well as induce DNA damage that can result in mutations and eventually lead to the emergence of chronic illnesses like cancer (Kaefer and Milner 2008). Numerous spices and herbs are the subject of continuing research on their antioxidant qualities and health. According to epidemiological research, there is a link between consuming more antioxidants in your diet and having a reduced incidence of illness and death (Devasagayam et al. 2004). Approximately 500 newly diagnosed gastric adenocarcinoma patients and approximately 1100 control participants participated in a population-based, case-control study in Sweden that revealed the total antioxidant capacity of many plant-based dietary components to be inversely linked with stomach cancer risk (Serafini et al. 2002). Herbs may also protect against oxidative stress. The formation of reactive oxygen species (such as hydroxyl radicals and superoxide radical anion) and cellular antioxidant capacity are out of balance, which leads to oxidative stress. Herbs and spices contain a variety of water-soluble phenolic acids, flavonoids, tocopherols, lipid-soluble tocopherols, carotenoids and sterols that are naturally occurring. These substances have the ability to remove reactive oxygen species (free radicals) from the environment and may guard against the formation of genotoxic lipid peroxidation products like trans-4-hydroxy-2-nonenalUsing DPPH• and ABTS+• radical scavenging techniques, as shown in a study that conducted research using various plants to determine their antioxidant activity. In the DPPH• and ABTS+• techniques, marjoram and thyme were discovered to be the most effective antiradical scavengers. In scavenging the DPPH• radical, rosemary was shown to have double the activity of thyme (Gramza-Michaowska et al. 2008 and Wojdyo et al. 2007).

Figure 12.1 depicts the antioxidant qualities of numerous plants in modern times as a result of their possession of diverse potent antioxidant components (Embuscado 2015). Numerous research suggests the antioxidant power of rosemary leaves. Posadas et al. (2009) claim that rosemary extract contains 20% carnosic acid, an antioxidant that reduces oxidative stress in older rats. Over 90% of their antioxidant action is probably accounted for by carnosic acid and carnosol (Jiang 2019). Carnosic acid and carnosol reduced membrane damage by 40–50% and lowered lipid peroxidation by 88–100 and 38–89%, respectively, under oxidative stress conditions in cell culture studies. The maximum antioxidant capacity was found in rosemary and sage (Cosio et al. 2006). This is probably because of the unique composition of rosmarinic acid, carnosic acid and its metabolites. According to other research, several leafy herbs, such as sage, rosemary, oregano and thyme, have a significant antioxidant potential (Schwarz and Ternes 1992a, 1992b). The antioxidant capacity of many culinary and medicinal plants is shown in Table 12.3.

12.5 ANTIMICROBIAL PROPERTIES

Many disease-causing microbes can be killed off by herbs having antimicrobial properties such as simple phenols and phenolic acids, coumarins, terpenoids and alkaloids (Kaefer and Milner 2008). There have been investigations on the relationship between herbs and spices and gram-negative bacteria such as Escherichia coli, Salmonella and Vibrio cholera. Liu and Nakano (1996) discovered that the growth of *S. aureus*, *E. coli*, *Salmonella*, *V. cholera*, and *Bacillus* was suppressed at three distinct concentrations (0.5%, 0.2% and 0.1%) utilising five herbs and 17 spices. Overall, it was discovered that herbs and spices more effectively suppressed gram-positive bacteria than gram-negative bacteria. All of the following streptococcal strains have been found to be resistant to the hydroalcoholic extract of *Rosmarinus*

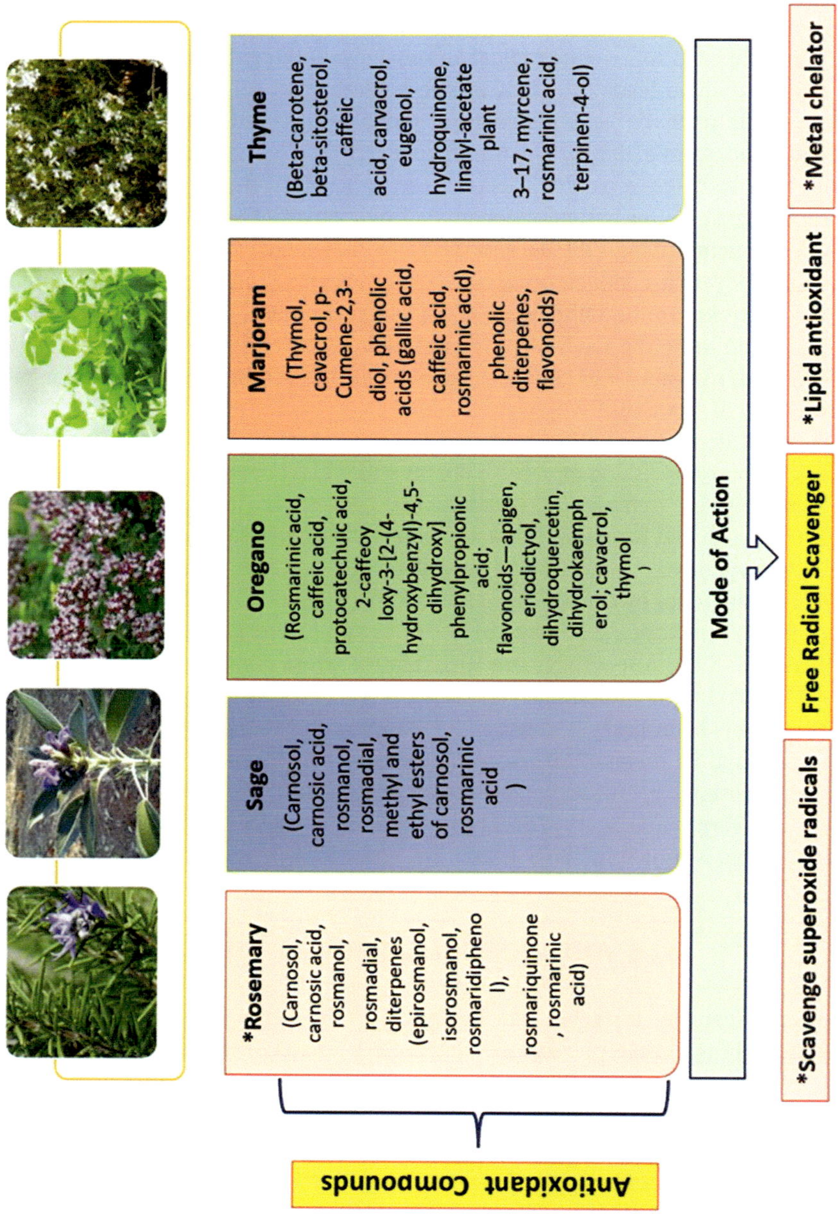

Figure 12.1 Different antioxidant compounds isolated from several herbs

Table 12.3 Antioxidant capacity (ORAC) and DPPH from various herbs extract

General name	Botanical name	Types	Antioxidant capacity (ORAC) μmol of TE/g of fresh weight	DPPH (mg AAE/g)
Creeping thyme	*Thymus praecox ssp. arcticus*	medicinal	13.40	32.5–79.3
Garden thyme	*Thymus vulgaris*	medicinal	19.49	159.59 μg/ml
Garden sage	*Salvia officinalis*	medicinal	13.28	14.1–64.5
Lemon balm	*Melissa officinalis*	culinary	9.54	171.5
Lovage	*Levisticum officinale*	culinary	21.54	34.1
Peppermint	*Mentha × piperita*	medicinal	15.84	203.8
Greek mountain oregano	*Origanum vulgare* ssp. *hirtum*	culinary	64.71	209.1
Cuban oregano	*Plectranthus amboinicus*	culinary	4.71	281.26 μg/mL
Hard sweet marjoram	*Origanum × majoricum*	culinary	71.64	42.1
Lemon thyme	*Thymus × citriodorus*	culinary	13.28	0.32 mg/ml
Mexican oregano	*Poliomintha longiflora*	culinary	92.18	83.70 μg/mL
Orange mint	*Mentha aquatica*	culinary	19.80	46.05 μg/mL
Parsley	*Petroselinum crispum*	culinary	11.03	3310 μg/mL
Purple amaranth	*Amaranthus cruentus*	culinary	28.92	56 - 423 μg/mL
Rosemary	*Rosmarinus officinalis*	culinary	19.15	44.7
Spearmint	*Mentha spicata*	culinary	8.10	12 μg/mL
Sweet basil	*Ocimum basilicum*	culinary	14.27	55.5
Vietnamese Coriander	*Polygonum odoratum*	culinary	22.30	190.19 μg/mL
Sweet bay	*Laurus nobilis*	culinary	31.70	0.46 mg/mL

Source: Zheng and Wang 2001; Chrpová et al. 2010; Ulewicz-magulska and Wesolowski 2019

officinalis Linn: *Streptococcus mitis*, *Streptococcus sanguinis*, *Streptococcus mutans*, *Streptococcus sobrinus* and *Lactobacillus casei* (rosemary). It can be demonstrated that every test, with the exception of the one against *S. mitis*, had antibacterial activity (Silva et al. 2008). A few pathogen strains, such as Listeria monocytogenes, S. *aureus*, *E. coli* and *Salmonella enteritidis*, isolated from food, were tested using essential oils from the leaves of oregano, thyme, basil, marjoram, lemongrass, ginger rhizomes and clove flower buds (Barbosa et al. 2009). The results showed that gram-positive microbes were susceptible to oils, with lemongrass oil being the most effective against them. Against two strains of *L. monocytogenes* and

Salmonella enteritidis in meat products, it has been demonstrated that the essential oils from oregano (*Origanum vulgare*), thyme (*Thymus vulgaris*), marjoram (*Origanum majorana*) and basil (*Ocimun basilicum*) exhibit antibacterial properties. It was demonstrated that all oils showed antibacterial effects, especially against gram-positive bacteria (Costa et al. 2009). Multi-resistant bacteria including *E. coli*, E. faecalis, Acinetobacter baumannii, K. pneumoniae and P. aeruginosa are susceptible to the antibacterial effects of oregano essential oils (*Origanum vulgare*). With the exception of P. aeruginosa, the concentration that produced the maximum inhibition was 0.125%. (2009). Oregano oil has an antimicrobial effect because it contains carvacrol and thymol (Sakkas & Papadopoulou, 2017), and its antimicrobial spectrum includes parasites, fungi (*Aspergillus* spp. and *Candida* spp.), bacteria (*L. monocytogenes, E. coli, E. coli O157:H7, Bacillus subtilis, B. cereus* and *Y. entero*), and parasite (*Blastocystis hominis, Entamoeba hartmanni* and *Endolimax nana*). The parsley essential oil was ineffective against all of the tested bacteria, but it worked particularly well against Salmonella enterica, Listeria monocytogenes, and *Staphylococcus aureus* (Linde et al. 2016). We investigated the presence of any antibacterial activity against two bacterial strains, including Escherichia coli and Bacillus subtilis, in the hydromethanolic extract of freeze-dried and radioactively irradiated parsley (*Petroselinum crispum*) and cilantro (Coriandrum sativum) leaves. When comparing methanol extract to water extract, the former's antibacterial effect is more pronounced. Thyme oil contains antibacterial properties that may be used to treat a wide range of pathogens, such as bacteria (*Aeromonas* spp., *B. cereus, B. subtilis, E. faecalis, L. monocytogenes, methicillin-resistant S. aureus, S. enteritidis, S. Typhimurium, Helicobacter pylori, E. coli* and *E. coli* (Sakkas and Papadopoulou, 2017). A test against a fungus that was identified from bread products involved the evaluation of 16 essential oils from diverse plants. Thyme's essential oil has the strongest inhibitory effects (Guynot et al. 2003). Thyme essential oil was one of the most potent antifungals tested against *Aspergillus flavus, Aspergillus parasiticus, Aspergillus ochraceus* and *Fusarium moniliforme* out of 12 essential oils (Soliman et al. 2002). Another research discovered significant antifungal activity of oregano and thyme oils against *Candida albicans, Candida parapsilosis, Candida tropicalis* and *Candida glabrata*. Further research discovered that oregano and thyme oils significantly reduced the growth of clinical strains of *Candida albicans, Candida parapsilosis, Candida tropicalis* and *Candida glabrata* (Sakkas et al. 2016). All of the fungi that were studied displayed fungistatic activity when exposed to parsley essential oil, however *Penicillium ochrochloron* and *Trichoderma viride* were particularly susceptible (Linde et al. 2016).

12.6 ANTI-CANCER ACTIVITY

Researchers and people are becoming increasingly interested in the numerous health benefits associated with food products like herbs, including lowering cancer risk and altering tumour behaviour, as a result of the rising prevalence of chronic diseases and the corresponding rise in healthcare costs on a global scale (Kaefer and Milner

2008). According to a study, herbs may also be able to fight against oxidative stress and inflammation, two variables that increase the risk of cancer development and progression as well as other pathological disorders (Michael 2006). The formation of reactive oxygen species (such as hydroxyl radicals and superoxide radical anion) and cellular antioxidant capacity are out of balance, which leads to oxidative stress. Herbs and spices naturally include lipid-soluble compounds including tocopherols, carotenoids and sterols as well as water-soluble phenolic acids, flavonoids (including quercetin and caffeic acid) and flavonoids. The production of genotoxic lipid peroxidation products like trans-4-hydroxy-2-nonenal may be prevented by these substances, which have the ability to eliminate reactive oxygen species from the environment. Some Helicobacter species strains, particularly those from the intestine, gallbladder, liver and stomach (*H. pylori*, *H. cholecystus*, *H. pullorum*, *H. bilis* and *H. hepaticus*), may encourage the invasion and growth of different cancers. Parsley, dill, sage and oregano were among the herbs that showed bactericidal effects, particularly against H. pylori species (O'Mahony et al. 2005). The culinary herb parsley leaf oil is often consumed and is extensively used as a flavouring and garnish. Myristicin, the primary volatile fragrance compound in parsley, showed high action as an inducer of the phase 2 enzyme glutathione S-transferase in the liver and small intestinal mucosa of strain A/Jalbino mice and a 65% inhibition of tumour multiplicity (Michael 2006). Male Fischer 344 rats were subjected to the early hepatocarcinogenesis-inhibitory effects of lemon grass extract after beginning with diethylnitrosamine (Puatanachokchai et al. 2002). Lemon grass (*Cymbopogon citratus*) contains the acyclic monoterpene alcohol geraniol, which inhibits colon cancer cell growth and polyamine synthesis. Amagase et al. (1996) found that rosemary's active ingredients may be able to prevent breast cancer by preventing the binding of DMBA metabolites to mammary cell DNA in rats. As shown in Figure 12.2, a study with few limits demonstrated that herbs and spices' bioactive components, with their antibacterial, antioxidant, anti-tumourigenic and carcinogen

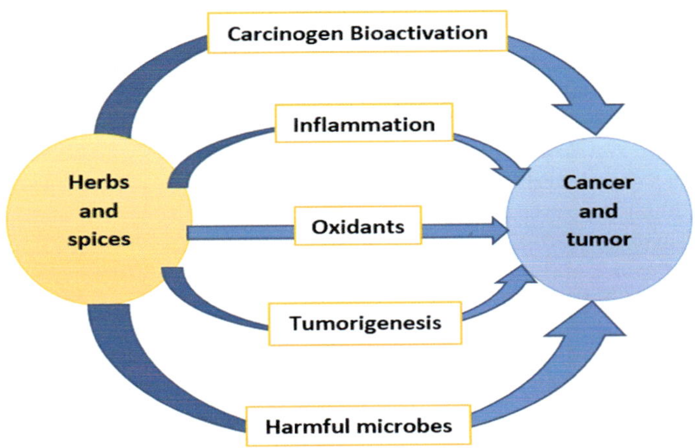

Figure 12.2 Cancer prevention by herbs and spices

bioactivation properties, occasionally function alone or in combination to lessen cancer risk (Kaefer and Milner 2008).

12.7 CONCLUSION

Herbs are used as functional foods due to their nutraceutical potential. It has been seen that there are several herbs that have the potential to be included in our daily diets due to their nutritional composition, alongside carbohydrates, protein, lipids, energy minerals and vitamins, which are necessary for the growth and development of humans. Herbs are important due to their therapeutic effect in curing diseases and improving immunity. Herbs and spices contain a variety of water-soluble phenolic acids, flavonoids, tocopherols, lipid-soluble tocopherols, carotenoids and sterols, which are naturally occurring. These substances have the ability to remove reactive oxygen species. Many disease-causing microbes can be killed off by herbs having antimicrobial properties, such as simple phenols and phenolic acids, coumarins, terpenoids and alkaloids. Researchers and people are becoming increasingly interested in the numerous health benefits associated with food products like herbs, including lowering cancer risk and altering tumour behaviour.

REFERENCES

Amagase, H., Sakamoto, K., Segal, E. R., & Milner, J. A. (1996). Dietary rosemary suppresses 7,12-dimethylbenz(a)anthracene binding to rat mammary cell DNA. *The Journal of Nutrition*, 126(5), 1475–1480. https://doi.org/10.1093/jn/126.5.1475

Ballali, S., & Lanciai, F. (2012). Functional food and diabetes: A natural way in diabetes prevention? *International Journal of Food Sciences and Nutrition*, 63(supl), 51–61. https://doi.org/10.3109/09637486.2011.637487

Barbosa, L. N., Rall, V. L. M., Fernandes, A. A. H., Ushimaru, P. I., da Silva Probst, I., & Fernandes Jr, A. (2009). Essential oils against foodborne pathogens and spoilage bacteria in minced meat. *Foodborne pathogens and disease*, *6*(6), 725–728.

Bishnoi, S. (2016). Herbs as functional foods. In D. Mudgil, S. Barak (Eds.) *Functional Food: Sources and Health Benefits*, 141–165.

Bown, D. (1995). *The Royal Horticultural Society Encyclopedia of Herbs & Their Uses*. D. Mudgil, S. Barak. (Eds.), 141–172. Dorling Kindersley Limited.

Brower, V. (2005). A nutraceutical a day may keep the doctor away: Consumers are turning increasingly to food supplements to improve well-being when pharmaceuticals fail. *EMBO Reports*, 6(8), 708–711. https://doi.org/10.1038/sj.embor.7400498

Buckenhueskes, H. J. (2009). Herbs and spices as functional food ingredients? *Zeitschrift für Arznei-& Gewürzpflanzen*, 14(2), 62–68.

Chadha, Y. R. (1976). The wealth of India: A dictionary of Indian raw materials and industrial products. *Publication Information Directorate*, CSIR, New Delhi, 10, 522–524.

Chauhan, B., Kumar, G., Kalam, N., & Ansari, S. H. (2013). Current concepts and prospects of herbal nutraceutical: A review. *Journal of Advanced Pharmaceutical Technology and Research*, 4(1), 4. https://doi.org/10.4103/2231-4040.107494

Chintale Ashwini, G., Kadam Vaishali, S., Sakhare Ram, S., Birajdar Ganesh, O., & Nalwad Digambar, N. (2013). Role of nutraceuticals in various diseases: A comprehensive review. *International Journal of Research in Pharmacy and Chemistry*, 3, 290–299.

Choudhary, R., & Tandon, R. V. (2009). Consumption of functional food and our health concerns. *Pakistan Journal of Physiology*, 30(5).

Chrpová, D., Kourimská, L., Gordon, M. H., Hermanová, V., Roubícková, I., & Pánek, J. (2010). Antioxidant activity of selected phenols and herbs used in diets for medical conditions. *Czech Journal of Food Sciences*, 28(4), 317–325. https://doi.org/10.17221/129/2010-CJFS

Cosio, M. S., Buratti, S., Mannino, S., & Benedetti, S. (2006). *Use of an electrochemical method to evaluate the antioxidant activity of herb extracts from the Labiatae family*, 97(4), 725–731. https://doi.org/10.1016/j.foodchem.2005.05.043

Costa, A. C. D., Santos, B. H. C. D., Santos Filho, L., & Lima, E. D. O. (2009). Antibacterial activity of the essential oil of Origanum vulgare L.(Lamiaceae) against bacterial multiresistant strains isolated from nosocomial patients. *Revista Brasileira de Farmacognosia*, 19(1b), 236–241. https://doi.org/10.1590/S0102-695X2009000200010

Devasagayam, T. P., Tilak, J. C., Boloor, K. K., Sane, K. S., Ghaskadbi, S. S., & Lele, R. D. (2004). Free radicals and antioxidants in human health: Current status and future prospects. *The Journal of the Association of Physicians of India*, 52, 794–804.

Devendra, P., Patel, S. S., Birwal, P., Basu, S., Deshmukh, G., & Datir, R. (2018). Brahmi (Bacopa monnieri) as functional food ingredient in food processing industry. *Journal of Pharmacognosy and Phytochemistry*, 7(3), 189–194. http://doi.org/10.54085/ap.2022.11.1.14

Dureja, H., Kaushik, D., & Kumar, V. (2003). Developments in nutraceuticals. *Indian Journal of Pharmacology*, 35(6), 363–372.

Embuscado, M. E. (2015). Herbs and spices as antioxidants for food preservation. In *Handbook of Antioxidants for Food Preservation*, pp. 251–283. Woodhead Publishing. Elsevier Ltd.

Frewer, L. J., Scholderer, J., & Bredahl, L. (2003). Communicating about the risks and benefits of genetically modified foods: The mediating role of trust. *Risk Analysis: An International Journal*, 23(6), 1117–1133. https://doi.org/10.1111/j.0272-4332.2003.00385.x

Gramza-Michałowska, A., Abramowski, Z., Eduardo, J., & Marzanna, H. (2007). Antioxiant potential of herbs extracts and impact on HEPG2 cells viability. *Acta Scientiarum Polonorum, Technologia Alimentaria*, 7(4), 61–72.

Guynot, M. E., Ramos, A. J., Seto, L., Purroy, P., Sanchis, V., & Marin, S. (2003). Antifungal activity of volatile compounds generated by essential oils against fungi commonly causing deterioration of bakery products. *Journal of Applied Microbiology*, *94*(5), 893–899.

Jiang, T. A. (2019). Health benefits of culinary herbs and spices. *Journal of AOAC International*, 102(2), 395–411. https://doi.org/10.5740/jaoacint.18-0418

Kaefer, C. M., & Milner, J. A. (2008). The role of herbs and spices in cancer prevention Christine. *The Journal of Nutritional Biochemistry*, 19(6), 347–361. https://doi.org/10.1016/j.jnutbio.2007.11.003

Khajuria, A., Thusu, N., & Zutshi, U. (2002). Piperine modulates permeability characteristics of intestine by inducing alterations in membrane dynamics: Influence on brush border membrane fluidity, ultrastructure and enzyme kinetics. *Phytomedicine*, 9(3), 224–231. https://doi.org/10.1078/0944-7113-00114

Knight, J. A. (2000). Review: Free radicals, antioxidants, and the immune system. *Annals of Clinical and Laboratory Science*, 30(2), 145–158.

Kris-Etherton, P. M., Hecker, K. D., Bonanome, A., Coval, S. M., Binkoski, A. E., Hilpert, K. F., Griel, A. E., & Etherton, T. D. (2002). Bioactive compounds in foods: Their role in the prevention of cardiovascular disease and cancer. *The American Journal of Medicine*, 113(9), 71–88. https://doi.org/10.1016/S0002-9343(01)00995-0

Linde, G. A., Gazim, Z. C., Cardoso, B. K., Jorge, L. F., Tešević, V., Glamoćlija, J., Soković, M., & Colauto, N. B. (2016). Antifungal and antibacterial activities of Petroselinum crispum essential oil. *Genetics and Molecular Research*, 15(3), 1–11. http://doi.org/10.4238/gmr.15038538

Liu, Z. H., & Nakano, H. (1996). Antibacterial activity of spice extracts against food-related bacteria. *Journal of the Faculty of Applied Biological Science, Hiroshima University*, 35(2), 181–190.

Mahendra, P., & Bisht, S. (2011). *Coriandrum sativum*: A daily use spice with great medicinal effect. *Pharmacognosy Journal*, 3(21), 84–88. https://doi.org/10.5530/pj.2011.21.16

Michael, F. (2006). Cancer in: The health benefits of herbs and spices: How strong is the evidence? *The Medical Journal of Australia*, 185(4), S7–S12. https://doi.org/10.3109/9781420067064

O'Mahony, R., Al-Khtheeri, H., Weerasekera, D., Fernando, N., Vaira, D., Holton, J., & Basset, C. (2005). Bactericidal and anti-adhesive properties of culinary and medicinal plants against Helicobacter pylori. *World Journal of Gastroenterology*, 11(47), 7499–7507. https://doi.org/10.3748/wjg.v11.i47.7499

Peter, K. V. (2004). *Handbook of Herbs and Spices 2*. New York: CRC Press. https://doi.org/10.1533/9780857095688.193

Posadas, S. J., Caz, V., Largo, C., De la Gándara, B., Matallanas, B., Reglero, G., & De Miguel, E. (2009). Protective effect of supercritical fluid rosemary extract, Rosmarinus officinalis, on antioxidants of major organs of aged rats. *Experimental Gerontology*, 44(6–7), 383–389. https://doi.org/10.1016/j.exger.2009.02.015

Puatanachokchai, R., Kishida, H., Denda, A., Murata, N., Konishi, Y., Vinitketkumnuen, U., & Nakae, D. (2002). Inhibitory effects of lemon grass (Cymbopogon citratus, Stapf) extract on the early phase of hepatocarcinogenesis after initiation with diethylnitrosamine in male Fischer 344 rats. *Cancer Letters*, 183(1), 9–15. https://doi.org/10.1016/S0304-3835(02)00111-8

Rezai, G., Teng, P.K., Shamsudin, M.N., Mohamed, Z., & Stanton, J.L. 2017 Effect of perceptual differences on consumer purchase intention of natural functional food. *J. Agribus. Dev. Emerg. Econ;7,* 153–173. doi: 10.1108/JADEE-02-2015-0014.

Sakkas, H., & Papadopoulou, C. (2017). Antimicrobial activity of basil, oregano, and thyme essential oils.

Sakkas, H., Gousia, P., Economou, V., Petsios, S., & Papadopoulou, C. (2016). Antifungal activity of four essential oils against Candida clinical isolates. *Asian Journal of Ethnopharmacology and Medicinal Foods*, 2, 22–25.

Schwarz, K., & Ternes, W. (1992a). Antioxidative constituents of Rosmarinus officinalis and Salvia officinalis. I. *Zeitschrift für Lebensmittel-Untersuchung und -Forschung*, 195(2), 95–98. https://doi.org/10.1007/bf01201767

Schwarz, K., & Ternes, W. (1992b). Antioxidative constituents of Rosmarinus officinalis and Salvia officinalis II. *Zeitschrift für Lebensmittel-Untersuchung und -Forschung*, 195(2), 99–103. https://doi.org/10.1007/bf01201765

Sendker, J., & Sheridan, H. (2017). History and current status of herbal medicines. In *Toxicology of Herbal Products*, pp. 11–27. Cham: Springer. https://doi.org/10.1007/978-3-319-43806-1_2

Serafini, M., Bellocco, R., Wolk, A., & Ekström, A. M. (2002). Total antioxidant potential of fruit and vegetables and risk of gastric cancer. *Gastroenterology*, 123(4), 985–991. https://doi.org/10.1053/gast.2002.35957

Shahidi, F. (2012). Nutraceuticals, functional foods and dietary supplements in health and disease. *Journal of Food and Drug Analysis*, 20(1), 226–230.

Silva, M. S. A., Silva, M. A. R., Higino, J. S., Pereira, M. S. V. and Carvalho, A. A. T. 2008. In vitro antimicrobial activity and antiadherence of Rosmarinus officinal is Linn. against oral planktonic bacteria. *Rev. Bras. Pharmacogn.*, *18*(2), 236–40.

Singh, H. K., & Dhawan, B. N. (1997). Neuropsychopharmacological effects of the Ayurvedic nootropic *Bacopa monniera* Linn.(Brahmi). *Indian Journal of Pharmacology*, 29(5), 359.

Smith, T., Elwood, P., Keating, C., Rothwell, P., Detering, E., Freedman, A., Langley, R., Logan, R., Phillips, C., & DeCensi, A. (2014). The aspirin foundation scientific conference: The history, the present state and the future of aspirin prophylaxis. *Ecancermedicalscience*, 8. https://doi.org/10.3332/ecancer.2014.388

Soliman, K. M., & Badeaa, R. I. (2002). Effect of oil extracted from some medicinal plants of different mycotoxigenic fungi. *Food and Chemical Toxicology*, 40(11), 1669–1675. https://doi.org/10.1016/S0278-6915(02)00120-5

Srinivasan, K. (2005). Plant foods in the management of diabetes mellitus: Spices as beneficial antidiabetic food adjuncts. *International Journal of Food Sciences and Nutrition*, 56(6), 399–414. https://doi.org/10.1080/09637480500512872

Srinivasan, K. (2007). Black pepper and its pungent principle-piperine: A review of diverse physiological effects. *Critical Reviews in Food Science and Nutrition*, 47(8), 735–748. https://doi.org/10.1080/10408390601062054

Suneetha, W. J., & Krishnakantha, T. P. (2005). Antiplatelet activity of coriander and curry leaf spices. *Pharmaceutical Biology*, 43(3), 230–233. https://doi.org/10.1080/13880200590928807

Ulewicz-magulska, B., & Wesolowski, M. (2019). Total phenolic contents and antioxidant potential of herbs used for medical and culinary purposes. *Plant Foods for Human Nutrition*, 74(1), 61–67. https://doi.org/10.1007/s11130-018-0699-5

Vallette, L., Rabadeaux, C., Sirdaarta, J., Davis, C., & Cock, I. E. (2016). An upscaled extraction protocol for Tasmannia lanceolata (Poir.) AC Sm.: Anti-bacterial, anti-Giardial and anticancer activity. *Pharmacognosy Communications*, 6(4). https://doi.org/10.5530/pc.2016.4.7

Wachtel-Galor, S., & Benzie, I. F. (2011). Herbal medicine. Lester packer, Ph.D., 1.

Wani, S. A., Naik, H. R., Wagay, J., Ganie, N. A., Mulla, M. Z., & Dar, B. N. (2022). Mentha: A review on its bioactive compounds and potential health benefits. *Quality Assurance and Safety of Crops and Foods*, 14(4), 154–168.

Wojdyło, A., Oszmiański, J., & Czemerys, R. (2007). Antioxidant activity and phenolic compounds in 32 selected herbs. *Food Chemistry*, 105(3), 940–949.https://doi.org/10.1016/j.foodchem.2007.04.038

World Health Organization. (2005). *National Policy on Traditional Medicine and Regulation of Herbal Medicines: Report of a WHO Global Survey*. World Health Organization.

Wulan, I. A. D. P., Priyanto, S. H., & Ihalauw, J. J. (2020). Model of herbal functional food adoption. *Calitatea*, 21(174), 135–138.

Zhao, J. (2007). Nutraceuticals, nutritional therapy, phytonutrients, and phytotherapy for improvement of human health: A perspective on plant biotechnology application. *Recent Patents on Biotechnology*, 1(1), 75–97. https://doi.org/10.2174/187220807779813893

Zheng, W., & Wang, S. Y. (2001). Antioxidant activity and phenolic compounds in 32 selected herbs. *Journal of Agricultural and Food Chemistry*, 49(11), 5165–5170. https://doi.org/10.1021/jf010697n

CHAPTER 13

Potential Role of Bioactive Compounds from Foods and Current Trends

Nur Alim Bahmid*, **Andi Rahmayanti**, **Heru Pitria Hastuti**, **Muhpidah** and **Ikawati Karim**

13.1 INTRODUCTION

Food is one of the fundamental requirements for living organisms to fulfil their functions optimally. Foods contain various macro and micro nutritional components that contribute to human health (Ngoma et al. 2020). In addition to having nutritional components, food materials also contain bioactive ingredients. In food products, bioactive components are present in small quantities. Many studies describe food items including bioactive compounds with medical benefits for human health. Several food types, including vegetables, fruits, nuts, whole grains and oils, have been identified to contain bioactive compounds and offer nutraceutical benefits (Kris-Etherton et al. 2002), as shown in Table 13.1.

Functional foods from natural substances with health-promoting impacts and nutritional claims have been accelerated by current consumer trends for the adoption of better lifestyles (Fogliano and Vitaglione 2005). Consuming bioactive substances, including the phenolic chemicals present in fruits and vegetables, has been related to a reduction in the risk of chronic non-communicable illnesses. According to (WHO 2022), of all deaths worldwide (41 million per year), 71% are attributable to chronic non-communicable diseases, which include cancer, diabetes, chronic respiratory diseases, neurodegenerative illnesses and cardiovascular diseases. Almost three-quarters (32 million) of these deaths take place in low- and middle-income nations (WHO 2022). In addition, mortality from chronic non-communicable illnesses affects people between the ages of 30 and 69, meaning they do not just affect the elderly (WHO 2021). The scientific community has concentrated its efforts on looking for

* Corresponding author E-mail: nur.alim.bahmid@brin.go.id

DOI: 10.1201/9781003315100-13

Table 13.1 Natural bioactive compounds sourced from food materials

Bioactive compounds	Food source
Carotenoid	Margarine, vegetables, carrots, oranges, apricots, spinach
Vitamin E	Sunflower, soya, corn, olive oil, rapeseed oil
Ascorbate acid	Oranges, citrus, lemons, strawberries
Riboflavin	Dairy milk, cheese, eggs, chicken breast, salmon
Selenium (Se)	Seafood, Brazil nuts, chicken, brown rice
Copper (Cu)	Potatoes, grains, beans,
Zinc (Zn)	Seafood, legumes, eggs, chicken
Protein	Animal-based foods and plant-based foods
Biogenic amine	Cheese, sausage, fermented vegetables, wine, fish
Phenolic compounds (tyrosol, vanillic acid, thymol, carvacrol, gingerol, isothiocyanate)	Olive oil, vanillin, thyme oil, ginger oil, broccoli, mustard seeds
Polyphenolic compounds (Flavonoid, lignans, stilbenes)	Berries, herbs, cocoa powder, flaxseeds
Tannin (catechin, gallic acid)	Tea, fruits, vegetables
Omega–3 fatty acids	Salmon, other seafood, flaxseeds, chia seeds, canola oil
Peptides	Fish meat, milk, maize, animal, egg, soybeans, wheat

bioactive substances in the plant kingdom that might prevent or lessen the incidence of chronic non-communicable illnesses in light of this alarming situation. Bioactive substances are able to modify metabolic functions and exhibit advantageous traits such as antioxidant activity, receptor activity inhibition, enzyme induction or inhibition, and induction and inhibition of gene expression. The bioactive compounds provide health advantages that surpass their fundamental nutritional worth discussed in this chapter.

The bioavailability and biological qualities of bioactive substances are influenced by the variety of chemical structures, but the bioavailability of certain compounds can be reduced by anti-nutritional factors that also stop digestive enzymes (Santos et al. 2019). The digestive process impacts bioactive compounds and their stability, which in turn affects the bioavailability and possible health benefits. Potential health effects of bioactive compounds and nutrients are therefore reliant on the digestion process (Acquah et al. 2021; M. Wang et al. 2022). Although many active compounds are water-insoluble and have limited mucosal penetration and cell absorption capacities, they may nonetheless have low bioavailability. Highly hydrophobic chemicals can be made more bioavailable and bioactive by being encapsulated in hydrophilic polymer nanoparticles, which increases their solubility in aqueous solution

(Gómez-Guillén and Montero 2021). This chapter shows the development and roles of encapsulation in improving stability, bioavailability and solubility so that bioactive compounds can have better potential health effects on human health.

The creation of functional meals can therefore be aided by an awareness of the roles that influence the variables at play in the modification of bioactive substances while also optimising their physiological usage to support possible health advantages. This chapter presents the role of bioactive compounds from food in health management and recent advancement in bioactive compounds for the advancement of functional foods as a current and future market trend. New technology in controlled delivery for bioactive compounds with nanoparticles or nanoencapsulation is also discussed in this chapter by outlining the role, potential and challenges for new functional foods for human health.

13.2 ROLE OF BIOACTIVE COMPOUNDS FROM FOOD IN HEALTH MANAGEMENT

Bioactive constituents are one of the ingredients suitable for dietary interventions to prevent non-communicable illnesses. Bioactive components that are reported to be able to prevent chronic illnesses, like non-communicable diseases, include flavonoids (Tung et al. 2020), organosulfur (Catanzaro et al. 2022), dietary fibre (Nepali et al. 2022) and plant sterol (Khandelwal et al. 2013). Non-communicable diseases are caused by a confluence of environmental, genetic, behavioural and physiological variables. Diabetes mellitus, chronic respiratory diseases, cardiovascular illnesses and malignancies are the most common forms of non-communicable diseases (NCD) (WHO 2021). A healthy diet is essential in preventing and treating NCDs because it plays a role in these diseases' genesis, development, and therapeutic interventions (Ribeiro et al. 2019). Based on these findings, the current population's consumption of foods containing bioactive components is beginning to become a trend. As a result, we gathered information about bioactive constituents from food and their relationship to health in this section, which summarises the most recent progress in bioactive compounds for disease prevention and clinical trials.

13.2.1 Bioactive Compounds in Disease Prevention and Health Promotion

Consumers, healthcare professionals, regulatory agencies, and food manufacturers and distributors are all affected in some way by the role that bioactive components play in human nutrition, which is one of the most important fields of inquiry currently being conducted (Visen et al. 2022). Nutritionally valuable and physiologically or pharmacologically active compounds can be found in foods obtained from animals or plants. These compounds will be only helpful for preventing and treating diseases after their safety, bioavailability and health effects have been validated. Eating a diet high in vegetables and fruits is strongly related to a lower risk of developing chronic diseases (R. H. Liu 2013). This confirms that alterations in dietary practices and lifestyle choices, such as elevating one's intake of fruits and vegetables and engaging in more physical activity, are effective methods for lowering the prevalence of chronic diseases.

The bioactive components from fruits and vegetables are reported to have nutraceutical activity. Various types of fruits and vegetables contain a variety of bioactive components and nutrients, such as phytochemicals (carotenoids, flavonoids, alkaloids and phenolic acids), minerals (magnesium, calcium and potassium), vitamins (pro-vitamin A, folate and vitamin C) and fibre (R. H. Liu 2004). Bioactive compounds are found not only in fruits and vegetables but also in other types of food like spices, fermented food, fish product, cereals, honey, herbs, tea, coffee, chocolate, legumes, hydrocolloid and oils. Fermented foods, employed as carriers for probiotic bacteria, contain bioactive peptides, playing a significant part in the metabolic activities of living humans due to their diverse biological functions (Marco et al. 2017a). Herbal teas and beverages contain high concentrations of naturally occurring bioactive chemicals, including saponins, terpenoids, alkaloids, coumarins, phenolic acids, polyacetylenes, carotenoids and flavonoids (Chandrasekara and Shahidi 2018). Other products that also contain bioactive compounds are spices (Wani et al. 2022; Yashin et al. 2017), grains (Saini A et al. 2019) and legumes (Conti et al. 2021). The roles of those compounds in disease prevention and health promotion are described in Figure 13.1 and explained in the following sections.

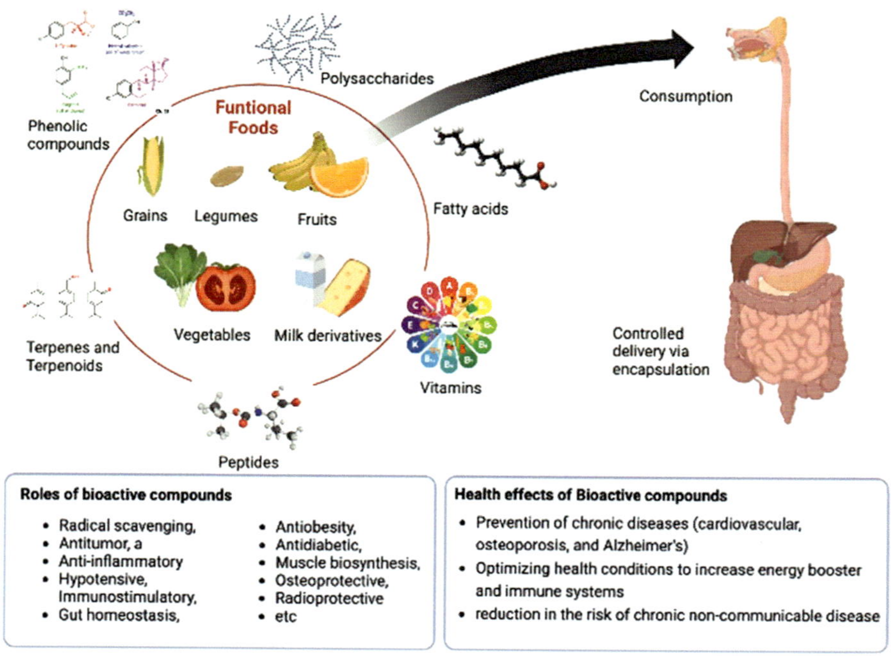

Figure 13.1 Roles and health effects of consumption of functional foods containing bioactive compounds (this figure was created with BioRender.com)

13.2.1.1 Phenolic Compounds

A variety of subgroups can be assigned to phenolic compounds based on the chemical structures of these compounds; however, the major categories include flavonoids, stilbenes, tannins, phenolic acids, carotenoids and lignans, mostly found at high concentrations in vegetables, tea, red wine, fruits, coffee, chocolate and grains (Schmidt et al. 2019). Phenolic acids are regarded as one of the most vital classes of bioactive molecules belonging to phenolic compounds. Phenolic acids are secondary aromatic metabolites that give foods their colour, flavour, astringency, harshness and other typical organoleptic characteristics. These chemicals might have a particular phenolic molecule or a complex high-molecular-mass polymer as their structure (Banwo et al. 2021).

Flavonoids and carotenoids are the most reported phenolic compounds with evidence from epidemiological studies (Andarwulan et al. 2021). According to the properties of the aglycones, flavonoids have five major subclasses: anthocyanidins, flavones, flavanones, flavonols and flavan-3-ols (Banwo et al. 2021). At high concentrations, flavonoids have effects on human health, due to their capability to scavenge free radicals, their capacity to fight oxidation to ward off critical diseases, such as cancer, and their capability to reduce viral infections (Natalello et al. 2020). Carotenoids are a class of lipid-soluble phytochemicals that are found in high concentrations in algae, microorganisms and plants. Examples of carotenoids include alpha-carotene, beta-cryptoxanthin and beta-carotene. Each of these carotenoids performs the role of pro-vitamin A and acts as a precursor for vitamin A in the human body. Recent years have seen a surge in research into the health benefits of carotenoids, which have shown promising results in studies of their effects on a variety of systemic ailments and eye problems, as well as in the protection of the retina from the effects of phototoxic light (Álvarez-Barrios et al. 2021). Carotenoids may protect against certain types of cancer by inhibiting the uncontrolled proliferation of cells and improving communication across gap junctions (Tanaka, Shnimizu and Moriwaki 2012). Carotenoids may also help prevent heart disease by preventing the production and oxidation of low-density lipoprotein, causing heart disease (Yoshida et al. 2010). In general, the consumption of high phenolic compounds in food is related to chronic disease prevention because of their antioxidant activity with a variety of mechanisms: gene expression, cellular signalling and enzymatic activity (Milenkovic et al. 2011). Phenolic acids also have important functional properties with health benefits such as cardioprotective, anti-thrombotic, anti-cancer, anti-microbial, anti-atherogenic, immunoregulatory, anti-inflammatory, anti-oxidantant and anti-allergenic properties (Rashmi and Negi 2020).

13.2.1.2 Terpenes and Terpenoids

Terpenes are organic compounds formed from five-carbon isoprene building blocks, resulting in cyclic molecules. Terpenes are mostly used in cosmetics, synthetic chemicals, rubber products, beverages, perfume and the biofuel industry. Terpenoids are secondary metabolites derived from terpenes, containing oxygen and cyclic groups (Banwo et al. 2021). Terpenes have a reputation for being volatile, but terpenoids are often either non-volatile or volatile. Because of their volatile nature, terpenes and terpenoids are frequently used in the flavouring and scent industries. Terpenes and terpenoids can be found in various foods, including aromatic plant resources, such as basil, cannabis, green tea, mangoes, pepper, lemon and orange. Tetrahydrocannabinol (THC), cannabigerol (CBG) and cannabidiol (CBD) are the important oils that are produced by the crucial components known as terpenes and terpenoids. The CB groups play a role in blocking hepatic carcinogenesis with aflatoxin, helping in retaining memory loss by inhibiting acetylcholinesterase, and having analgesic and anti-inflammatory activity. The THC groups increase gastro-oesophageal reflux, upregulate the gamma (γ)-aminobutyric acid compositions, can be a local anaesthetic or sedative, and have anti-dermatophy properties (Banwo et al. 2021). Terpenes and terpenoids are the largest class of chemical compounds formed in the essential oils of different plants and play an important part in reducing the risk of developing cancer. Terpenes and terpenoids are also known to inhibit the growth of cancerous cells (Masyita et al. 2022; Silva et al. 2021).

13.2.1.3 Vitamins

Vitamins are a type of micronutrient found in food and are known to play a crucial role in retaining human health. Vitamins are divided into two categories: those that are fat-soluble and those that are water-soluble. Both types of vitamins are essential to the body and serve various purposes. Vitamins dissolved in fat are absorbed and distributed throughout the whole body, which is essential for proper cell membrane function. On the other hand, vitamins dissolved in water are known as coenzymes in metabolic reactions carrying electrons and chemical groups (Morris and Mohiuddin 2022; Freeland-Graves and Bavik 2003). Vitamin A is essential to many bodily processes, including sight, the expression of genes, reproduction, the development of embryos, growth and immunological function.

The enhanced intestinal calcium absorption that results from vitamin D's contribution to bone production and resorption is the key role that vitamin D plays in the process. In the small intestine, vitamin D from food is readily absorbed alongside the fat that comes from food. Vitamin D deficiency is associated with the risk of hypertension or cardiovascular events, a greater incidence of malignancies, more musculoskeletal pain or migraines, and neuropsychiatric diseases such as schizophrenia, dementia or depression, as proven by observational studies (Weydert 2014).

Vitamin E is a dietary antioxidant that is fat-soluble and has a role in healing plasma membranes. The majority of vitamin E can be found in cooking oils, with the greatest concentration in plants with high total lipid content. Tocopherols and tocotrienols are both components of vitamin E, which belong to a class of vitamers known collectively as tocols. Tocopherol and tocotrienol are two compounds that are considered to be powerful antioxidants due to their ability to use as scavengers for lipid peroxy radicals. Tocols perform the role of scavengers, inhibiting the production of chemical and singlet oxygen species, which in turn disrupts the process of lipid peroxidation, allowing them to perform an antioxidant function (Jain et al. 2022). Additionally, tocopherol and tocotrienol have a part in the prevention of obesity. Those who are obese have lower amounts of the antioxidant tocopherol than those who are of a healthy weight, indicating a vitamin E deficit.

Vitamin C is one of the water-soluble vitamins that exist in many foods like fruits and vegetables. There are two primary forms of vitamin C: ascorbic acid and dehydroascorbic acid. L-ascorbic acid can also be produced by eukaryotic cells, plants, fungi and most animal species. A recent comprehensive study on the connection between vitamin C and heart health concluded that further vitamin C supplementation does not correlate with an increased risk of cardiovascular disease (CVD) in communities where there is already an appropriate consumption of vitamin C (Moser and Chun 2016). However, if the nutritional supply was below par, there was an increased risk of cardiovascular disease (Moser and Chun 2016). In contrast to the beneficial effects that were observed, the risk of cardiovascular disease mortality increased in postmenopausal women who had diabetes when they used vitamin C supplements (D. H. Lee et al. 2004). This suggests that cases with specific pathologies should be considered separate from the general population. Numerous clinical studies have also used vitamin C interventions concerning specific outcomes in patient groups, such as HIV, type 2 diabetes, depressive disorders and end-stage renal diseases. These intervention trials frequently used vitamin C administration in combination with other antioxidant vitamins (most frequently vitamin E and beta-carotene) or as part of a multivitamin–mineral mix, both of which might confuse the interpretation of good findings (Granger and Eck 2018).

13.2.1.4 Peptides

Bioactive peptides produced from food proteins are another example of naturally occurring compounds that have the ability to prevent or alleviate the symptoms of chronic diseases (Cicero, Fogacci and Colletti 2017). Biologically active peptides that participate in a variety of biological processes have been isolated from various dietary proteins, including those found in milk, eggs, fish, poultry and legumes (X. Cao, Liao and Wang 2022). Bioactive peptides are beneficial ingredients to develop functional and nutraceutical food products and have the potential to be used in preventing chronic diseases (X. Cao, Liao and Wang 2022). This potential is related to chronic disease enzymes, antihypertensive antioxidant activity and anti-inflammatory, osteoprotective, antibacterial and opioid properties (Chakrabarti, Guha and Majumder 2018). Furthermore, food-derived peptides can inhibit α-glucosidase, an enzyme that plays a function in the metabolism of carbohydrates, so food-isolated peptides have anti-diabetic properties (Yutang Wang et al. 2020).

13.2.1.5 Fatty Acids

Bioactive lipids are lipid molecules derived from plants, animals or microbes with functional and nutraceutical properties. These lipids can be found in foods. Alterations to the production process might be made in order to increase the concentration of a certain fatty acid, get rid of certain components or enhance particular physicochemical properties (Espero, Mendoza-sa and Garcı 2022). Omega-3 fatty acids (FAs) are polyunsaturated fatty acids whose first double bond is positioned at carbon 3 from the hydrocarbon chain's methylene group. The FAs of the n-3 and n-6 families are on the list of necessary nutrients for humans, compounds that cannot be produced and must be consumed (B. Chen, McClements and Decker 2013).

Cholesterol and triglyceride levels can be lowered by n-3 fatty acids, while high-density lipoprotein cholesterol levels can be raised. Since n-3 FAs play various roles in brain function, their protective effects on the cognitive level of older persons, notably those with Alzheimer's disease (AD), have also been evaluated. Alzheimer's disease (AD), depression (depression), schizophrenia (schizophrenia), hyperreactivity (hyperreactivity), multiple sclerosis (MS), ischaemia (brain damage) and peroxisomal disorders (brain cell dysfunction) have all been linked to decreased DHA levels in brain tissue, and DHA supplementation has been shown to ameliorate these conditions (Espero, Mendoza-sa and Garcı 2022).

Conjugated linoleic acid (CLA) is the name given to a group of at least 28 fatty acids that can be found in the meat and dairy products of ruminants, such as cows, goats and sheep (Espero, Mendoza-sa and Garcı 2022). In experimental animals, CLA treatment normalises dyslipidemias and lowers the number of atheromatous plaques. Accordingly, it has been observed that administering 1% CLA as a dietary supplement to hypercholesterolemic rats lowers the amount of cholesterol found in the lipoprotein fractions and prevents the development of atheroma in the aorta (Espero, Mendoza-sa and Garcı 2022). The usage of CLA in individuals with type 2 diabetes indicated that when the intake was 6 g/day for eight weeks, it generated a substantial drop in the concentration of fasting glucose, leptin, body mass index and weight. This was discovered through the study's findings on the antidiabetogenic effects of CLA. Consumption of CLA during childhood has been demonstrated to improve glucose intolerance, which is connected with diabetes. The glucose tolerance is normalised and the hyperinsulinemia is reduced owing to the CLA supplement (Espero, Mendoza-sa and Garcı 2022).

Saturated fatty acids with 6–12 carbon atoms are known as medium-chain fatty acids (MCFAs) (Espero, Mendoza-sa and Garcı 2022). These fatty acids include hexanoic or caproic acid (C6:0), octanoic or caprylic acid (CA) (C8:0), decanoic or capric acid (C10:0), and dodecanoic or lauric acid (C12:0) (Espero, Mendoza-sa and Garcı 2022). Patients who suffer from fat malabsorption, poor digestion and metabolic difficulties related to cystic fibrosis, Crohn's disease, colitis and enteritis may benefit from the use of medium-chain fatty acids (MCFAs) and medium-chain triacylglycerides (MCTs) in enteral and parenteral nutrition (Espero, Mendoza-sa and Garcı 2022). MCFAs can lead to a decrease in total body weight and total body fat, as well as a reduction in lipoprotein production and a dampening of the postprandial response to TAGs. In addition, its consumption, even in modest amounts and in conjunction with low-fat diets, brings about a faster and more significant decrease in serum lipid levels than the consumption of monounsaturated or polyunsaturated fatty acids (Esperón Rojas et al. 2017).

One category of fatty acids called short-chain fatty acids (SCFAs) includes the acids acetate (C2), propionate (C3) and butyrate (C4). These fatty acids have been demonstrated to exhibit functional activity. SCFAs could be obtained through the consumption of foods that contain SCFAs. The fermentation process that takes place during the manufacturing of some foods and beverages, such as yoghurt and cheese, may cause the presence of short-chain fatty acids in the final product (Espero, Mendoza-sa and Garcı 2022). The blend of SCFAs found in bovine milk, which contains roughly 5–10% butyric acid (BA) and 3–5% caproic acid, makes bovine milk a valuable component of a healthy diet (K. Lee and Akoh 1998). SCFAs are essential to the upkeep of a healthy colon and are thought to offer protection against colorectal cancer. Nevertheless, they can also modulate immunological responses, which is the primary reason for their significance in various chronic inflammatory illnesses (Espero, Mendoza-sa and Garcı 2022). Butyrate is the short-chain fatty acid (SCFA) that is produced by the intestinal microbiota. It has received the most attention and has been the subject of the most research because it appears to exert broad anti-inflammatory activity (Espero, Mendoza-sa and Garcı 2022). It does this by affecting immune cell migration, adhesion and cytokine expression, including IL-8, IL-17, IL-1b, IL-6 and TNF-a, as well as cellular processes such as proliferation, activation and apoptosis, with particular emphasis on its significance in protecting against colon cancer and colitis (Gill et al. 2018).

13.2.1.6 Polysaccharides

A polysaccharide is a type of significant biomacromolecule that can be found in the cells of a wide variety of organisms, including plants, animals and microorganisms. With the advancement of research in molecular biology, it has gradually been found that polysaccharides from various sources have a wide range of complex biological activities. As a result, this discovery has attracted a great deal of attention in medicine and has become one of the most active areas of life research (Siddiq et al. 2013). Polysaccharides have been proven to have a variety of physiological roles, including anti-tumour, anti-virus, hypoglycaemic, anti-oxidation and immunological regulatory effects, according to research (Zhou, Huang and Huang 2022). Onion polysaccharide and its derivatives had strong antioxidant activity, as measured by their ability to scavenge superoxide anion, hydroxyl radicals and lipid peroxide, as well as their capacity to reduce lipid peroxidation (Zhou, Huang and Huang 2022).

Several studies have been done to look at the potential bioactivity of polysaccharides in Ginkgo biloba leaves. The polysaccharide in Ginkgo biloba leaves demonstrated a substantial DPPH radical and OH radical scavenging ability, which could be attributed to glucose, mannose, rhamnose and xylose (Yun Wang et al. 2018). The use of many purified herbal antioxidant polysaccharides as functional antioxidant components and nutraceuticals in numerous meals and medicinal items is becoming increasingly common (Mohanta et al. 2022). They eliminate the potentially harmful effects of oxidative stress by acting as free radical scavengers in living organisms (Mohanta et al. 2022). The antioxidants that were produced by the leaves were shown to have protective properties against free radicals. Antioxidant polysaccharides derived from herbal sources have already earned particular acclaim due to their ability to scavenge free radicals, inhibit the peroxidation of lipids and prevent the induction of brain damage (Mohanta et al. 2022).

Polysaccharides are also extracted from the hull of flaxseeds (*Linum usitatissimum* L.) (Tian et al. 2020). Mannose, rhamnose, galactose, glucose, galactose, xylose, arabinose and fucose were discovered as monosaccharide compositions in flaxseed polysaccharides (Mohanta et al. 2022). Some flaxseed polysaccharide fractions exhibited high antioxidant activity, elevated levels of nitric oxide and enhanced cytokine production (TNF-α and IL-6) (Mohanta et al. 2022). There is potential for flaxseed polysaccharides to be used as natural antioxidants and immune-boosting dietary supplements (Mohanta et al. 2022).

The antioxidant potential of polysaccharides derived from Chinese watermelon was examined. It has been claimed that Chinese watermelon polysaccharides contain six monosaccharides: glucose, galactose, mannose, xylose, arabinose and rhamnose (Liu et al., 2018). The polysaccharides possessed the β-glycosidic bond. In addition, the isolated polysaccharides were discovered to have a high ability to scavenge superoxide anions (Mohanta et al. 2022). Additionally, a polysaccharide isolated from guava leaves exhibited anti-diabetic action (Luo et al. 2019).

13.2.2 Clinical Evidence of Bioactive Compound from Food in Health Management

13.2.2.1 Clinical Evidence of Bioactive Compounds in Cardiovascular Disease

Consumption of bioactive compounds has been linked to a reduced risk of cardiovascular death and new events. The clinical data has already been gathered to understand more about how a bioactive molecule could reduce the risk of cardiovascular disease (CVD). Thus, we have collected some evidence of the clinical trial of bioactive compounds in cardiovascular disease. Related to the bioactive components of phytochemicals, epidemiological studies showed that a higher intake of dietary (poly)phenols, particularly flavonoids, is related to a lower risk of mortality from all causes and cardiovascular disease (X. Wang et al. 2014).

Phenolic chemicals can potentially suppress the activity of digestive enzymes (amylases and lipases), resulting in decreased absorption of dietary carbohydrates and fats in the intestines. This activity may explain why (poly)phenols influence glucose and lipid metabolism, particularly after a meal in the postprandial phase (Vetrania et al. 2020). In addition, it has been demonstrated that (poly)phenols can influence lipoprotein composition by regulating important enzymes that are involved in the dynamic exchange of lipids between the various lipoproteins found in plasma (LPL, CEPT), as well as those enzymes that are related to the metabolic processes that take place in the liver (fat oxidation, lipoprotein secretion, etc.) (Pepa et al. 2020). On the other hand, it has been found that ingestion of (poly)phenols, notably anthocyanins and flavan-3-ols, results in an increase in insulin sensitivity. This is the conclusion that has been drawn from a study. According to in vitro studies, this effect may be related to (1) the reduction of cell stress, which might translate into β-cell preservation, and (2) the enhanced peroxisome proliferator-activated receptor gamma (PPAR) expression and activated glucose transporter protein 4 (GLUT-4) expression, which might yield to an improvement of adipose tissue insulin sensitivity, the modulation of adipogenesis and increased glucose uptake in adipose tissue and muscle (B Scazzocchio et al. 2011).

A meta-analysis of prospective cohort studies discovered that the intakes of anthocyanidins, proanthocyanidins, flavones, flavanones and flavan-3-ols were inversely associated with the risk of cardiovascular disease (CVD) when comparing the highest and lowest categories of intake (X. Wang et al. 2014). This was the case for all of the flavonoids as well. The meta-analysis of 13 prospective cohort studies with a total of 344,488 participants and 12,445 occurrences of cardiovascular disease demonstrated a correlation between flavonol consumption and the risk of cardiovascular disease. According to the findings of the dose-response study, increasing one's daily flavonol consumption by an average of 10 milligrams was connected with a 5% lower risk of cardiovascular disease (X. Wang et al. 2014). These findings lend credence to the recommendations that higher consumption of foods rich in flavonoids be made to lower the risk of cardiovascular disease (X. Wang et al. 2014).

A vitamin C shortage has been connected to an increase in plasma levels of LDL-cholesterol and its oxidation, both of which are well-known for the antioxidant ability that vitamin C possesses and its involvement in oxidoreductive and hydroxylation reactions that take place in the body (Chambial et al. 2013). Meanwhile, it has been demonstrated that taking in a suitable amount of it or supplementing with between 500 and 1000 milligrams per day can lower amounts of oxidised LDL and lipid peroxidation (Chambial et al. 2013). As a result, it contributes to a reduction in the process of atherosclerosis. Consuming a proper amount of it also strengthens the vascular system by increasing the production of collagen, a structural protein that protects the integrity of the vascular endothelium and other tissues (Chambial et al. 2013).

13.2.2.2 Clinical Evidence of Bioactive Compounds in Cancer

The bioactive molecule curcumin has been proven in several different clinical investigations to have the potential to boost the sensitivity of cancer cells to the chemotherapeutic medications that are being used to treat them. Beneficial effects have been observed in pancreatic and breast cancer patients when doses ranging from 6000 to 8000 milligrams were administered. These benefits include a reduction in the inflammatory state (caused by a lower level of NF- κ β expression), a reduction in tumour size and an increase in the profile of anti-inflammatory cytokines (Bayet Robert et al. 2010). In addition, supplementation with 360 milligrams three times daily increased the expression of the p53 protein, an essential protein in regulating the cell cycle and induction of programmed cell death (Z. He et al. 2011). This led to an increase in the amount of apoptosis that was induced in patients with colon cancer (Z. He et al. 2011). In addition to this, it decreased TNF-α levels by enhancing the patients' overall inflammatory condition (Z. He et al. 2011).

Clinical trials utilising dietary bioactive compounds against breast cancer are still in the early stages; however, an overwhelmingly large number of breast cancer-fighting compounds are currently in the process of being tested in humans. The efficacy of numerous bioactive chemicals found in foods, such as berberine, curcumin, EGCG, genistein, green tea, lycopene, quercetin, resveratrol and sulforaphane, has been demonstrated in preclinical research (Ganesan, Du and Chen 2022). The consumption of approximately ten cups of green tea per day has been shown through epidemiological research to have chemopreventive benefits for women who have breast cancer (Yuan 2013). The intake of resveratrol (50 mg/ twice a day, for 12 weeks) reduced methylation of Ras association domain family 1 isoform A, a gene linked with breast cancer. The participants in the study were 39 women with a high risk of developing breast cancer. The results showed that increased trans-resveratrol and resveratrol-glucuronide levels in the blood circulation suppressed the expression of cancer-promoting PGE2 in the breast tissues (Zhu et al. 2012).

13.3 RECENT ADVANCEMENTS IN BIOACTIVE COMPOUNDS FOR THE DEVELOPMENT OF FUNCTIONAL FOODS

Functional foods have benefits for human health due to their high nutritional content and bioactive compounds. Bioactive compounds are substances with the power to influence calorie intake while reducing metabolic disorders (Niwano et al. 2022), oxidative stress and proinflammatory states. Epidemiological studies demonstrate that eating foods high in bioactive components with antioxidant activity, such as phenolic compounds, phytochemicals and vitamins, can prevent clinical diseases such as stroke, cancer, age-related functional decline, cataracts and heart disease (Samtiya et al. 2021). One of the most common and well-studied phytochemicals in the plant world is flavonoids, which are also known as polyphenols. In vitro and in vivo studies have connected long-term intake of dietary phenolic acid and flavonoids to a decreased risk of acquiring chronic non-communicable illnesses. Some phenolic acids and flavonoid subclasses have been associated with the ability to inhibit cancer cell growth in vitro, reduce vascularisation, protect neurons, promote vasodilation and increase insulin production (Sánchez et al. 2018; G. Li et al. 2020; Batra and Sharma 2013). The development of functional foods containing bioactive compounds can be seen in Table 13.2.

13.3.1 Fermented Food Products

Foods that have been fermented are those produced through regulated microbial growth and enzymatic transformations of major and minor food components (Marco et al. 2017b). Fermentation is possible with a wide variety of foods, such as cereal-based goods, dairy, fruits and vegetables, plants, and meat and fish, and it can take place in any environment. During the process of fermenting these foods, several different biochemical reactions are triggered by multiple microorganisms. These reactions result in the release of various substances, some of which may be beneficial to

Table 13.2 Recent advancement of bioactive compounds and their application in functional foods

Food products	Applied functional foods or media	Study type	Bioactive compounds	Functions for health	Reference
Fermented products	Wheat bran fermented with Eurotium cristatum (*E. cristatum*)	In vivo	Polyphenol, anthocyanin, soluble dietary fibres	Enhancement of antioxidant and anti-obesity activities	(Lu et al. 2022)
	Fermented milk with *Lacticaseibacillus casei* SJRP38 and buriti pulp	In vivo	Phenolic compounds	Increase in fermented milk's nutritional value	(Borgonovi, Casarotti, and Penna 2021)
	Fermented dairy yoghurt	-	Bioactive peptides	Higher bioactivity	(Nielsen et al. 2022)
Plant-based products	Cooked beans adzuki	In vitro	p-courmaric acid, sinapic acid, and syringic acid	Digestible fraction showing higher health-promoting activities	(Sangsukiam and Duangmal 2022)
	Soybean *chhurpi*	In vitro	Glycinin-derived peptide	*In silico* antihypertensive property, potential ACE-inhibitory and antioxidant activity	(Chourasia et al. 2022)
	Virgin olive oil	In vivo	Polyphenols and triterpenic acids	Improvement in Matsuda index of insulin sensitivity, insulinogenic index and gastric inhibitory polypeptide (GIP)	(Sanchez-Rodriguez et al. 2021)
	Melon peels		Flavonoids, hydroxybenzoic acids, β-Carotene and β-cryptoxanthin	A high antioxidant activity	(Gómez-García et al. 2021)
	Coffee (Coffea arabica L.) pulp aqueous extract	In vitro	Phenolic compounds, chlorogenic acid, caffeine and anthocyanin	Higher antioxidant and anti-microbial properties	(Khochapong et al. 2021)
	Eugenia pyriformis fruit	In vitro	Phenolic acid and flavonoids	A higher antioxidant capacity of the seed extract in the intestinal phase	(de Paulo Farias et al. 2021)
	Food matrices (fish soup apple juice)	In vitro	Fish protein hydrolysates	Improved bioactivity to reduce health concerns, e.g., cardiovascular risk problems, due to the antihypertensive and antioxidant activity	(Rivero-Pino, Espejo-Carpio, and Guadix 2020b)
	Sea cucumbers	In vitro	*H. forskali* hydrolysate protein, Selenium, Iodine, total polifenols and antioxidants	A high antioxidant and antihypertensive activity	(García et al. 2019)
	Eggs	In vitro	linolenic fatty acid and omega-3 fatty acid	A high bioaccessibility of bioactive molecules in the human digestive system	(Nolasco et al. 2021)

(*Continued*)

Table 13.2 (Continued)

Food products	Applied functional foods or media	Study type	Bioactive compounds	Functions for health	Reference
Processed food products	Mayonnaise	-	Lipophilic extracts from cranberry bush and sea buckthorn containing fatty acids, triacylglycerols, tocols and phytosterols	Improvement of oxidative stability	(Dienaitė, Baranauskienė, and Rimantas Venskutonis 2021)
	Dense foods, such as pasta; enriched with moringa seed powder	In vivo	Thiamine, riboflavin, γ- aminobutyric acid, glucosinolates	Improvement in nutritional and functional properties	(Coello et al. 2021)
	Resistant starch	In vitro	Ferulic acid	A high anti-cancer and antioxidant potential	(Noor et al. 2022)
	Sardine hydrolysates	In vitro	Peptides	Antihypertensive and antioxidant capacity	(Rivero-Pino, Espejo-Carpio, and Guadix 2020a)
Algae-based products	Marine algae	In vitro	Phenolic compounds	Showing bioactivities such as antioxidant, antitumoral or anti-microbial.	(Jimenez-Lopez et al. 2021)
	Microalgae	In vitro	Selenium	Alternative selenised food for humans, due to their high Se bioavailability	(Gómez-Jacinto et al. 2020)

human health, such as vitamins, amino acids, enzymes, exopolysaccharides (EPS), short-chain fatty acids (SCFAs), organic acids, phenolic compounds and bioactive peptides (Diez-Ozaeta and Astiazaran 2022). Furthermore, fermented foods can modify the intestinal microbiota composition due to the presence of microorganisms and the raw material, which is a source of bioactive substances. The intestinal microbiota is one of the components of the intestinal barrier and plays a role in the process of shaping the integrity of the barrier. Therefore, it is reasonable to expect that including fermented goods in one's diet will assist in reducing the negative effects of the Western diet (dysbiosis), as well as being a potential factor in the prevention and treatment of inflammatory disorders (Kocot and Wr 2021).

Different nutritional and medicinal benefits may be derived from fermented foods and drinks. The beneficial health effects of fermented milks and associated products are mostly determined by lactic acid bacteria (LAB) (Mathur, Beresford and Cotter 2020). It is well known that probiotic dairy foods include *Lactobacilus acidophilus* and *Bifidobacteria spp.* Gokavi et al. (2005) discussed a fermented oat beverage as an example of a functional food, and its classification may be due to the beverage's interactions with dependent probiotic starter cultures and the prebiotic fibre β-glucan. Angelov, Yaneva-Marinova and Gotcheva (2018) found that the amount of β-glucan, which gives oats their high level of stickiness, has been fundamentally linked to the health advantages of oats. Both the blood cholesterol levels and the rate at which the intestines absorb glucose can be reduced by the β-glucan with reasonable success. Aside from milk or other aqueous formulations, oats can be processed to create functional beverages because of their high dietary fibre and protein content. This mixture is advertised in the market as a supplement food that is quick, easy and healthy for people to use. Food fermentation has further advantages, including reduced energy consumption during matrix processing, desirable biochemical changes for nutrient enhancement, food safety, increased product shelf life and better sensory qualities.

Yoghurt is now viewed as a probiotic food carrier. Numerous food powders, including yoghurt powder and curd powder, are created while taking into account the number of organisms that will still be present in the final product after drying. These meals, drinks and powders have a high level of consumer acceptance due to their flavour, scent and nutritional content. Since the starting bacteria's cell wall is linked to antitumor activity, the activity endures even after drying (Shiby and Mishra 2013a). Along with lowering blood cholesterol and preventing gastrointestinal infections, fermented milks also have antimutagenic properties (Shiby and Mishra 2013b). Therefore, patients with atherosclerosis and lactose intolerance are advised to consume fermented food.

Due to their oestrogen-like properties, soybeans are one of the richest sources of isoflavones, often known as phytoestrogens. It has been shown that soybean fermentation increases the aglycone isoflavone content and hence the bioavailability of isoflavone (Z. H. Cao et al. 2019). Other fermented dishes like sufu, miso, douchi and soy sauce have been staples in Asian cuisine for a long time. Many locations in Asia have similar products to these. Consumption of these traditional fermented soy foods has been shown to promote bone health, reduce the risk of cancer and delay the advancement of diabetes, according to a number of studies (F. Jin He and Chen 2013).

However, a variety of contaminated pathways, ranging from containers used to collect raw milk ingredients to acidophilic microorganisms, pose significant health concerns (Aliyo and Teklemariam 2022). As a result, the advantages and hazards associated with fermentative and pathogenic bacteria should be thoroughly explored. Bacterial toxins are a source of worry in this area, particularly in tropical developing regions such as South Asia and Africa (Ashbolt 2004). Furthermore, oligosaccharides, tannins, saponins, protein inhibitors, lectins, cyanogenic glycosides and phytate are inherent anti-nutritional components found in the majority of raw materials used in bacterial fermentation of foods and drinks (Samtiya, Aluko and Dhewa 2020). As a result, fermented foods should be adequately checked for nutritional content and safety. Most historically fermented drinks have not been properly examined for their alcohol content levels, which are of interest to today's consumers. This information can be used to track a person's health.

13.3.2 Plant-Based Products

Plant products are generally considered safe by consumers and industry due to their natural nature. As a result, the food industry is presently focusing on the creation of plant-based food items fortified with bioactive compounds that have high nutritional value, are low in saturated fat and are loaded with useful bioactive chemicals to minimise the risk of chronic illnesses, which are known as functional foods. For functional food, a variety of food resources can be employed and included in food composition. Garlic (*Allium sativum*) is a good source of active organ sulfur contents that lower blood pressure (e.g., allicin, allylic sulfides) (Ried and Fakler 2014). Curcumin, a valuable medication with several pharmacological applications, was isolated from *Curcuma sp*, which has been utilised as an old-fashioned medicine to treat sickness in underdeveloped nations (Hatcher et al. 2008). In European countries, golden milk made from pure curcumin extract is commercially available (Idowu-Adebayo, Fogliano and Linnemann 2022). Other plant-based sources, such as maize, wheat, rice, millets and oats are excellent sources of micronutrients such as folates, vitamin E, zinc, phenolic and dietary fibre. The oxidative degradation of bioactive compounds can result in rancidity, which reduces nutritional and functional quality (F. Li, Wu and Wu 2021). The existing physical (heating and chilling processes) and chemical (salts, organic acid and ammonia) methods may have a negative impact on nutritional and overall quality (Abraha et al. 2018), while synthetic preservatives have a harmful influence on food quality, human health and the environment. Recent research findings showed the deleterious influence of synthetic preservatives on the environment, human health and food quality (Fernandes et al. 2020). Plant compounds with significant antibacterial and antioxidant activity, such as plant anti-microbial peptides (pAMPs), terpenoids, phenylpropanoids, essential oils, flavonoids, alkaloids and polyphenolics could be employed as safer alternatives compared to conventional preservatives. Furthermore, essential oils derived from aromatic plants have piqued the interest of the food industry due to their robust antibacterial and antioxidant properties (Dawidowicz and Olszowy 2014). Angelica, basil, citrus peels, lemongrass, thyme, ylang-ylang, carvone, cinnamaldehyde, citral, p-cymene, eugenol, limonene, menthol, linalool and other plant products have been designated as GRAS (Generally Recognised as Safe).

Functional foods can also be made from by-products of food processing containing high bioactive chemicals. Several articles mention that bioactive compounds can also be produced from food waste (Šeregelj et al. 2020) and by-products (Ben-Othman et al. 2020). There are various examples: fibres in melon juice peels (Gómez-García et al. 2021), coffee pulp extracts (Khochapong et al. 2021), soup apple juices (Rivero-Pino, Espejo-Carpio and Guadix 2020b) and sea cucumbers (García et al. 2019), as shown in Table 13.2. Gorinstein et al. (2001) state that the non-edible parts of fruit contain as high amounts of bioactive compounds as the edible parts. Carrots also produce approximately 30–50% of residue after processing. Carrot waste consists of carrot pomace and peels, and several researchers found through an appropriate extraction process that this produces bioactive components such as carotenoids, phenolics and flavonoids (Šeregelj et al. 2020). If using industrial waste as raw material for bioactive components, the health and safety and health of the food end-product must be considered. Waste from fresh materials such as tomatoes, carrots, apples, oranges and some dairy products and meat is often contaminated with contaminants such as pesticides, bacteria, mould and heavy metals (grape skin) (Vilas-Boas, Pintado and Oliveira 2021). Furthermore, the food and beverage business is one of the greatest producers of trash or by-products. Non-edible components of fruit and vegetable waste are produced by the beverage sector or fruit and vegetable processing, for example.

13.3.3 Processed Food Products

Minimally processed functional food items, as opposed to typical plant-based food products, are a novel type of food that combines usefulness with freshness. Functional foods with little processing are an example of sophisticated manufacturing. The functional meal idea combines ready-to-eat meal convenience with freshness and utility. In general, functional meals outperform regular functional foods in terms of nutrients. For example, mayonnaise contains a lipophilic extract from cranberry bush and sea buckthorn and contains fatty acids, triacylglycerols, tocols and phytosterols, maintaining the oxidative stability of the products (Dienaitė, Baranauskienė and Rimantas Venskutonis 2021). For dense foods such as pasta, including thiamine, riboflavin, γ- aminobutyric acid and glucosinolates can improve the nutritional and functional properties of the product and thereby have an effect on human health (Coello et al. 2021).

Resistant starch, as well as foods high in dietary fibre, are among the food components that are widely used as functional foods (Bede and Zaixiang 2021; Repo et al., 2011). These functional ingredients are used to prevent many diseases or pathological conditions such as colon cancer, diabetes, hypertension and obesity. Resistant starch is part of starch and carbohydrates in general, and resistant starch has a health function in the human body when consumed. The structural component of resistant starch is the same as raw starch, which consists of amylose and amylopectin fractions, but resistant starch has a crystalline structure that distinguishes it from raw starch.

In general, starch digestion occurs in the digestive tract from the mouth to the small intestine, but resistant starch is called resistant because it cannot be hydrolysed by digestive enzymes such as amylase, sucrose isomaltose and glucoamylase. The result of starch digestion is glucose, which can be used by the body as an energy source, but in some cases, excess glucose in the blood can have a negative effect on health, therefore resistant starch is needed to reduce the amount of glucose produced from other food sources. There are three reasons why resistant starch is said to be resistant to degradation by digestive enzymes (Bede and Zaixiang 2021). The first is the structure of resistant starch, which is more compact than other starches, resulting in digestive enzymes not having access to enter the starch granules. The second is that the design of starch granule structures are in such a way that the compounds in the enzymes cannot react with the molecules present in the starch granules; examples of resistant starch are found in natural foodstuffs, namely corn starch high amylose, potato starch and raw banana starch. The third is that starch structures that are rearranged after being heated, commonly referred to as retrograded starch, result in digestive enzymes being unable to find suitable substrates in the starch granule structure, thereby preventing enzymatic hydrolysis.

In practice, functional foods should be consumed on a regular basis to ensure the necessary daily consumption. Functional foods could be designed in modest amounts for specialised sectors such as institutional and commercial industrial catering, schools and the food services industry. The functional foods would provide benefits to specific medical problems, e.g. patients with a concern of adult obesity, liver illness, diabetes and healthcare for older people.

13.3.4 Algae-Based Products

The use of microalgae as a bulk protein source when designing functional foods has potential. Microalgae, as with plants, lack cysteine, methionine and sulphur-containing amino acids, and protein bioavailability becomes high after feeding (Caporgno and Mathys 2018). At this phase, three concepts must be mentioned: bioaccessibility, bioavailability and bioactivity. The fraction of the medication freed from the food matrix that is available for absorption is reflected in bioaccessibility, which is frequently measured using in vitro tests (Akter et al. 2021). Following that, the chemicals may enter the systemic circulation and be used, which is known as bioavailability. In vivo testing determines bioavailability. Finally, a compound's bioactivity explains its physiological reaction, such as its antioxidative, antihypertensive or anticancerogenic properties (Călinoiu and Vodnar 2018). In vitro, ex vivo and in vivo bioactivity can all be assessed. The bioavailability of protein from entire microalgae cells might be improved by disrupting cell walls, which impedes breakdown.

The antioxidant capabilities of microalgae-derived chemicals are likely the most attractive for commercial applications (Batista et al. 2013; Camacho, Macedo and Malcata 2019; Xia et al. 2021). Lipids and protein in foods are susceptible to oxidation during storage and processing, resulting in the loss of vital nutrients and the formation of potentially hazardous chemicals (Xia et al. 2021). Consumer acceptance is influenced by low molecular weight, which forms off-flavour chemicals caused by oxidation (Camacho, Macedo and Malcata 2019). Food products can deliver bioactive substances with health advantages in addition to supplying energy and nutrients for physical activity, maintenance and development. Peptides and protein are two of the most common types of molecules with industrial and biological functions (Batista et al. 2013; Xia et al. 2021). Peptides comprise around 20–30 amino acids per molecule and are inert inside the fundamental protein structures until the peptides are released into the gastrointestinal system during food processing or digestion. These unique protein fragments have the potential to improve biological functioning or circumstances, and hence influence health. Some macro and microalgae-peptides exhibit anticoagulant, hepatoprotective, immunomodulatory, antihypertensive, antioxidative and anticancerogenic properties.

13.4 NANOPARTICLES AND CONTROLLED DELIVERY FOR BIOACTIVE COMPOUNDS: OUTLINING CHALLENGES FOR NEW SMART-FOODS FOR HEALTH

Many bioactive chemicals have limited bioaccessibility and are vulnerable to harsh environmental variables such as pH, oxygen, light or enzymes, as well as low mucosal penetration and cell absorption capacity (Bazana, Codevilla and de Menezes 2019). Bioactive compounds have poor water solubility and are vulnerable to chemical breakdown and metabolism in foods and the human gut. Encapsulation is a method that traps an active component in a polymeric matrix, generating microscopic particles with the primary goal of protecting bioactive substances from environmental conditions during manufacturing and storage, hence increasing their useful life. Furthermore, these particles enhance release qualities and serve as delivery vehicles. Micro and nanoencapsulation are viable solutions to phenolic compounds' physical and chemical restrictions (Rahaiee et al. 2020). Phenolic compounds have been encapsulated into biopolymeric material to enhance their bioaccessibility after digestion and maintain their stability by protecting the compounds from external factors released in the gastrointestinal tract. Nanoencapsulation is described as a method to encapsulate bioactive compounds in polymer material at the nanoscale range (Pateiro et al. 2021). Encapsulation of bioactive compounds in hydrophilic polymer nanoparticles (NPs) is a well-recognised method for boosting the solubility of very hydrophobic substances in aqueous media and thereby improving their bioavailability and bioactivity (Martínez-Ballesta et al. 2018).

Controlled delivery of nanoparticles has demonstrated potential results for targeted bioactive compounds using polymeric formulations and liposomes and to enhance bioavailability and pharmacokinetics (Martínez-Ballesta et al. 2018). It is also useful for protecting hydrophilic compounds with high molecular weight, e.g., peptides and proteins, against proteolytic degradation in the gastrointestinal (GI) tract (Renukuntla et al. 2013). The permeability of the compounds using the mucosal barrier needs also to be improved because of the large number and large molecular size of reactive groups. Encapsulation, from a technical standpoint, also provides stability against physical degradation or the loaded chemicals against chemicals within food preparation and storage, as well as aiding in the masking of undesirable flavours. Mucoadhesive polymer nanoparticles are especially intended to avoid early leakage in a hostile stomach environment, as well as to offer prolonged targeted and release mechanisms of both hydrophilic and hydrophobic chemicals at intestinal absorption sites (Shaikh et al. 2011). Table 13.3 shows details of the controlled delivery of bioactive compounds from different nanoparticle polymers.

For controlled delivery, encapsulation can be done via nanomaterials or other materials that are able to trap and protect the bioactive compounds from external factors, as well as control compound release, as shown in Figure 13.1. Table 13.3 shows various polymers that control the delivery of bioactive compounds, zein nanoparticles, alginate, κ-carrageenan, chitosan, etc. Compared to other polymer materials, chitosan is a widely used material to encapsulate bioactive compounds for the pharmaceutical and medical industries due to lower immunogenicity, better biodegradability and greater biocompatibility (Brodaczewska et al. 2017). Chitosan with additional active sites of hydroxyl and amine groups has a high solubility in acidic solutions but is insoluble in alkaline and neutral solutions. However, this intrinsic feature makes chitosan not able to be used as a pharmacological adjuvant or carrier in vivo (Scherließ et al. 2013). To improve the chitosan properties in pharmacology, hydrogen bonding and minimal deacetylation can be applied. The deacetylated chitosan has a possible biosafety issue, amplified as the molecular weight of chitosan increases. The application of encapsulation of compounds using polymer material such as chitosan can control the delivery of encapsulated compounds in an antagonistic gastrointestinal environment (Kaushalya and Gunathilake 2022), in simulated gastric fluids and undigested protein (Zou et al. 2016) and acidic environments (Q. Chen et al. 2022).

Curcumin is a well-studied bioactive chemical for encapsulation (Zou et al. 2016; Dou, Lu and Nitin 2021; Yang et al. 2022). Curcumin has been found in clinical studies to be particularly effective against colorectal and pancreatic cancer due to its bioactive effects by disrupting multiple metabolic pathways. The effective effects of curcumin as a bioactive agent allow a potential application in pharmaceuticals, supplements and functional foods. Nonetheless, curcumin has low oral bioavailability, low chemical instability and poor solubility in water, limiting its inclusion in commercial products aimed to benefit human health. Encapsulating curcumin into colloidal delivery methods may help to overcome the hurdles. Casein micelles, zein nanoparticles, curcumin nanoparticles, colloidosomes, liposomes, nanoemulsions and emulsions have all been researched as ways to improve the oral bioavailability, chemical stability and water dispersibility of curcumin (Zou et al. 2016). Combining a curcumin-rich material with an excipient nanoemulsion constituted of dispersed lipid droplets in an aqueous medium is an alternative strategy for improving curcumin bioavailability (Gonçalves et al. 2021). To boost the bioavailability of hydrophobic bioactives, researchers have developed novel delivery technologies and food

Table 13.3 Controlled delivery of various bioactive compounds from different nanoparticle polymers

Materials	Bioactive compounds	Delivery systems	Effects on health	Challenges	Objective	Reference
Zein and Mesona chinensis polysaccharide (MCP)	Curcumin	Control resistance of digestive enzymes in the gastrointestinal tract,	High stability, antioxidant activity, vitro antitumor activity and bioaccessibility	Significantly toxic to HepG-2 cells in a dose-dependent manner	Dietary supplement	(Yang et al. 2022)
κ-carrageenan	Limonene (DL, R-(+)-Limonene)	Control secretion of DL within the gastric phase, but freely released at neutral pH within the intestinal phase	Improved photostability and thermostability of DL	pH-dependent release manner in vitro	Formulation of pharma/food products	(Fani et al. 2022)
Yeast cell microcarriers	Curcumin	Control diffusion by high viscosity of cell-based formulations	A novel cell-based carrier for high affinity binding to skin surface and enhanced permeation to the dermis section	Potential risks for human health due to nanoscale carriers may also penetrate into the skin through various pathways	Skincare	(Dou, Lu, and Nitin 2021)
Zein, polyethylene glycol (PEG), and ethylene cellulose	Curcumin and resveratrol	Controlled shell formation with ethyl cellulose and additional hydrophobic coating over the zein core	Increased bioaccessibility and intestinal permeability	Poor aqueous solubility, high sensitivity to light and heat, and constraints reducing bioavailability and nutraceutical effect	Nutraceutical carriers	(Dou, Lu, and Nitin 2021)
Bioactive oleanolic acid (OA)	β-carotene (Car)	Controlled release medium and degradation in simulated gastric fluid (SGF)	Improved water dispersibility, stability against UV, radiation, heat, ionic strength, and acidic conditions, as well as inherent hepatoprotective effects	Rapid degradation of Car/OA NPs and not suitable for long-term storage	Nutritional components in the food industry	(S. Liu et al. 2022)
Alginate/chitosan-coated zein nanoparticles	Phenolic compounds	Suitability of the polysaccharide layers to increase the encapsulation efficiency	Improving the encapsulation efficiency and bioaccessibility	No significant bioaccessibility increase was shown for flavonoid aglycones	Decreased risk of Chronic non-communicable diseases	(Carrasco-Sandoval et al. 2021)

(*Continued*)

Table 13.3 (Continued)

Materials	Bioactive compounds	Delivery systems	Effects on health	Challenges	Objective	Reference
Chitosan	Cinnamaldehyde	Control exposed abundant amino groups and an acidic environment	Overcoming tumour resistance by decreasing intracellular levels of ATP and PARP-1, thus strengthening antitumor efficiency in vivo	The complexity of nanosystem is conducive to mass production and not practical application	Chemoresistance in breast cancer	(Q. Chen et al. 2022)
Zein nanoparticles	Curcumin	Controls the presence of some pepsin in the simulated gastric fluids and undigested protein	Increasing bioaccessibility and solubilisation capacity	Low solubility in gastrointestinal fluids and necessary to include additives to prevent chemical degradation or metabolic enzymes	Clinical trials	(Zou et al. 2016)
Chitosan	Brown seaweed phlorotannins	Control active compounds in the antagonistic gastrointestinal environment	High antioxidant activities	The anti-diabetic potential of phlorotannins seemed to be not upheld by encapsulation	Functional food and nutraceutical formulations	(Kaushalya and Gunathilake 2022)

matrix design approaches. Emulsion-based delivery methods, for example, have been identified as a very effective carrier for both hydrophobic and hydrophilic bioactives.

However, nanoparticle-based drug delivery systems have a number of drawbacks and challenges (Desai 2012). For starters, because they are completely "foreign," the immune system swiftly clears them, resulting in a short circulation time duration and inadequate availability at the target place. Second, unlike native cells, these nanoparticles are incapable of detecting the disease microenvironment, resulting in decreased drug accumulation at the target site. One of the key hurdles to overcome for optimal therapeutic agent delivery to the target location is the quick clearance of nanoparticles from circulation (Zelepukin et al. 2020).

13.5 CURRENT MARKET INTEREST IN FUNCTIONAL FOODS

With the increasing consumer knowledge about the importance of health and well-being (Lahouar, Achour and Latiri 2021), the demand for functional food will also grow. Globally, the functional food market is forecasted to increase by 2.71 % in the period 2022–2027 (Mordor Intelligence 2022). IndustryARC (2022) forecasted that a $7,815 million functional food market will be reached by 2026. In the current situation, the growth of the functional food market is supported by the demands for a healthy lifestyle among consumers. Food companies will need to focus on food products and develop functional products and modifications (Khan et al. 2014; Moons, Barbarossa and de Pelsmacker 2018). Functional foods do not only aim to satisfy hunger but, more importantly, they provide nutrients in food to prevent dangerous diseases so that the physical and mental health of customers improves. The European Food Information Council (EUFIC) (2022) stated that functional foods contain active components that offer specific health benefits, including bone health, heart health, eye health and cholesterol management, through the daily intake of nutrients. Fortifying food additives and increasing demand for nutrition are anticipated to drive market growth. Food manufacturers add nutritional additives such as fibre, vitamins, minerals and omega-3 fatty acids into their product offerings to increase the nutritional content of foodstuffs impact that are marketed ("Dietary Reference Values for Nutrients Summary Report" 2017).

A desire for plant protein in new food products is driven by consumers that demand healthy, sustainable products and non-animal proteins (Tan, Nawaz and Buckow 2021). For example, spirulina as microalgae-based plant food is a source of plant protein offering health benefits (Moons, Barbarossa and de Pelsmacker 2018). Moreover, whey protein in functional food filtered from cheese whey is a possible source of nutrients because it contains healthy components such as bioactive peptides, immunome-potentiators, antioxidants and essential amino acids. Whey protein has been verified for its anti-obesity, gut homeostasis, immunostimulatory, radical scavenging, hypotensive, anti-cancer and anti-inflammatory properties. As a result, customers are increasingly finding protein components in food items, which is likely to boost the growth of the functional food industry. Healthy food promotion and environmental concerns in functional food drive marketers' willingness to promote

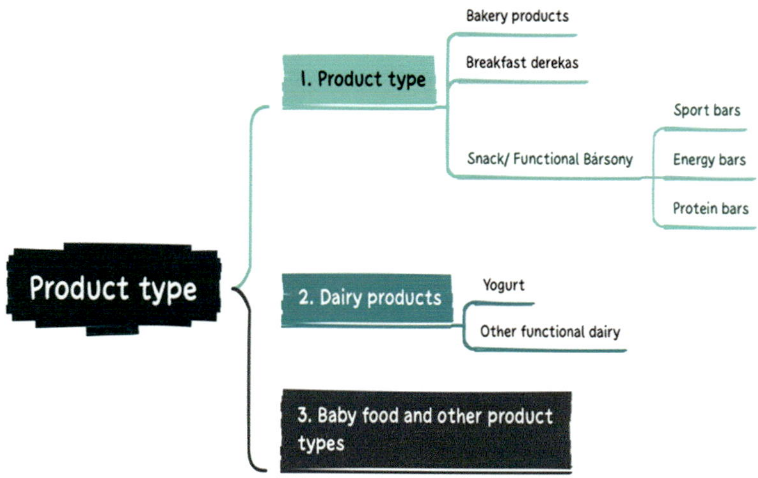

Figure 13.2 Functional food market interest based on product types

these items. As shown in Figure 13.2, the functional food market can be separated based on product type. Each product as a functional food has a different market. The functional food market based on product type can be further divided into bakery products, breakfast cereals, snack or functional bars, dairy products and baby foods. All types of functional foods are provided for the health and physiological benefits of humans (Banerjee 2019). The motivation of consumers in purchasing different types of functional foods is based on the properties of the foods in preventing chronic diseases, such as cardiovascular disease, osteoporosis and Alzheimer's disease, and also optimising health conditions to increase energy and the immune system (Khan et al. 2013). Current marketing strategies can use these insights to promote eco-friendly functional foods for consumers that favour healthy alternatives in food consumption.

The functional food market is an example of a business area founded by the intersection of food and medicines. According to market experts, functional foods will achieve double-digit growth in the future. Functional food factories need to address concerns about whether a functional food actually qualifies as such when dealing with its suppliers (Lamberti and Lettieri 2011). Most countries in the world have required the classification of functional foods to be more specific, for example in the United States, the European Union (EU), Asia Pacific, the Middle East and Africa. Functional foods were first recognised in Japan in 1980. To improve the health nation's ageing population, the Ministry of Health and Welfare arranged a regulatory system to approve certain foods with documented health benefits in Food Specified Health Use (FOSHU) (Hasler 2002). In the United States, the Committee's Dietary Goals for The United States, The National Research Council's Diet and Health, Healthy People from US Public Service emphasised the importance of consuming functional foods with fruits, high in vegetables, low saturated fat and other functional foods to reduce of chronic diseases risks (Hasler 2002). Food functional promotion in an appropriate manner will ensure that people are informed and can see the value of functional foods, which would create significant demand and opportunities to ensure public trust.

China and Brazil have large markets that revised the regulations regarding the market entry of functional foods. Chinese regulation, designed by China's Food and Drug Administration (CFDA) in 2016, requires the listing of vitamins and minerals in health foods (Shao 2017). In Brazil, the regulatory framework of functional foods published a series of six consultations in mid-2018. By having these rules and regulations, customers can trust that functional foods are safe and protected products. Functional foods are suggested in North Africa, the Middle East and Southeast Asia as a healthy way of food consumption that reduces stress and depression. Functional foods are an effective and safe strategy for reducing chronic diseases (Lahouar, Achour and Latiri 2021). In 2022, after the Covid-19 crisis, functional foods and drinks were forecasted to reach US $173.7 billion in the global market. In the United States, these numbers were estimated at US $59.8 billion, and in China, the estimated numbers were approximately US $29.8 billion.

13.6 CURRENT AND FUTURE DEVELOPMENT IN BIOACTIVE COMPOUNDS

Bioactive compounds used as functional food additives for health enhancement are being developed. Bioactive compounds generally come from secondary metabolites of natural resources, e.g., lactic acid, carotenoids, phenolic acids, flavonoids, ascorbic acid, cinnamaldehyde and vanillin (Pai, Hebbar and Selvaraj 2022). The use of bioactive compounds in functional foods is intended to be established by the inclusion of active extracts into polymer material. The current development of bioactive compounds in the food industry is the process of using more effective bioactive compound methods, such as the nanoencapsulation process to increase the effectiveness and bioactivity of bioactive components (Guía-García et al. 2022). Bioactive compound characteristics and bioactivities depend strongly on their nature. Bioactive compounds are grouped, in general, into phenolic compounds, terpenes and nitrogen compounds. The availability of natural resources makes bioactive compounds very easy to extract from source. Furthermore, raw materials especially those from food or industrial waste, can also be sources of bioactive compounds. The development of bioactive compounds from sources continues, especially the use of bioactive compounds in other sectors, such as pharmaceutical and therapeutic sectors, bioremediation and chemical sectors. It is important to note that the supply chain plays an important role in bringing functional food to the market.

Technological innovations in functional food are not only for convenience but also for modified functional food to bring various choices to reach all market segmentations. Conventional functional foods are composed of natural or whole-food ingredients that preserve functional substances, while modified functional are food products with additional ingredients for specific health functions. Plant-based food, fermented foods, processed food, algae-based products and drinks contain bioactive compounds that play an essential role in the human body against chronic illness. These product types and geographical expansions are key strategies adopted by players in the functional food market. In addition, improving the diet of the public is an essential public policy goal and drives continuous innovation in the food industry.

REFERENCES

Abraha, Bereket, Habtamu Admassu, Abdu Mahmud, Negasi Tsighe, Xia Wen Shui, and Yang Fang. 2018. "Effect of Processing Methods on Nutritional and Physico-chemical Composition of Fish: A Review." *MOJ Food Processing & Technology* 6(4). https://doi.org/10.15406/MOJFPT.2018.06.00191.

Acquah, Caleb, Godfred Ohemeng-Boahen, Krista A. Power, and Susan M. Tosh. 2021. "The Effect of Processing on Bioactive Compounds and Nutritional Qualities of Pulses in Meeting the Sustainable Development Goal 2." *Frontiers in Sustainable Food Systems* 5(May): 166. https://doi.org/10.3389/FSUFS.2021.681662/XML/NLM.

Akter, Saleha, Rama Addepalli, Michael Netzel, Ujang Tinggi, Mary Fletcher, Yasmina Sultanbawa, and Simone Osborne. 2021. "In Vitro Bioaccessibility and Intestinal Absorption of Selected Bioactive Compounds in Terminalia ferdinandiana." *Frontiers in Nutrition* 8(January). https://doi.org/10.3389/FNUT.2021.818195.

Aliyo, Alqeer, and Zelalem Teklemariam. 2022. "Assessment of Milk Contamination, Associated Risk Factors, and Drug Sensitivity Patterns among Isolated Bacteria from Raw Milk of Borena Zone, Ethiopia." *Journal of Tropical Medicine* (June): 1–9. https://doi.org/10.1155/2022/3577715.

Álvarez-Barrios, Ana, Lydia Álvarez, Montserrat García, Enol Artime, Rosario Pereiro, and Héctor González-Iglesias. 2021. "Antioxidant Defenses in the Human Eye: A Focus on Metallothioneins." *Antioxidants* 10(1): 1–33. https://doi.org/10.3390/ANTIOX10010089.

Andarwulan, Nuri, Niken Cahyarani Puspita, Saraswati, and Dominika Średnicka-Tober. 2021. "Antioxidants Such as Flavonoids and Carotenoids in the Diet of Bogor, Indonesia Residents." *Antioxidants* 10(4). https://doi.org/10.3390/ANTIOX10040587/S1.

Angelov, Angel, Teodora Yaneva-Marinova, and Velitchka Gotcheva. 2018. "Oats as a Matrix of Choice for Developing Fermented Functional Beverages." *Journal of Food Science and Technology* 55(7): 2351–60. https://doi.org/10.1007/S13197-018-3186-Y/TABLES/1.

Ashbolt, Nicholas John. 2004. "Microbial Contamination of Drinking Water and Disease Outcomes in Developing Regions." *Toxicology* 198(1): 229. https://doi.org/10.1016/J.TOX.2004.01.030.

Banerjee, Pradipta. 2019. "Functional Food: A Brief Overview." *International Journal of Bioresource Science* 6(2). https://doi.org/10.30954/2347-9655.02.2019.2.

Banwo, Kolawole, Ayoyinka Olufunke Olojede, Adekemi Titilayo Adesulu-Dahunsi, Deepak Kumar Verma, Mamta Thakur, Soubhagya Tripathy, Smita Singh, et al. 2021. "Functional Importance of Bioactive Compounds of Foods with Potential Health Benefits: A Review on Recent Trends." *Food Bioscience* 43(October): 101320. https://doi.org/10.1016/J.FBIO.2021.101320.

Batista, Ana Paula, Luísa Gouveia, Narcisa M. Bandarra, José M. Franco, and Anabela Raymundo. 2013. "Comparison of Microalgal Biomass Profiles as Novel Functional Ingredient for Food Products." *Algal Research* 2(2): 164. https://doi.org/10.1016/J.ALGAL.2013.01.004.

Batra, Priya, and Anil K. Sharma. 2013. "Anti-cancer Potential of Flavonoids: Recent Trends and Future Perspectives." *3 Biotech* 3(6): 439. https://doi.org/10.1007/S13205-013-0117-5.

Bayet-Robert, Mathilde, Fabrice Kwiatowski Mathilde, Marianne Leheurteur, Eloïse Planchat, Catherine Abrial, Xavier Durando, Chantal Barthomeuf, X. Durando, C. Barthomeuf, and P. Chollet. 2010. "Phase I Dose Escalation Trial of Docetaxelplus Curcumin in Patients with Advanced and Metastatic Breast Cancer." *Cancer Biology & Therapy*. https://doi.org/10.4161/cbt.9.1.10392.

Bazana, Maiara Taís, Cristiane Franco Codevilla, and Cristiano Ragagnin de Menezes. 2019. "Nanoencapsulation of Bioactive Compounds: Challenges and Perspectives." *Current Opinion in Food Science* 26(April): 47. https://doi.org/10.1016/J.COFS.2019.03.005.

Bede, Dery, and Lou Zaixiang. 2021. "Recent Developments in Resistant Starch as a Functional Food." *Starch - Stärke* 73(3–4): 2000139. https://doi.org/10.1002/STAR.202000139.

Ben-Othman, Sana, Ivi Jõudu, Rajeev Bhat, Maria Beatriz, Pinto Oliveira, and Rita Carneiro Alves. 2020. "Bioactives from Agri-Food Wastes: Present Insights and Future Challenges." *Molecules* 25(3): 510. https://doi.org/10.3390/MOLECULES25030510.

Borgonovi, Tais Fernanda, Sabrina Neves Casarotti, and Ana Lúcia Barretto Penna. 2021. "Lacticaseibacillus Casei SJRP38 and Buriti Pulp Increased Bioactive Compounds and Probiotic Potential of Fermented Milk." *LWT* 143(May): 111124. https://doi.org/10.1016/J.LWT.2021.111124.

Brodaczewska, Klaudia, Natalia Wolaniuk, Katarzyna Lewandowska, Katarzyna Donskow-Łysoniewska, and Maria Doligalska. 2017. "Biodegradable Chitosan Decreases the Immune Response to Trichinella Spiralis in Mice." *Molecules : A Journal of Synthetic Chemistry and Natural Product Chemistry* 22(11). https://doi.org/10.3390/MOLECULES22112008.

Călinoiu, Lavinia Florina, and Dan Cristian Vodnar. 2018. "Whole Grains and Phenolic Acids: A Review on Bioactivity, Functionality, Health Benefits and Bioavailability." *Nutrients* 10(11). https://doi.org/10.3390/NU10111615.

Camacho, Franciele, Angela Macedo, and Francisco Malcata. 2019. "Potential Industrial Applications and Commercialization of Microalgae in the Functional Food and Feed Industries: A Short Review." *Marine Drugs* 17(6). https://doi.org/10.3390/MD17060312.

Cao, Xinyi, Wang Liao, and Shaokang Wang. 2022. "Food Protein-Derived Bioactive Peptides for the Management of Nutrition Related Chronic Diseases." *Advances in Food and Nutrition Research* 101(January): 277–307. https://doi.org/10.1016/BS.AFNR.2022.04.004.

Cao, Zhen Hui, M. Julia, Johnson Green, D. Nicole Buckley, and Qiu Ye Lin. 2019. "Bioactivity of Soy-Based Fermented Foods: A Review." *Biotechnology Advances* 37: 223–38. https://doi.org/10.1016/j.biotechadv.2018.12.001.

Caporgno, Martín P., and Alexander Mathys. 2018. "Trends in Microalgae Incorporation Into Innovative Food Products With Potential Health Benefits." *Frontiers in Nutrition* 5(July): 58. https://doi.org/10.3389/FNUT.2018.00058.

Carrasco-Sandoval, Jonathan, Mario Aranda-Bustos, Karem Henríquez-Aedo, Amparo López-Rubio, and María José Fabra. 2021. "Bioaccessibility of Different Types of Phenolic Compounds Co-encapsulated in Alginate/Chitosan-Coated Zein Nanoparticles." *LWT* 149(September): 112024. https://doi.org/10.1016/J.LWT.2021.112024.

Catanzaro, Elena, Donatella Canistro, Valentina Pellicioni, Fabio Vivarelli, and Carmela Fimognari. 2022. "Anticancer Potential of Allicin: A Review." *Pharmacological Research* 177(March): 106118. https://doi.org/10.1016/J.PHRS.2022.106118.

Chakrabarti, Subhadeep, Snigdha Guha, and Kaustav Majumder. 2018. "Food-Derived Bioactive Peptides in Human Health: Challenges and Opportunities." *Nutrients* 10(11): 1738. https://doi.org/10.3390/NU10111738.

Chambial, S., S. Dwivedi, K.K. Shukla, P.J. John, and P. Sharma. 2013. "Vitamin C in Disease Prevention and Cure: An Overview." *Indian Journal of Clinical Biochemistry* 28(4): 314–28. https://doi.org/10.1007/s12291-013-0375-3.

Chandrasekara, Anoma, and Fereidoon Shahidi. 2018. "Herbal Beverages: Bioactive Compounds and Their Role in Disease Risk Reduction - A Review." *Journal of Traditional and Complementary Medicine* 8(4): 451–58. https://doi.org/10.1016/J.JTCME.2017.08.006.

Chen, B., D.J. McClements, and E.A. Decker. 2013. "Design of Foods with Bioactive Lipids for Improved Health." *Annual Review of Food Science and Technology* 4(1): 35–56. https://doi.org/10.1146/annurev-food-032112-135808.

Chen, Qiang, Chaochao Jia, Yingran Xu, Zhuanzhuan Jiang, Ting Hu, Conghu Li, and Xu Cheng. 2022. "Dual-PH Responsive Chitosan Nanoparticles for Improving In Vivo Drugs Delivery and Chemoresistance in Breast Cancer." *Carbohydrate Polymers* 290(August): 119518. https://doi.org/10.1016/J.CARBPOL.2022.119518.

Chourasia, Rounak, Loreni Chiring Phukon, Md. Minhajul Abedin, Dinabandhu Sahoo, and Amit Kumar Rai. 2022. "Production and Characterization of Bioactive Peptides in Novel Functional Soybean Chhurpi Produced Using Lactobacillus delbrueckii WS4." *Food Chemistry* 387(September): 132889. https://doi.org/10.1016/J.FOODCHEM.2022.132889.

Cicero, Arrigo F.G., Federica Fogacci, and Alessandro Colletti. 2017. "Potential Role of Bioactive Peptides in Prevention and Treatment of Chronic Diseases: A Narrative Review." *British Journal of Pharmacology* 174(11): 1378–94. https://doi.org/10.1111/BPH.13608.

Coello, Karín E., Elena Peñas, Cristina Martinez-Villaluenga, María Elena Cartea, Pablo Velasco, and Juana Frias. 2021. "Pasta Products Enriched with Moringa Sprout Powder as Nutritive Dense Foods with Bioactive Potential." *Food Chemistry* 360(October): 130032. https://doi.org/10.1016/J.FOODCHEM.2021.130032.

Conti, Maria Vittoria, Lorenzo Guzzetti, Davide Panzeri, Rachele de Giuseppe, Paola Coccetti, Massimo Labra, and Hellas Cena. 2021. "Bioactive Compounds in Legumes: Implications for Sustainable Nutrition and Health in the Elderly Population." *Trends in Food Science and Technology* 117(November): 139–47. https://doi.org/10.1016/J.TIFS.2021.02.072.

Dawidowicz, Andrzej L., and Małgorzata Olszowy. 2014. "Does Antioxidant Properties of the Main Component of Essential Oil Reflect Its Antioxidant Properties? The Comparison of Antioxidant Properties of Essential Oils and Their Main Components." *Natural Product Research* 28(22): 1952–63. https://doi.org/10.1080/14786419.2014.918121.

Desai, Neil. 2012. "Challenges in Development of Nanoparticle-Based Therapeutics." *The AAPS Journal* 14(2): 282. https://doi.org/10.1208/S12248-012-9339-4.

Dienaitė, Lijana, Renata Baranauskienė, and Petras Rimantas Venskutonis. 2021. "Lipophilic Extracts Isolated from European Cranberry Bush (Viburnum opulus) and Sea Buckthorn (Hippophae rhamnoides) Berry Pomace by Supercritical CO2 – Promising Bioactive Ingredients for Foods and Nutraceuticals." *Food Chemistry* 348(June): 129047. https://doi.org/10.1016/J.FOODCHEM.2021.129047.

"Dietary Reference Values for Nutrients Summary Report." 2017. *EFSA Supporting Publications* 14(12). https://doi.org/10.2903/SP.EFSA.2017.E15121.

Diez-Ozaeta, Iñaki, and Oihana Juaristi Astiazaran. 2022. "Fermented Foods: An Update on Evidence-Based Health Benefits and Future Perspectives." *Food Research International* 156(June): 111133. https://doi.org/10.1016/J.FOODRES.2022.111133.

Dou, Fang, Yixing Lu, and Nitin Nitin. 2021. "Yeast Cell Microcarriers for Delivery of a Model Bioactive Compound in Skin." *International Journal of Pharmaceutics* 609(November): 121123. https://doi.org/10.1016/J.IJPHARM.2021.121123.

Espero, Alaina Alessa, Liliana Gabriela Mendoza-Sa, and Hugo Sergio Garcı. 2022. "Structured Acylglycerides Emulsifiers with Bioactive Fatty Acids as Food Ingredients." *Value-Addition in Food Products and Processing Through Enzyme Technology*: 221–33. https://doi.org/10.1016/B978-0-323-89929-1.00004-4.

Esperón Rojas, A.A., R. Baeza Jiménez, C. Cano Sarmiento, and H.S. García. 2017. "Structured Mono- and Diacylglycerols with a High Content of Medium Chain Fatty Acids." *Journal of Oleo Science* 6(9): 991–96. https://doi.org/10.5650/jos.ess17010.

Eufic. 2022. "Functional Foods | Eufic." 2022. https://www.eufic.org/en/food-production/category/functional-foods.

Fani, Niloofar, M.H. Enayati, Hadis Rostamabadi, and Seid Reza Falsafi. 2022. "Encapsulation of Bioactives within Electrosprayed κ-Carrageenan Nanoparticles." *Carbohydrate Polymers* 294(October): 119761. https://doi.org/10.1016/J.CARBPOL.2022.119761.

Fernandes, Filipa A., Márcio Carocho, Sandrina A. Heleno, Paula Rodrigues, Maria Inês Dias, José Pinela, Miguel A. Prieto, Jesus Simal-Gandara, Lillian Barros, and Isabel C.F.R. Ferreira. 2020. "Effect of Natural Preservatives on the Nutritional Profile, Chemical Composition, Bioactivity and Stability of a Nutraceutical Preparation of Aloe arborescens." *Antioxidants* 9(4). https://doi.org/10.3390/ANTIOX9040281.

Fogliano, Vincenzo, and Paola Vitaglione. 2005. "Functional Foods: Planning and Development." *Molecular Nutrition and Food Research* 49(3): 256–62. https://doi.org/10.1002/MNFR.200400067.

Freeland-Graves, J.H., and C. Bavik. 2003. "Coenzymes." *Encyclopedia of Food Sciences and Nutrition*, January, 1475–81. https://doi.org/10.1016/B0-12-227055-X/00267-4.

Ganesan, Kumar, Bing Du, and Jianping Chen. 2022. "Effects and Mechanisms of Dietary Bioactive Compounds on Breast Cancer Prevention." *Pharmacological Research* 178 (October 2021): 105974. https://doi.org/10.1016/j.phrs.2021.105974.

García, J., D. Méndez, M. Álvarez, B. Sanmartin, Rebeca Vazquez Sobrado, L. Regueiro, and M. Atanassova. 2019. "Design of Novel Functional Food Products Enriched with Bioactive Extracts from Holothurians for Meeting the Nutritional Needs of the Elderly." *LWT* 109(July): 55–62. https://doi.org/10.1016/J.LWT.2019.03.097.

Gill, P.A., M.C. van Zelm, J.G. Muir, and P.R. Gibson. 2018. "Review Article: Short Chain Fatty Acids as Potential Therapeutic Agents in Human Gastrointestinal and Inflammatory Disorders." *Alimentary Pharmacology and Therapeutics* 48(1): 15–34. https://doi.org/10.1111/apt.14689.

Gokavi, Sumangala, Lanwei Zhang, Huang Ming Kuei, Xin Zhao, and Mingruo Guo. 2005. "Oat-Based Symbiotic Beverage Fermented by Lactobacillus plantarum, Lactobacillus paracasei ssp. Casei, and Lactobacillus acidophilus." *Journal of Food Science* 70(4): M216–23. https://doi.org/10.1111/J.1365-2621.2005.TB07191.X.

Gómez-García, Ricardo, Débora A. Campos, Ana Oliveira, Cristóbal N. Aguilar, Ana R. Madureira, and Manuela Pintado. 2021. "A Chemical Valorisation of Melon Peels Towards Functional Food Ingredients: Bioactives Profile and Antioxidant Properties." *Food Chemistry* 335(January): 127579. https://doi.org/10.1016/J.FOODCHEM.2020.127579.

Gómez-Guillén, M. Carmen, and M. Pilar Montero. 2021. "Enhancement of Oral Bioavailability of Natural Compounds and Probiotics by Mucoadhesive Tailored Biopolymer-Based Nanoparticles: A Review." *Food Hydrocolloids* 118(September): 106772. https://doi.org/10.1016/J.FOODHYD.2021.106772.

Gómez-Jacinto, Verónica, Francisco Navarro-Roldán, Inés Garbayo-Nores, Carlos Vílchez-Lobato, Ana Arias Borrego, and Tamara García-Barrera. 2020. "In Vitro Selenium Bioaccessibility Combined with In Vivo Bioavailability and Bioactivity in Se-Enriched Microalga (Chlorella sorokiniana) to Be Used as Functional Food." *Journal of Functional Foods* 66(March): 103817. https://doi.org/10.1016/J.JFF.2020.103817.

Gonçalves, Raquel F.S., Joana T. Martins, Luís Abrunhosa, António A. Vicente, and Ana C. Pinheiro. 2021. "Nanoemulsions for Enhancement of Curcumin Bioavailability and Their Safety Evaluation: Effect of Emulsifier Type." *Nanomaterials* 11(3). https://doi.org/10.3390/NANO11030815.

Gorinstein, Shela, Olga Martín-Belloso, Yong Seo Park, Ratiporn Haruenkit, Antonin Lojek, Milan Ĉí ž, Abraham Caspi, Imanuel Libman, and Simon Trakhtenberg. 2001. "Comparison of Some Biochemical Characteristics of Different Citrus Fruits." *Food Chemistry* 74(3): 309–15. https://doi.org/10.1016/S0308-8146(01)00157-1.

Granger, Matthew, and Peter Eck. 2018. "Dietary Vitamin C in Human Health." *Advances in Food and Nutrition Research* 83(January): 281. https://doi.org/10.1016/BS.AFNR.2017.11.006.

Guía-García, Jorge L., Ana Charles, Francisca Ramírez-Godina, Armando Robledo, Hermila Trinidad García-Osuna, Miguel Cerqueira, and María Liliana Flores-López. 2022. "Micro and Nanoencapsulation of Bioactive Compounds for Agri-Food Applications: A Review." *Industrial Crops and Products* 186(October): 115198. https://doi.org/10.1016/J.INDCROP.2022.115198.

Hasler, Clare M. 2002. "Functional Foods: Benefits, Concerns and Challenges—A Position Paper from the American Council on Science and Health." *The Journal of Nutrition* 132(12): 3772–81. https://doi.org/10.1093/JN/132.12.3772.

Hatcher, H., R. Planalp, J. Cho, F.M. Torti, and S. V. Torti. 2008. "Curcumin: From Ancient Medicine to Current Clinical Trials." *Cellular and Molecular Life Sciences : CMLS* 65(11): 1631. https://doi.org/10.1007/S00018-008-7452-4.

He, Fen jin, and Jin qiang Chen. 2013. "Consumption of Soybean , Soy Foods , Soy Isoflavones and Breast Cancer Incidence : Differences between Chinese Women and Women in Western Countries and Possible Mechanisms." *Food Science and Human Wellness* 2(3–4): 146–61. https://doi.org/10.1016/j.fshw.2013.08.002.

He, Z.Y., C.B. Shi, H. Wen, F.L. Li, B.L. Wang, and J. Wang. 2011. "Upregulation of P53 Expression in Patients with Colorectal Cancer by Administration of Curcumin." *Cancer Investigation* 29(3): 208. https://doi.org/10.3109/07357907.2010.550592.

Idowu-Adebayo, Folake, Vincenzo Fogliano, and Anita Linnemann. 2022. "Turmeric-Fortified Cow and Soya Milk: Golden Milk as a Street Food to Support Consumer Health." *Foods* 11(4): 558. https://doi.org/10.3390/FOODS11040558.

IndustryARC. 2022. "Global Functional Foods Market Share, Size and Industry Growth Analysis: 2020–2025." https://www.industryarc.com/Research/Global-Functional-Foods-Market-Research-513806?https://www. industryarc.com/Research/Global-Functional-Foods-Market-Research-.

Jain, Priti, Inderbir Singh, Sanjay J. Surana, and Atul A. Shirkhedkar. 2022. "Tocopherols and Tocotrienols: The Essential Vitamin E." *Bioactive Food Components Activity in Mechanistic Approach*, January: 139–54. https://doi.org/10.1016/B978-0-12-823569-0.00009-6.

Jimenez-Lopez, C., A.G. Pereira, C. Lourenço-Lopes, P. Garcia-Oliveira, L. Cassani, M. Fraga-Corral, M.A. Prieto, and J. Simal-Gandara. 2021. "Main Bioactive Phenolic Compounds in Marine Algae and Their Mechanisms of Action Supporting Potential Health Benefits." *Food Chemistry* 341(March): 128262. https://doi.org/10.1016/J.FOODCHEM.2020.128262.

Kaushalya, K.G.D., and K.D.P.P. Gunathilake. 2022. "Encapsulation of Phlorotannins from Edible Brown Seaweed in Chitosan: Effect of Fortification on Bioactivity and Stability in Functional Foods." *Food Chemistry* 377(May): 132012. https://doi.org/10.1016/J.FOODCHEM.2021.132012.

Khan, Rao Sanaullah, John Vincent Grigor, Alan G. Win, and Mike Boland. 2014. "Differentiating Aspects of Product Innovation Processes in the Food Industry: An Exploratory Study on New Zealand." *British Food Journal* 116(8): 1346–68. https://doi.org/10.1108/BFJ-04-2013-0094/FULL/PDF.

Khan, Rao Sanaullah, John Grigor, Ray Winger, and Alan Win. 2013. "Functional Food Product Development – Opportunities and Challenges for Food Manufacturers." *Trends in Food Science & Technology* 30(1): 27–37. https://doi.org/10.1016/J.TIFS.2012.11.004.

Khandelwal, Shweta, Rahul Shidhaye, Isabelle Demonty, Ramakrishnan Lakshmy, Ruby Gupta, Dorairaj Prabhakaran, and Srinath Reddy. 2013. "Impact of Omega-3 Fatty Acids and/or Plant Sterol Supplementation on Non-HDL Cholesterol Levels of Dyslipidemic Indian Adults." *Journal of Functional Foods* 5(1): 36–43. https://doi.org/10.1016/J.JFF.2012.08.001.

Khochapong, Wiriya, Sunantha Ketnawa, Yukiharu Ogawa, and Niramol Punbusayakul. 2021. "Effect of In Vitro Digestion on Bioactive Compounds, Antioxidant and Antimicrobial Activities of Coffee (Coffea arabica L.) Pulp Aqueous Extract." *Food Chemistry* 348(June): 129094. https://doi.org/10.1016/J.FOODCHEM.2021.129094.

Kocot, Aleksandra Maria, and Barbara Wróblewska. 2021. "Fermented Products and Bioactive Food Compounds as a Tool to Activate Autophagy and Promote the Maintenance of the Intestinal Barrier Function." *Trends in Food Science & Technology* 118(July): 905–19. https://doi.org/10.1016/j.tifs.2021.11.014.

Kris-Etherton, Penny M., Kari D. Hecker, Andrea Bonanome, Stacie M. Coval, Amy E. Binkoski, Kirsten F. Hilpert, Amy E. Griel, and Terry D. Etherton. 2002. "Bioactive Compounds in Foods: Their Role in the Prevention of Cardiovascular Disease and Cancer." *The American Journal of Medicine* 113(9): 71–88. https://doi.org/10.1016/S0002-9343(01)00995-0.

Lahouar, Lamia, Lotfi Achour, and Imed Latiri. 2021. "Talbina as a Functional Food and a Source of Health-Beneficial Ingredients: A Narrative Review." *The North African Journal of Food and Nutrition Research* 5(12): 139–51. https://doi.org/10.51745/NAJFNR.5.12.139-151.

Lamberti, Lucio, and Emanuele Lettieri. 2011. "Gaining Legitimacy in Converging Industries: Evidence from the Emerging Market of Functional Food." *European Management Journal* 29(6): 462. https://doi.org/10.1016/J.EMJ.2011.08.002.

Lee, Duk Hee, Aaron R. Folsom, Lisa Harnack, Barry Halliwell, and David R. Jacobs. 2004. "Does Supplemental Vitamin C Increase Cardiovascular Disease Risk in Women with Diabetes?" *The American Journal of Clinical Nutrition* 80(5): 1194–1200. https://doi.org/10.1093/AJCN/80.5.1194.

Lee, K.T., and C.C. Akoh. 1998. "Structured Lipids: Synthesis and Applications." *Food Reviews International* 14(1): 17–34. https://doi.org/10.1080/87559129809541148.

Li, Fang, Xiaojuan Wu, and Wei Wu. 2021. "Effects of Oxidative Modification by Malondialdehyde on the In Vitro Digestion Properties of Rice Bran Protein." *Journal of Cereal Science* 97(January): 103158. https://doi.org/10.1016/J.JCS.2020.103158.

Li, Guangxing, Kaiyue Ding, Yanling Qiao, Liu Zhang, Luping Zheng, Taowen Pan, and Lin Zhang. 2020. "Flavonoids Regulate Inflammation and Oxidative Stress in Cancer." *Molecules* 25(23). https://doi.org/10.3390/MOLECULES25235628.

Liu, Rui Hai. 2004. "Potential Synergy of Phytochemicals in Cancer Prevention: Mechanism of Action." *The Journal of Nutrition* 134(12): 3479S-3485S. https://doi.org/10.1093/JN/134.12.3479S.

Liu, Rui Hai. 2013. "Dietary Bioactive Compounds and Their Health Implications." *Journal of Food Science* 78(s1): A18–25. https://doi.org/10.1111/1750-3841.12101.

Liu, Shiqi, Jing Zhang, Rao Fu, Hao Feng, Yibing Chu, Dong Huang, Han Liu, Chaonan Li, Chao Ma, and A.M. Abd El-Aty. 2022. "Improved Stability and Aqueous Solubility of β-Carotene via Encapsulation in Self-Assembled Bioactive Oleanolic Acid Nanoparticles." *Food Chemistry* 373(March): 131498. https://doi.org/10.1016/J.FOODCHEM.2021.131498.

Liu, Y., G. Huang, and J. Hu. 2018. "Extraction, Characterisation, and Antioxidant Activities of Polysaccharides from Chinese Watermelon." *International Journal Biology Macromolecule* 111: 1304–1307.

Lu, Xiaojie, Yue Jing, Yanyi Li, Naisheng Zhang, and Yongguo Cao. 2022. "Eurotium Cristatum Produced β-Hydroxy Acid Metabolite of Monacolin K and Improved Bioactive Compound Contents as Well as Functional Properties in Fermented Wheat Bran." *LWT* 158(March): 113088. https://doi.org/10.1016/J.LWT.2022.113088.

Luo, You, Bin Peng, Weiqian Wei, Xiaofei Tian, and Z. Wu. 2019. "Antioxidant and Anti-diabetic Activities of Polysaccharides from Guava Leaves." *Molecules*: 1–14. https://doi.org/10.3390/molecules24071343.

Marco, Maria L., Dustin Heeney, Sylvie Binda, Christopher J. Cifelli, Paul D. Cotter, Benoit Foligné, Michael Gänzle, et al. 2017a. "Health Benefits of Fermented Foods: Microbiota and Beyond." *Current Opinion in Biotechnology* 44(April): 94–102. https://doi.org/10.1016/J.COPBIO.2016.11.010.

Marco, Maria L., Dustin Heeney, Sylvie Binda, Christopher J. Cifelli, Paul D. Cotter, Benoit Foligné, Michael Gänzle, et al. 2017b. "Health Benefits of Fermented Foods: Microbiota and Beyond." *Current Opinion in Biotechnology* 44(April): 94–102. https://doi.org/10.1016/J.COPBIO.2016.11.010.

Martínez-Ballesta, M. Carment, Ángel Gil-Izquierdo, Cristina García-Viguera, and Raúl Domínguez-Perles. 2018. "Nanoparticles and Controlled Delivery for Bioactive Compounds: Outlining Challenges for New 'Smart-Foods' for Health." *Foods* 7(5): 72. https://doi.org/10.3390/FOODS7050072.

Masyita, Ayu, Reka Mustika Sari, Ayun Dwi Astuti, Budiman Yasir, Nur Rahma Rumata, Talha bin Emran, Firzan Nainu, and Jesus Simal-Gandara. 2022. "Terpenes and Terpenoids as Main Bioactive Compounds of Essential Oils, Their Roles in Human Health and Potential Application as Natural Food Preservatives." *Food Chemistry: X* 13(March): 100217. https://doi.org/10.1016/J.FOCHX.2022.100217.

Mathur, Harsh, Tom P. Beresford, and Paul D. Cotter. 2020. "Health Benefits of Lactic Acid Bacteria (LAB) Fermentates." *Nutrients* 12(6): 1–16. https://doi.org/10.3390/NU12061679.

Milenkovic, Dragan, Christiane Deval, Claude Dubray, Andrzej Mazur, and Christine Morand. 2011. “Hesperidin Displays Relevant Role in the Nutrigenomic Effect of Orange Juice on Blood Leukocytes in Human Volunteers: A Randomized Controlled Cross-Over Study.” *PLOS ONE* 6(11): e26669. https://doi.org/10.1371/JOURNAL.PONE.0026669.

Mohanta, Bulu, Dhrubo Jyoti, Beduin Mahanti, and Amit Kumar. 2022. “Antioxidant Potential of Herbal Polysaccharides : An Overview on Recent Researches.” *Sensors International* 3(December 2021): 100158. https://doi.org/10.1016/j.sintl.2022.100158.

Moons, Ingrid, Camilla Barbarossa, and Patrick de Pelsmacker. 2018. “The Determinants of the Adoption Intention of Eco-friendly Functional Food in Different Market Segments.” *Ecological Economics* 151(September): 151–61. https://doi.org/10.1016/J.ECOLECON.2018.05.012.

Mordor Intelligence. 2022. “Global Functional Food Market | 2022 - 27 | Industry Share, Size, Growth - Mordor Intelligence.” https://www.mordorintelligence.com/industry-reports/global-functional-food-market.

Morris, Alyssa L., and Shamim S. Mohiuddin. 2022. “Biochemistry, Nutrients.” *StatPearls*, May. https://www.ncbi.nlm.nih.gov/books/NBK554545/.

Moser, Melissa A., and Ock K. Chun. 2016. “Vitamin C and Heart Health: A Review Based on Findings from Epidemiologic Studies.” *International Journal of Molecular Sciences* 17(8): 1328. https://doi.org/10.3390/IJMS17081328.

Natalello, Antonio, Gonzalo Hervás, Pablo G. Toral, Giuseppe Luciano, Bernardo Valenti, Alejandro G. Mendoza, Mariano Pauselli, Alessandro Priolo, and Pilar Frutos. 2020. “Bioactive Compounds from Pomegranate By-Products Increase the In Vitro Ruminal Accumulation of Potentially Health Promoting Fatty Acids.” *Animal Feed Science and Technology* 259(January): 114355. https://doi.org/10.1016/J.ANIFEEDSCI.2019.114355.

Nepali, Prakash, Surya Suresh, Gauri Pikale, Sharan Jhaveri, Chaithanya Avanthika, Mridul Bansal, Rabia Islam, and Aditya Chanpura. 2022. “Hypertension and the Role of Dietary Fiber.” *Current Problems in Cardiology* 47(7): 101203. https://doi.org/10.1016/J.CPCARDIOL.2022.101203.

Ngoma, Catherine M., Janetta Roos, Seter Siziya, Abeer M. Alharbi, Thamer Alduraywish, Andrew Luk Leung, C. Rabindran, et al. 2020. “Challenges in Disease and Health Research Vol. 2.” September https://doi.org/10.9734/BPI/CDHR/V2.

Nielsen, Søren D., Louise M.A. Jakobsen, Nina R.W. Geiker, and Hanne Christine Bertram. 2022. “Chemically Acidified, Live and Heat-Inactivated Fermented Dairy Yoghurt Show Distinct Bioactive Peptides, Free Amino Acids and Small Compounds Profiles.” *Food Chemistry* 376(May): 131919. https://doi.org/10.1016/J.FOODCHEM.2021.131919.

Niwano, Yoshimi, Hidetsugu Kohzaki, Midori Shirato, Shunichi Shishido, and Keisuke Nakamura. 2022. “Putative Mechanisms Underlying the Beneficial Effects of Polyphenols in Murine Models of Metabolic Disorders in Relation to Gut Microbiota.” *Current Issues in Molecular Biology* 44(3): 1353. https://doi.org/10.3390/CIMB44030091.

Nolasco, Emerson, Junsi Yang, Ozan Ciftci, Danh C. Vu, Sophie Alvarez, Sheila Purdum, and Kaustav Majumder. 2021. “Evaluating the Effect of Cooking and Gastrointestinal Digestion in Modulating the Bio-accessibility of Different Bioactive Compounds of Eggs.” *Food Chemistry* 344(May): 128623. https://doi.org/10.1016/J.FOODCHEM.2020.128623.

Noor, Nairah, Adil Gani, Faiza Jhan, Mohammad Ashraf Shah, and Zanoor ul Ashraf. 2022. "Ferulic Acid Loaded Pickering Emulsions Stabilized by Resistant Starch Nanoparticles Using Ultrasonication: Characterization, In Vitro Release and Nutraceutical Potential." *Ultrasonics Sonochemistry* 84(March): 105967. https://doi.org/10.1016/J.ULTSONCH.2022.105967.

Pai, Sanidhya, Akshatha Hebbar, and Subbalaxmi Selvaraj. 2022. "A Critical Look at Challenges and Future Scopes of Bioactive Compounds and Their Incorporations in the Food, Energy, and Pharmaceutical Sector." *Environmental Science and Pollution Research International* 29(24): 35518. https://doi.org/10.1007/S11356-022-19423-4.

Pateiro, Mirian, Belén Gómez, Paulo E.S. Munekata, Francisco J. Barba, Predrag Putnik. Danijela Bursać Kovačević, and José M. Lorenzo. 2021. "Nanoencapsulation of Promising Bioactive Compounds to Improve Their Absorption, Stability, Functionality and the Appearance of the Final Food Products." *Molecules* 26(6): 1547. https://doi.org/10.3390/MOLECULES26061547.

Farias, Paulo, Fábio Fernandes de David, Iramaia Angélica de Araújo, Flávia Luísa Dias-Audibert Neri-Numa, Rodrigo Ramos Catharino Jeany Delafiori, Glaucia Maria Pastore, and G.M. Pastore. 2021. "Effect of In Vitro Digestion on the Bioaccessibility and Bioactivity of Phenolic Compounds in Fractions of Eugenia pyriformis Fruit." *Food Research International* 150(December): 110767. https://doi.org/10.1016/J.FOODRES.2021.110767.

Pepa, Giuseppe Della, Claudia Vetrani, Marilena Vitale, Lutgarda Bozzetto, Giuseppina Costabile, Paola Cipriano, Anna Mangione, et al. 2020. "Effects of a Diet Naturally Rich in (Poly)Phenols on Lipid Composition of Postprandial Lipoproteins in High Cardiometabolic Risk Individuals: An Ancillary Analysis of a Randomized Controlled Trial." *European Journal of Clinical Nutrition* 74(1): 183. https://doi.org/10.1038/s41430-019-0459-0.

Rahaiee, Somayeh, Elham Assadpour, Afshin Faridi Esfanjani, Ana Sanches Silva, and Seid Mahdi Jafari. 2020. "Application of Nano/Microencapsulated Phenolic Compounds against Cancer." *Advances in Colloid and Interface Science* 279(May): 102153. https://doi.org/10.1016/J.CIS.2020.102153.

Rashmi, Havalli Bommegowda, and Pradeep Singh Negi. 2020. "Phenolic Acids from Vegetables: A Review on Processing Stability and Health Benefits." *Food Research International* 136(October): 109298. https://doi.org/10.1016/J.FOODRES.2020.109298.

Renukuntla, Jwala, Aswani Dutt Vadlapudi, Ashaben Patel, Sai H.S. Boddu, and Ashim K. Mitra. 2013. "Approaches for Enhancing Oral Bioavailability of Peptides and Proteins." *International Journal of Pharmaceutics* 447(0): 75. https://doi.org/10.1016/J.IJPHARM.2013.02.030.

Repo, Carrasco-Valencia, Ritva Ann Mari, and Lesli Astuhuaman Serna. 2011. "Quinoa (Chenopodium Quinoa, Willd.) as a Source of Dietary Fiber and Other Functional Components." *Food Science and Technology* 31(1): 225. https://doi.org/10.1590/S0101-20612011000100035.

Ribeiro, Priscila Vaz de Melo, Patrícia Amaro Andrade, Helen Hermana Miranda Hermsdorff, Carolina Araújo dos Santos, Rosângela Minardi Mitre Cotta, Juliana de Assis Silva Gomes Estanislau, Aline Aparecida de Oliveira Campos, and Carla de Oliveira Barbosa Rosa. 2019. "Dietary Non-Nutrients in the Prevention of Non-Communicable Diseases: Potentially Related Mechanisms." *Nutrition* 66(October): 22–28. https://doi.org/10.1016/J.NUT.2019.03.016.

Ried, Karin, and Peter Fakler. 2014. "Potential of Garlic (Allium sativum) in Lowering High Blood Pressure: Mechanisms of Action and Clinical Relevance." *Integrated Blood Pressure Control* 7(December): 71. https://doi.org/10.2147/IBPC.S51434.

Rivero-Pino, Fernando, F. Javier Espejo-Carpio, and Emilia M. Guadix. 2020a. "Bioactive Fish Hydrolysates Resistance to Food Processing." *LWT* 117(January): 108670. https://doi.org/10.1016/J.LWT.2019.108670.

Rivero-Pino, Fernando, F. Javier Espejo-Carpio, and Emilia M. Guadix. 2020b. "Evaluation of the Bioactive Potential of Foods Fortified with Fish Protein Hydrolysates." *Food Research International* 137(November): 109572. https://doi.org/10.1016/J.FOODRES.2020.109572.

Saini, A., D. Panwar, P.S. Panesar, and M.B. Bera. 2019. "Bioactive Compounds from Cereal and Pulse Processing Byproducts and Their Potential Health Benefits." *Journal of Nutrition and Metabolism* 2(6). https://austinpublishinggroup.com/nutrition-metabolism/fulltext/ajnm-v6-id1068.php.

Samtiya, Mrinal, Rotimi E. Aluko, and Tejpal Dhewa. 2020. "Plant Food Anti-nutritional Factors and Their Reduction Strategies: An Overview." *Food Production, Processing and Nutrition* 2(1): 1–14. https://doi.org/10.1186/S43014-020-0020-5.

Samtiya, Mrinal, Rotimi E. Aluko, Tejpal Dhewa, and José Manuel Moreno-Rojas. 2021. "Potential Health Benefits of Plant Food-Derived Bioactive Components: An Overview." *Foods* 10(4). https://doi.org/10.3390/FOODS10040839.

Sánchez, Manuel, Miguel Romero, Manuel Gómez-Guzmán, Juan Tamargo, Francisco Pérez-Vizcaino, and Juan Duarte. 2018. "Cardiovascular Effects of Flavonoids." *Current Medicinal Chemistry* 26(39): 6991–7034. https://doi.org/10.2174/0929867326666181220094721.

Sanchez-Rodriguez, Estefania, Laura Alejandra Vazquez-Aguilar, Sara Biel-Glesson, Jose Ramon Fernandez-Navarro, Juan Antonio Espejo-Calvo, Jose Maria Olmo-Peinado, Rafael de la Torre, et al. 2021. "May Bioactive Compounds from the Olive Fruit Improve the Postprandial Insulin Response in Healthy Adults?." *Journal of Functional Foods* 83(August): 104561. https://doi.org/10.1016/J.JFF.2021.104561.

Sangsukiam, Thasanporn, and Kiattisak Duangmal. 2022. "Changes in Bioactive Compounds and Health-Promoting Activities in Adzuki Bean: Effect of Cooking Conditions and In Vitro Simulated Gastrointestinal Digestion." *Food Research International* 157(July): 111371. https://doi.org/10.1016/J.FOODRES.2022.111371.

Santos, Diana I., Jorge Manuel Alexandre Saraiva, António A. Vicente, and Margarida Moldão-Martins. 2019. "Methods for Determining Bioavailability and Bioaccessibility of Bioactive Compounds and Nutrients." *Innovative Thermal and Non-Thermal Processing, Bioaccessibility and Bioavailability of Nutrients and Bioactive Compounds*, January: 23–54. https://doi.org/10.1016/B978-0-12-814174-8.00002-0.

Scazzocchio, B., Varì, R., Filesi, C., D'Archivio, M., Santangelo, C., Giovannini, C., Iacovelli, A., Silecchia, G., Volti, G. L., Galvano, F., & Masella, R. (2011). "Cyanidin-3-O-β-Glucoside and Protocatechuic Acid Exert Insulin-Like Effects by Upregulating PPARγ Activity in Human Omental Adipocytes." *Diabetes 60*(9): 2234–2244. https://doi.org/10.2337/DB10-1461

Scherließ, Regina, Simon Buske, Katherine Young, Benjamin Weber, Thomas Rades, and Sarah Hook. 2013. "In Vivo Evaluation of Chitosan as an Adjuvant in Subcutaneous Vaccine Formulations." *Vaccine* 31(42): 4812–19. https://doi.org/10.1016/J.VACCINE.2013.07.081.

Schmidt, Helena de Oliveira, Fernanda C. Rockett, Carlos H. Pagno, Jucelaine Possa, Renato Q. Assis, Viviani R. de Oliveira, Vanuska L. da Silva, Simone H. Flôres, and Alessandro de Oliveira Rios. 2019. "Vitamin and Bioactive Compound Diversity of Seven Fruit Species from South Brazil." *Journal of the Science of Food and Agriculture* 99(7): 3307–17. https://doi.org/10.1002/JSFA.9544.

Šeregelj, Vanja, Jelena Vulić, Gordana Ćetković, Jasna Čanadanovć-Brunet, Vesna Tumbas Šaponjac, and Slađana Stajčić. 2020. "Natural Bioactive Compounds in Carrot Waste for Food Applications and Health Benefits." *Studies in Natural Products Chemistry* 67(January): 307. https://doi.org/10.1016/B978-0-12-819483-6.00009-6.

Shaikh, Rahamatullah, Thakur Raj Singh, Martin Garland, A. Woolfson, and Ryan Donnelly. 2011. "Mucoadhesive Drug Delivery Systems." *Journal of Pharmacy and Bioallied Sciences* 3(1): 89. https://doi.org/10.4103/0975-7406.76478.

Shao, Andrew. 2017. "Global Market Entry Regulations for Nutraceuticals, Functional Foods, Dietary/Food/Health Supplements." *Developing New Functional Food and Nutraceutical Products*, January: 279–90. https://doi.org/10.1016/B978-0-12-802780-6.00015-8.

Shiby, V.K., and H.N. Mishra. 2013a. "Fermented Milks and Milk Products as Functional Foods--A Review." *Critical Reviews in Food Science and Nutrition* 53(5): 482–96. https://doi.org/10.1080/10408398.2010.547398.

Shiby, V.K., and H.N. Mishra. 2013b. "Fermented Milks and Milk Products as Functional Foods—A Review." *Critical Reviews in Food Science and Nutrition* 53(5): 482–96. https://doi.org/10.1080/10408398.2010.547398.

Siddiq, M., S. Roidoung, D.S. Sogi, and K.D. Dolan. 2013. "Total Phenolics, Antioxidant Properties and Quality of Fresh-Cut Onions (Allium cepa L.) Treated with Mild-Heat." *Food Chemistry* 136(2): 803–806. https://doi.org/10.1016/j.foodchem.2012.09.023.

Silva, Bruno I.M., Erika A. Nascimento, Cleber J. Silva, Teresinha G. Silva, and Jaciana S. Aguiar. 2021. "Anticancer Activity of Monoterpenes: A Systematic Review." *Molecular Biology Reports* 48(7): 5775–5785. https://doi.org/10.1007/S11033-021-06578-5/FIGURES/3.

Tan, Melvin, Malik Adil Nawaz, and Roman Buckow. 2021. "Functional and Food Application of Plant Proteins – A Review." *Food Reviews International*. https://doi.org/10.1080/87559129.2021.1955918.

Tanaka, Takuji, Masahito Shnimizu, and Hisataka Moriwaki. 2012. "Cancer Chemoprevention by Carotenoids." *Molecules* 17(3): 3202–3242. https://doi.org/10.3390/MOLECULES17033202.

Tian, Wenni, Nan Xiao, Yunyun Yang, Jie Xiao, Ruiping Zeng, Lanhua Xie, Ziyou Qiu, Pan Li, and Bing Du. 2020. "International Journal of Biological Macromolecules Structure, Antioxidant and Immunomodulatory Activity of a Polysaccharide Extracted from Sacha Inchi Seeds." *International Journal of Biological Macromolecules* 162: 116–26. https://doi.org/10.1016/j.ijbiomac.2020.06.150.

Tung, Wei-Cheng, Bryan Rizzo, Yusef Dabbagh, Suraj Saraswat, Mark Romanczyk, Rocio Rebollido-rios, Paul W. Needs, et al. 2020. "Polyphenols Bind to Low Density Lipoprotein at Biologically Relevant Concentrations That Are Protective for Heart Disease." *Archives of Biochemistry and Biophysics* 694(June). https://doi.org/10.1016/j.abb.2020.108589.

Vetrania, Claudia, Giuseppina Costabilea, Marilena Vitalea, and Rosalba Giacco. 2020. "(Poly)Phenols and Cardiovascular Diseases: Looking in to Move Forward." *Journal of Functional Foods* 71: 104013. https://doi.org/https. https://doi.org/10.1016/j.jff.2020.104013.

Vilas-Boas, Ana A., Manuela Pintado, and Ana L.S. Oliveira. 2021. "Natural Bioactive Compounds from Food Waste: Toxicity and Safety Concerns." *Foods* 10(7): 1564. https://doi.org/10.3390/FOODS10071564.

Visen, Aastha, Srishti Visen, Asha Sharma, and Pradeep K.S. Visen. 2022. "Nutraceuticals as a Natural Alternative for Preventive and Proactive Health Care." *Functional Foods and Nutraceuticals in Metabolic and Non-Communicable Diseases*, January: 603–18. https://doi.org/10.1016/B978-0-12-819815-5.00040-9.

Wang, Min, Jianjun Zhou, Joaquim Calvo-Lerma, Yixuan Liu, María Carmen Collado, and Francisco J. Barba. 2022. "Effects of Marine Bioactive Compounds on Gut Ecology Based on In Vitro Digestion and Colonic Fermentation Models." *Nutrients* 14(16): 3307 14 (16): 3307. https://doi.org/10.3390/NU14163307.

Wang, Xia, Ying Y. Ouyang, Jun Liu, and Gang Zhao. 2014. "Systematic Review with Meta-analysis Flavonoid Intake and Risk of CVD : A Systematic Review and Meta-analysis of Prospective Cohort Studies." *British Journal of Nutrition*: 1–11. https://doi.org/10.1017/S000711451300278X.

Wang, Yun, Xuelian Wei, Fuhou Wang, Jingjing Xu, Xiaozhen Tang, and Ningyang Li. 2018. "Structural Characterization and Antioxidant Activity of Polysaccharide from Ginger." *International Journal of Biological Macromolecules* 111: 862–69. https://doi.org/10.1016/j.ijbiomac.2018.01.087.

Wang, Yutang, Daniel Russo, Chang Liu, Qian Zhou, Hao Zhu, and Yinghua Zhang. 2020. "Functional Structure/Activity Relationships Predictive Modeling of Angiotensin I-Converting Enzyme (ACE) Inhibitory Peptides Using Various Machine Learning Approaches Predictive Modeling of Angiotensin I-Converting Enzyme (ACE) Inhibitory Peptides Using Various Machine Learning." https://doi.org/10.1021/acs.jafc.0c04624.

Wani, S.A., A. Singh, and P. Kumar (Eds.). 2022. *Spice Bioactive Compounds: Properties, Applications, and Health Benefits* (1st ed.). CRC Press. https://doi.org/10.1201/9781003215387.

Weydert, Joy A. 2014. "Vitamin D in Children's Health." *Children* 1(2): 208–226. https://doi.org/10.3390/CHILDREN1020208.

WHO. 2021. *Noncommunicable Diseases 2021.* https://www.who.int/news-room/fact-sheets/detail/noncommunicable-diseases.

WHO. 2022. *World Health Statistics 2022: Monitoring Health for the SDGs 2022.* https://repository.gheli.harvard.edu/repository/11008/.

Xia, Donghua, Wen Qiu, Xianxian Wang, and Junying Liu. 2021. "Recent Advancements and Future Perspectives of Microalgae-Derived Pharmaceuticals." *Marine Drugs* 19(12). https://doi.org/10.3390/MD19120703.

Yang, Jun, Jieqiong Lin, Xianxiang Chen, Liyuan Rong, Mingyue Shen, Yuanxing Wang, and Jianhua Xie. 2022. "Mesona chinensis Polysaccharide/Zein Nanoparticles to Improve the Bioaccesibility and In Vitro Bioactivities of Curcumin." *Carbohydrate Polymers* 295(November): 119875. https://doi.org/10.1016/J.CARBPOL.2022.119875.

Yashin, Alexander, Yakov Yashin, Xiaoyan Xia, and Boris Nemzer. 2017. "Antioxidant Activity of Spices and Their Impact on Human Health: A Review." *Antioxidants* 6(3): 70. https://doi.org/10.3390/ANTIOX6030070.

Yoshida, Hiroshi, Hidekatsu Yanai, Kumie Ito, Yoshiharu Tomono, Takashi Koikeda, Hiroki Tsukahara, and Norio Tada. 2010. "Administration of Natural Astaxanthin Increases Serum HDL-Cholesterol and Adiponectin in Subjects with Mild Hyperlipidemia." *Atherosclerosis* 209(2): 520–23. https://doi.org/10.1016/J.ATHEROSCLEROSIS.2009.10.012.

Yuan, Jian-min. 2013. "Cancer Prevention by Green Tea : Evidence from Epidemiologic Studies 1–4." *The American Journal of Clinical Nutrition* 98: 1676–1681. https://doi.org/10.3945/ajcn.113.058271.1.

Zelepukin, Ivan V., Alexey V. Yaremenko, Mikhail V. Yuryev, A.B. Mirkasymov, I.L. Sokolov, S.M. Deyev, P.I. Nikitin, and M.P. Nikitin. 2020. "Fast Processes of Nanoparticle Blood Clearance: Comprehensive Study." *Journal of Controlled Release* 326(October): 181–91. https://doi.org/10.1016/J.JCONREL.2020.07.014.

Zhou, Shiyang, Gangliang Huang, and Hualiang Huang. 2022. "Extraction, Derivatization and Antioxidant Activities of Onion Polysaccharide." *Food Chemistry* 388(December 2021): 133000. https://doi.org/10.1016/j.foodchem.2022.133000.

Zhu, Weizhu, Wenyi Qin, Ke Zhang, George E. Rottinghaus, Yin-Chieh Chen, Beth Kliethermes, and Edward R. Sauter. 2012. "Trans-Resveratrol Alters Mammary Promoter Hypermethylation in Women at Increased Risk for Breast Cancer." *Nutrition and Cancer* 64(3): 393–400. https://doi.org/10.1080/01635581.2012.654926.

Zou, Liqiang, Bingjing Zheng, Ruojie Zhang, Zipei Zhang, Wei Liu, Chengmei Liu, Hang Xiao, and David Julian McClements. 2016. "Enhancing the Bioaccessibility of Hydrophobic Bioactive Agents Using Mixed Colloidal Dispersions: Curcumin-Loaded Zein Nanoparticles plus Digestible Lipid Nanoparticles." *Food Research International* 81(March): 74–82. https://doi.org/10.1016/J.FOODRES.2015.12.035.

CHAPTER 14

Challenges and Solutions to the Addition of Nutraceuticals into the Diet

Monika Choudhary* and **Amarjeet Kaur**†

14.1 INTRODUCTION

Today's consumers are curious and well aware of the significance of a balanced diet and healthy foods. They search for a wide variety of food as well as beverage options in the market to aid them in fulfilling certain health requirements (Markets and Markets 2014). This is the reason that the market for nutraceuticals as well as functional foods across the world is influenced by increased consumer awareness about the nutritional advantages of foods for improving health and keeping diseases at bay (Shahidi et al. 2012). In addition, increasing medical or health care expenses are causing a liability on public funds. An aware and health-conscious population with sufficient income is the driving force behind the global market for nutraceuticals. The availability of various foods that allow people's well-being is an important way to improve living standards and endorse longevity (Andlauer and Furst 2002). The advantages of functional foods for health go beyond the essential nutrients within. Nutraceuticals can be a food or component of a food that is ingested to prevent and/or treat diseases along with providing the benefits of basic nutrition. The inclusion of nutraceuticals in food products in appropriate amounts may hinder the development of non-communicable diseases such as hypertension, obesity, cancer and heart disease (Del Rio et al. 2013).

Nutraceutical components such as phytochemicals, carotenoids, antioxidants, plant extracts, omega-3, prebiotics, probiotics, amino acids, peptides and proteins are incorporated into numerous categories of functional food groups such as superfoods or fortified foods, through delivery forms such as reconstitutable powders, liquids,

* Corresponding author E-mail: drmchoudhary@pau.edu;
† Corresponding author E-mail: sekhonamarjeet@pau.edu

DOI: 10.1201/9781003315100-14

Table 14.1 Selective nutraceuticals used in the food industry with their related health benefits

S. No.	Nutraceutical component	Health benefits	Food source
1.	**EPA, DHA,`α-linoleic acid**	Low risk of atherosclerosis and certain cancers Improved cardiovascular and cognition health	Fish, flaxseed, canola, genetically modified plants
2.	**CLA**	Low risk of atherosclerosis Immunomodulatory Anti-inflammatory and Anti-carcinogenic	Milk fat
3.	**Phenols, flavanols resveratrol, catechins curcuminoids**	Decreased oxidative stress Prevents cardiovascular, neurodegenerative and metabolic diseases	Grapes, tea, olives, pomegranates, wine, cocoa
4.	**Lycopene, lutein, β-carotene, astaxanthin**	Prevent age-related macular degeneration, heart disease, cancer	Tomatoes, carrots, marigolds, green leafy vegetables, microalgae
5.	**Peptides**	Antihypertensive properties	Milk, meat, fish, plants

(Bassaganya-Riera et al. 2002, Aboul-Enein et al. 2013, Copper 2004, Marques et al. 2012)

semisolids, solids, capsules or tablets (McClements et al. 2009). Functional foods or nutraceuticals are grouped into two major categories. The first group consists of naturally occurring foods high in nutraceutical components, and the second category consists of food products made with suitable amounts of added nutraceuticals, typically with a nutritional advantage or disease-preventive goal (Table 14.1). Food products in the second group have caught the attention of consumers seeking convenience foods that include additional doses of nutraceuticals. This has encouraged the involvement of the food industry involved in nutraceuticals. Consuming functional foods with additional nutraceuticals is a useful alternative to taking dietary supplements in the form of capsules or tablets. The difficulties and potential solutions in incorporating nutraceuticals into the diet are covered in this chapter.

14.2 OBSTACLES TO INCORPORATING DIFFERENT NUTRACEUTICALS IN THE DIET

Nutraceuticals can be delivered by using common foods and drinks that are consumed on a daily basis. Given that consumers are constantly asking for healthier food options, the food industry is very interested in the development of functional foods. The foremost concern for food processors or manufacturers is the availability of healthy foods that are inexpensive, easy to consume and tempting to consumers. The recognised efficiency of nutraceuticals that is associated with specific health benefits intensifies them as an alternative to dietary or health supplements. Scientists identified several nutraceuticals 20 years ago, but despite substantial research in this area, only a small number of nutraceuticals have successfully been used as food product constituents. Several marketing, regulatory and technical challenges such as age-specific products, products for a specific culture, religious or ethnic group, health claims allowed in the region, legal obligations in the encapsulation of nutraceuticals for solubilisation in a food product, and the form of the nutraceutical still need to be addressed for functional foods to flourish in the global market (Research and Markets 2013).

14.3 STRATEGIES FOR NUTRACEUTICAL INCORPORATION

The addition of nutraceuticals to food formulations is not an easy or simple task. Incorporation of nutraceutical ingredients in food products needs suitable food setups for appropriate delivery of nutraceuticals through various processes. These methods or strategies need to be identified for their successful application in the nutraceutical industry. Food delivery formats for nutraceutical incorporation in novel food products are solid, semisolid, liquid or powder form. Scientists and researchers with an interest in the nutraceutical industry take a variety of things into account when creating formulations and processing methods, which ultimately depend on the format of the finished product. These include the nutraceutical ingredients' chemical, biological and physical characteristics, as well as how they affect the texture, flavour and general acceptance of the finished product. Fortified cereals and grains, snack bars, confectioneries, spreads, energy drinks, functional beverages, fortified cereals and fortified grains are just a few examples of food products that have been given a nutraceutical boost (Auguston and Hemar 2009).

The process of nutraceutical incorporation into food through any technique is not simple as the addition of novel components to an already existing product, as this would influence the composition and physicochemical and sensory attributes of the original formulation. As a result, the new product's taste, flavour, texture and shelf life may all suffer. The design of the formulation and manufacturing process is influenced by the nature, form and solubility of the nutraceutical ingredient. With the addition of nutraceutical elements, this has an effect on the formulation's desirable properties. Nutraceuticals cannot typically be added directly to food since many of them are prone to degradation and may interrelate with other dietary ingredients. This could result in the beneficial characteristics of a nutraceutical component being lost and the quality of the finished product deteriorating. Therefore, a well-designed delivery mechanism to target specific food applications and protect the functional qualities of bioactive components is necessary for the successful incorporation of nutraceuticals into foods (Chen et al. 2013). Recently, methods for developing delivery systems using micro- or nano-encapsulation technologies have demonstrated potential for stabilisation, controlled discharge and targeted delivery of nutraceuticals (Yao et al. 2014). Different approaches for nutraceutical incorporation have been discussed below:

14.3.1 Encapsulation

A secondary material, such as a matrix, wall material or encapsulant, may be used to enclose an active ingredient through the process of encapsulation. Until an activation agent releases the encapsulated bioactive components at a desired location and time, they are shielded from the atmosphere (Ubbink et al. 2006). The intrinsic physicochemical properties of the active and matrix constituents, their stability in various atmospheres, the association between active ingredient and matrix, and methods utilised to develop an encapsulated ingredient are important concerns when designing an encapsulated agent with the desired arrangement and properties. In other words, encapsulation protects nutraceuticals when they are in the form of a component during processing, storage or consumption of the food until these bioactive compounds reach the target site in the body, where an activating agent, such as pH or enzymatic activity, releases the nutraceutical along the gastrointestinal tract (GIT). When developing delivery systems for synthetic chemicals, bioactive ingredients or nutraceuticals for use in food applications, only GRAS (Generally Recognised as Safe) components are utilised (Augustin and Sanguansri 2012). As a result, the food industry is limited to a small number of encapsulated materials, in contrast to the pharmaceutical industry, where the delivery of medications in medicines is accomplished by non-food-grade synthetic polymers (McClements 2012). There are certain constituents that are being used in the planning and development of encapsulated formulations containing nutraceuticals intended for use in the food sector (Table 14.1).

14.3.2 Various Techniques for Microencapsulating Ingredients

For the creation of the microencapsulated component, scientists can use more than one combination of encapsulating agents and more than one nutraceutical to capture. Nutraceutical delivery systems can be created using a variety of food-grade components and materials. These are made up of primary, double and multilayered emulsions, glassy matrices, inclusion complexes, biopolymer gels, lipospheres and liposomes (McClements et al. 2007, Augustin and Sanguansri 2008). The selected nutraceutical can be stabilised and shielded from the environment by encapsulant or matrix material in each of these systems until its release is prompted by a stimulus. To create encapsulated agents, a variety of procedures may be used, such as layer-by-layer depositions, supercritical anti-solvent precipitation, membrane emulsification, homogenisation, microfluidisation, coating, extrusion, coacervation and freeze-drying (Gouin 2004, McClements et al. 2007, Onwulata 2012, Ezhilarasi et al. 2013).

14.4 FORMULATION AND CHALLENGES: EMPHASIS ON STABILITY AND SAFETY EVALUATIONS

There is a dire need to tackle the hurdles involved in the execution of nutraceuticals in their pure form, i.e. non-encapsulated ingredients. To get beyond the limitations of using non-encapsulated nutraceuticals in health foods, strategies must be developed. These may involve creating bioactive components that are microencapsulated for use in food or including food ingredients as natural carriers for nutraceuticals. To successfully add nutraceuticals to a target food, a specific solution must be identified in a number of conditions. Some examples that are of major interest to the food sector are given below.

14.4.1 Omega-3 Fatty Acids

The health sector is becoming more interested in omega-3 fatty acids because of their many health-promoting qualities. These include reduced inflammatory symptoms and enhanced brain and cardiovascular health (Ruxton et al. 2007). Long-chain omega-3 or fatty acid-containing products, such as those containing eicosapentaenoic acid (EPA) and docosahexaenoic acid (DHA), are vulnerable to oxidation, which results in the production of unfavourable flavours and odours. Antioxidants can be added to items to slow down oxidation. For example, fish oils are stabilised with antioxidants to deliver significant amounts of EPA and DHA. Therefore, microencapsulation offers an alternative to increase omega-3 fatty acids' resistance to oxidation, allowing them to be added to a wide range of diets (Taneja and Singh 2012). Extrusion, coacervation, spray-drying emulsions, biopolymer gelation and fluidised bed coatings are some of the patented methods for encapsulating omega-3 fatty acids. Omega-3-rich beverages and milk powders can be prepared by blending n-3 oils in powder form. Omega-3 fatty acid powders also make it easier to mix oils into water-based solutions, which is normally challenging (Drusch and Mannino 2009). According to numerous human studies, fish oils taken as microencapsulated complex conjugates are equivalent in properties to oil taken as soft gel capsules (Barrow et al. 2009). Additionally, the bioavailability of fish oil supplied in the form of gelatine capsules is comparable to that of microencapsulated fish oils (stabilised by the Maillard reaction) added to different foods like cereal bars, yoghurt and orange juice (Sanguansri et al. 2013). The food carrier, however, changes the kinetics of the GIT transport of the microencapsulated fish oils (Sanguansri et al. 2013).

14.4.2 Dietary Bioactive Lipids

Dietary bioactive lipids such as fat-soluble vitamins, carotenoids, phytostanol and phytosterol have also piqued the interest of food manufacturers in the development of functional foods. Their lower water solubility and vulnerability to oxidation are the biggest obstacles to the assimilation of dietary bioactive lipids. Their low bioavailability is a result of their low water solubility. Recent studies have examined the use of different encapsulation techniques (emulsions, nano, micro-emulsions, liposomes and solid lipid nanoparticles) that efficaciously incorporate these lipids and fat-soluble bioactive components into functional foods, potentially enhancing their bioavailability (Chen et al. 2013, Yao et al. 2014).

14.4.3 Probiotics

The inclusion of probiotics in food is a topic that the food industry is highly interested in. Among their many health advantages, they are best recognised for promoting digestive health (Iannitti and Palmieri 2010). When probiotics are subjected to processing difficulties during food preparation, storage environments and transit through the gastrointestinal system with stomach acids and bile, they become less viable. Therefore, probiotics must be protected until they enter the intestinal tract where they are needed (Rokka and Rantamaki 2010, Heidebach et al. 2012). The most common technique for stabilising probiotic microorganisms is to embed them in biopolymeric gel matrices (e.g., carrageenan, gellan gum, alginate, milk protein gels). A hydrocolloid-probiotic cell mixture is extruded through a nozzle into a solution to harden cells. The size of capsules has an adverse influence on the sensory qualities of the meal if it is not managed, despite being excellent at protecting cells. Both bacterial freeze-drying and spray-drying work well for keeping live cells intact. Drying conditions and the probiotics' transport matrix have an impact on their viability. To increase their durability during drying, probiotics can be pre-adapted to a variety of stressors. In order to increase the stability of probiotic bacteria in the dried condition, sugars are usually utilised, either individually or in amalgamation with proteins or milk (Desmond et al. 2001, Zayed and Roos 2004, Meng et al. 2008). Probiotic bacteria are safeguarded in synbiotic formulations with protein-carbohydrate-oil matrices and prebiotics throughout non-refrigerated storage and while exposed to low pH, and they are also able to be released in simulated intestinal fluids (Crittenden et al. 2006). As different strains of Bifidobacterium and Lactobacillus can be exposed to varied environmental stressors, the efficiency of an encapsulant system must be examined focusing on the particular probiotic strain and food delivery system used (for example, temperature, oxygen, water activity and acidity). Foods, especially dairy products like yoghurt and cheese, have been utilised to deliver probiotics. Foods with encapsulated probiotics include ice cream, chocolate, pork products, nutrition bars, mayonnaise and orange juice (Burgain et al. 2011). Although low-pH juices are difficult for probiotics to thrive in, bacteria were able to withstand the low pH for extended periods of time when they were enclosed in whey protein matrices and then spray-dried. The ability of the whey protein to form a colloidal matrix that shielded the probiotics from the low-pH environment was associated with the probiotics' defence mechanisms (Ying et al. 2013). The impact of the food vehicle on probiotic functionality is poorly understood. On the other hand, higher probiotic doses might increase their effectiveness. There is still a dearth of human studies on probiotics' efficacy for particular health outcomes (Sanders and Marco 2010).

14.4.4 Polyphenols

Polyphenols are a diverse group of phenolic compounds present in a variety of fruits, vegetables and other plant products (e.g., tea, cocoa, grapeseed, olives). In addition to providing protection against heart disease, cancer and neurological problems, polyphenols have also been linked to better health. Epidemiological studies supporting their beneficial impact on health are convincing (Wani et al. 2023; Del Rio et al. 2013). After ingesting certain polyphenols, low levels of the parent chemical and some phase II metabolites can be seen in the bloodstream. The bacteria in the colon break down polyphenols into a series of metabolites, such as tiny phenolic acids and aromatic catabolites, which are then taken into the bloodstream (Borges et al. 2013). Epicatechins and epigallocatechins, two types of polyphenols, are generally absorbed in the small intestine, while anthocyanins, dihydrochalcones and chlorogenic acid are metabolised primarily in the large intestine (Scalbert and Williamson 2000, Visioli et al. 2014). It is impossible to develop a single encapsulation method that can provide all polyphenols. This is because the many structural variations of polyphenols and unique physicochemical characteristics each affect the metabolic processes in the body. Strategies for polyphenol encapsulation are covered in a recent review (Fang and Bhandari 2010).

14.4.5 Water-Soluble Polyphenols

Though they can be applied directly, water-soluble polyphenols (catechins in tea) should be safeguarded from high pH levels and oxidation. Epigallocatechin gallate (EGCG) may act as an EGCG carrier in functional food compositions by binding to naturally occurring food-based materials such as apple pomace, a byproduct of the production of apple juice (Wu et al. 2014). EGCG and catechin are more stable in the digestive system under alkaline conditions when they are encapsulated in chitosan-tripolyphosphate nanoparticles, according to in vitro experiments (Dube et al. 2010). It has also been demonstrated that anthocyanins can be protected from early digestion in the small intestine by being encapsulated in whey proteins or polysaccharides (pectin amide) (Oidtmann et al. 2012).

14.4.6 Polyphenols Moderately Soluble in Water and Oil

Few polyphenols, for instance, resveratrol and curcumin, are prone to degradation and are only slightly soluble in oils and aqueous solutions. These compounds need to be stabilised and rendered more soluble in aqueous media in order to be added to foods with a water base and delivered to the GI tract. Scarce water-soluble polyphenols can be administered in a variety of methods, including in polysaccharide (chitosan, pectinate) particles, protein, cyclodextrin and liposomal preparations (Augustin et al. 2013). Proteins are effective delivery systems because they can transport polyphenols (Bourassa et al. 2013). Using buttermilk, a byproduct of producing butter, is an additional choice. Occasionally, soluble materials can adhere to proteins in buttermilk and separate into an oil phase (Ye et al. 2013, Fu et al. 2014).

14.4.7 Peptides

There are several uses for bioactive peptides. They have the ability to increase mineral absorption and have immunomodulatory and antihypertensive opiate properties (Wada and Lonnerdal 2014). Bioactive peptides produced from diverse parent proteins, such as milk proteins and soy proteins, are offered as functional food components. It is frequently required to use the right encapsulation techniques to conceal the bitter taste of peptides and get around hygroscopicity problems in order to successfully incorporate them into food. For instance, casein hydrolysates' bitterness decreased through complex coacervation with soybean protein isolate and pectin due to the lower hygroscopic properties of encapsulated casein as compared to free hydrolysate (Mendanha et al. 2009). When a casein hydrolysate was spray-dried after being made with gelatine and soy proteins, similar benefits were attained (Favaro-Trindade et al. 2010).

14.5 SAFETY CONSIDERATIONS

Only a small percentage of the nutraceuticals available on the market are detrimental to human health; therefore the majority are not generally harmful when consumed by humans. When administered at the correct and recommended levels, studies have demonstrated that numerous frequently used nutraceuticals have many health advantages with relatively little harm. Anthocyanins, polyphenols and catechins are examples of common nutraceuticals that are safe for human consumption when employed under controlled conditions. Only a small number of studies have shown how these compounds are bad for human health.

However, research on nutraceuticals has demonstrated that the risks are outweighed by the benefits, and they are universally accepted for usage in humans when taken in the right amounts and dosages (Riccioni et al. 2018). Misuse and excessive usage of these products, however, could be harmful to people's health. The type, duration and quantity of use are the main determinants of these nutraceuticals' safety for customers. When certain nutraceuticals are used, especially by people who are taking medications, the interactions between the drugs and the nutraceutical substances might have extremely negative consequences on the body (Pirillo and Catapano 2015). As a result, for safe usage, they should only be used as directed by trained professionals, and only in the appropriate quality, quantity and time.

The most crucial characteristics of nutraceuticals are their affordability, extensive safety profiles in people and animals, acceptability and easy availability (Kim and Kim 2010). Despite having a high level of safety, a small number of nutraceuticals have been reported as having been contaminated with substances such as heavy metals, hazardous pesticides, addictive properties, toxic plants and fertilisers and mycotoxins (Anadon et al. 2016, Gul et al. 2016). Unfortunately, the safety profile of many nutraceuticals has not yet been fully investigated, and as a result, there is not enough documentation related to the safety of these agents. Toxico-kinetics, the start of use, necessary dose and dosing frequency may all be understood by having a thorough knowledge of the pharmacokinetic behaviour of each drug. Additionally, nutrient/food interactions are another critical component of safety assessments of nutraceuticals that evaluate the impact of interactions on efficiency, safety, half-life and consequent therapeutic response (Pirillo and Catapano 2015).

Nutraceuticals are classified as dietary supplements in many nations, including the United States, and may not be governed by the same rules and regulations governing allopathic pharmaceuticals' safety requirements. The "Dietary Supplements Health & Education Act," enacted by the US Congress in 1994, listed nutraceuticals as dietary supplements. Since they are not considered pharmaceuticals, their sale is allowed without any safety or effectiveness testing. However, in order to receive a licence for use in the general population, these herbal agents must demonstrate scientific proof of their quality, effectiveness and safety. In comparison to pure synthetic compound-based goods, nutraceuticals' toxicity and therapeutic evaluations are more challenging since they typically contain a complex blend of numerous components. Additionally, their chemical makeup changes depending on the region of plant collection, varying fertiliser/pesticide impacts, and the impact of stress (Gil et al. 2016). The limitations of nutraceuticals are the aforementioned factors, along with the absence of well-established methods for identification, chemical composition, extraction, potency, purity and protection of active pharmaceutical components, which are responsible for differences between batches and the difficulty in replicating therapeutic effects (Gupta et al. 2018). Nutraceuticals also contain a variety of substances that might cooperate or provoke an antagonistic reaction. High therapeutic results, toxicity and possibly a subtherapeutic response are all possible consequences of both of these actions (Gupta 2016).

In other words, the use of nutraceuticals has significantly expanded for both human and veterinary products. As there are no extensive analyses of their safety and effectiveness in scientific studies, the toxicity concerns related to their use are also higher. Additionally, for effective use of these substantial alternative agents, managing variances and the presence of toxins including pesticides, heavy metals and mycotoxins, as well as thorough pharmacokinetic and dynamic evaluations, are absolutely essential.

14.5.1 Association of Nutraceuticals with Genotoxicity and Carcinogenicity

A nutraceutical product's or component's propensity to create a DNA lesion that results in genetic mutations or cellular death that may cause different types of cancer is known as nutraceutical-associated genotoxicity. Although animal trials using oral therapy with aloe vera whole leaf extract for two years have shown a few carcinogenic results, aloe vera is still advised as a treatment for many ailments (Girdhar et al. 2017, Ayaz et al. 2019, Guo and Mei 2016,). In comparison to whole leaf extracts, aloe vera gel has been reported as less mutagenic and genotoxic (Pandiri et al. 2011), and in agreement with this finding, it was unable to significantly mutagenically alter the S. typhimurium TA100 strain (Shao et al. 2013). A favourable mutagenesis reaction was detected in *E. coli* repair-deficient mutant when subjected to aloe vera pulp extract (Williams et al. 2010). In human gastric cancer cells, danthron, a component of aloe, is known to produce DNA damage and trigger caspase-induced death via Bax-elicited and mitochondrial permeability transition pore pathways (Sehgal et al. 2013). In another investigation, mutagenic effects were discovered by the occurrence of hepatocellular cancer and hepatoblastoma in B6C3F1 mice after chronic *Ginkgo biloba* treatment both the presence and absence of exogenous S9 (liver extract from hamsters and rats) metabolic activation. Additionally, mutagenic effects were noted in *S. typhimurium* TA98 and TA strains as well as *E. coli* WP2 uvrA pKM101 strain

(Paes-Leme et al. 2005). *G. biloba* as well as its separated components, including quercetin, kaempferol-induced cytotoxicity and mutagenicity were demonstrated in a number of in vitro investigations utilising cell lines (Chiang et al. 2011). However, there was no evidence of genotoxicity in mice with gpt delta and androstane receptor deletion after an oral dose of 2000 mg/kg extract of G. biloba (Lin et al. 2014). Consequently, detecting *S. typhimurium* (TA98, TA100 1–10 mg/plate of goldenseal was detected in *E. coli* (WP2 uvrA/pKM101 strain) and other strains.

Despite the fact that root powder caused hepatocellular carcinoma and adenoma, it was not mutagenic with F344/N rats and B6C3F1 mice given the same dietary supplement orally for two years. However, increases were not seen when treatment lasted for three months. Frequencies in micro-nucleated erythrocytes were not statistically significant (Maeda et al. 2014). Goldenseal may damage DNA, as demonstrated by in vitro tests on HepG2 cells using commercial goldenseal products, as shown by the production of gamma-H2AX, a biomarker of DNA damage and breaks (Resende et al. 2012). The use of herbal products has significantly expanded globally despite conflicting scientific data. However, various studies have noted their positive effects on health. The findings of this research indicate that although nutraceuticals, herbs and dietary supplements are widely regarded as safe, comprehensive toxicological testing is still required. The amount of information currently available on the mutagenic and genotoxic effects of these products is not enough to draw firm conclusions on the safety of using all nutraceuticals.

14.5.2 Evaluation of Nutraceuticals in Terms of Efficiency, Safety, and Potential Toxicities

To assess the toxicity, efficacy and safety of nutraceutical products and their active ingredients, a number of models have been created. Numerous pharmacological effects, including sedative, antioxidant, immuno-modulatory, anti-inflammatory, hypnotic and adaptogenic ones, have been demonstrated by bioactive compounds found in nutraceuticals (Chen et al. 2013, Ajayi et al. 2017, Arslan et al. 2017, Gulati et al. 2016). Following application, these drugs' efficacy was evaluated by utilising in vitro and in vivo tests and omics technology (Sobrinho et al. 2017). Moreover, depending on the study design, several animal models have been used for invasive, non-invasive and in vivo studies (Wang 2016). For the purpose of assessing the safety and effectiveness of nutraceuticals, some invertebrate models have been created (Gonclaves et al. 2018, Peterson 2016, Barnett et al. 2016). Additionally, novel and mechanism-based predictive kinetic models have been created, and these models are incredibly helpful for learning about the pathways that contribute to toxicity and efficacy (Bian and Pei 2016, Krishna and Gopalakrishnan 2016).

14.5.3 Interactions of Nutraceuticals with Drugs

Any herbal or dietary supplement taken with medicine may have negative health effects that could be serious (Gonzalez-Suarez et al. 2016). Age, concurrent drug administration and a number of co-morbidities such as cancer, diabetes and infectious disorders all contribute to these undesirable interactions and predispose patients to the negative side effects of nutraceuticals. Drugs with a low therapeutic index, such as digitalis glycosides, anticoagulants, chemotherapeutics, psychoactive substances and immune modulators, are particularly important because their concentrations can cause life-threatening reactions if they exceed the safety range as a result of interactions (Gonzalez-Suarez et al. 2016). Some nutraceuticals act as cytochrome P450 (CYTP450) iso-enzyme inducers and inhibitors, inducing or inhibiting the liver enzymes necessary for these medicines' metabolism (Kadakkuzha et al. 2016, Mouly et al. 2017, Mooiman et al. 2014). For instance, a number of the phytochemicals found in G. biloba both activate and inhibit CYP450 (Oh et al. 2017, Zhang and Sparreboom 2017). Grapefruit juice is an inhibitor of CYP450, but ginseng is an inducer. Similarly, CYP1A2, CYP2C9, CYP2C19 and CYP3A4 are among the CYP450 isoforms that peppermint oil has been shown to stimulate in studies (Gaudineau et al. 2004). Additionally, the components of St. John's wort suppress the CYP3A4-induced metabolism of testosterone and midazolam. Studies showed that 14 days after St. John's wort administration, multiple-fold increases in the metabolic activity of hepatic CYP3A2 were seen (Shao et al. 2017, Unger and Frank 2004). Additionally, ginseng, an inhibitor of P-glycoproteins type efflux pumps, increases the action of many medicines three to four times (Shao et al. 2017). When polypharmacy is involved, St. John's wort administration has a significant impact on the kinetics and dynamics of many medicines (Durr et al. 2000, Goey et al. 2013). The majority of drug types can either directly or indirectly affect nutrition, while nutrition affects how drugs are disposed of and the ensuing therapeutic effects.

14.5.4 Chemicals Affecting Nutraceutical Quality

Numerous harmful contaminants or adulterants, including pesticides, heavy metals, mycotoxins and phytotoxicants can seriously impair the efficacy of nutraceuticals. High concentrations of these adulterants together can have fatal health effects (Dorman et al. 2016). For instance, pyrrolizidine alkaloids, which are found in a variety of plant species, are among the most deadly alkaloids. Numerous foods and nutraceuticals contain these alkaloids, which interact with proteins to cause aberrant mitosis, tissue necrosis and cellular dysfunctions (Herr et al. 2017, Yang et al. 2017, Merz and Schrenk 2016). These are potentially cytotoxic, carcinogenic and hepatotoxic (Dorman et al. 2016). In addition, microbial infection may cause dietary foods and supplements to lose some of their stability and quality (Panter et al. 2019). For instance, a Polish study revealed that, of the 152 goods and samples examined, 92.1% and 868.8%, respectively, showed varying degrees of bacterial and fungal contamination.

Mycotoxins are secondary metabolites produced by fungi and have a substantial impact on the nutritional value of foods. Often during cultivation, storage and transportation, these are mistakenly added to the foods and/or products. Citrinin, aflatoxins, ochratoxin and other moulds and their spores are some examples of frequent mycotoxins (Preliasco et al. 2017, Dlugaszewska et al. 2019). Similarly, contaminants like mercury, arsenic, cadmium and lead are added during harvesting or storage; as a result, the World Health Organization (WHO) highly suggests that all herbal products should be tested for the presence of heavy metals due to their serious adverse effects (Bugno et al. 2006, Prado et al. 2012).

14.5.5 Regulatory Status of Nutraceuticals

The public is now more aware of the benefits of using nutraceuticals and functional foods as a result of the promotion of these products by a number of organisations, including government bodies. There are some nations where there are no specific standards for the manufacture and consumption of nutraceuticals. The Dietary Supplement Health and Education Act (DSHEA), passed by Congress in 1994 for implementation, has several regulations in the United States. Similarly, Poland's Food and Nutrition Safety Act was approved in 2006, while Canada's Food and Drugs Act has set down particular requirements (Prabu et al. 2012, Sarma et al. 2011). As a result of these items' proper regulatory recognition, the pharmaceutical business will prosper with a plethora of new products and a corresponding increase in its market. The majority of nations have not developed specific regulatory criteria for nutraceuticals, which might potentially result in a rush of low-quality goods into the healthcare sector, particularly in developing nations. Therefore, regulatory rules for nutraceuticals are urgently needed by the global health community to guard against the arrival of fake and subpar medicines. Furthermore, to understand the precise mechanisms of action and potential side effects of functional foods and nutraceuticals, robust scientific research is required.

14.5.6 Processing Effect on Nutraceuticals

In order to increase the formation of secondary metabolites, various post-harvest abiotic elicitors have offered non-thermal processing procedures, including ultrasound, high-pressure processing (HPP) and pulsed electric fields (PEF). However, it has been demonstrated that these effects yield results that are strikingly comparable to those of plants under stress (Tong et al. 2017). These treatments stimulate the manufacture of nutraceuticals in plants. According to a different study, ultrasound therapy raises the amount of carotenoids in carrots (Hasler et al. 2005). The same treatment also improved Panax ginseng's phenylalanine ammonia-lyase activity (PAL) and phenolic compounds, and phenolic compounds in romaine lettuce increased (Del Cuellar-Villarreal et al. 2016). According to a different study, HHP boosted the levels of carotenoids, phenolic compounds and ascorbic acid in Ataulfo mango (Wu and Lin 2002). Additionally, PEF-specific reactions in potato metabolism were characterised by modifications in the hexose pool, which might lead to the breakdown of ascorbic acid and starch (Yu et al. 2016). When compared to foods that are consumed immediately after harvest, the unintentional creation of metabolites may further increase the health advantages that these foods exert. However, since this is a new field of study, little is known about the potential long-term advantages for human health.

14.6 CONCLUSION

Nutraceuticals are a cutting-edge and exciting research area for the development of novel health products with enormous potential for positive health effects, including safety, efficacy and affordability. Researchers have come to the conclusion that dietary supplements and good nutrition can both prevent and treat chronic diseases on a global scale. Numerous nutraceutical types have been extracted from foods, and they are manufactured in large quantities using biotechnology and genetic engineering techniques that have positive pharmaco-economic effects. In addition to their nutritional qualities, these items offer great health advantages by preventing a number of disorders. Numerous illnesses, such as cancer, rheumatoid arthritis, diabetes and other chronic diseases, have been successfully treated with nutraceuticals. It is certainly possible to improve health and avoid various diseases with the use of nutraceuticals, some of which have shown efficacy comparable to that of traditional medications.

Comparing nutraceuticals to both complementary treatments and traditional pharmaceuticals, there are generally fewer side effects, unfavourable effects and medication interactions. The absence of side effects, adverse effects and drug interactions does not mean that nutraceuticals lack these features, but risk–benefit analyses for the use of nutraceuticals have not yet been completed in the same way as for other traditional drugs. The pharmaceutical and food sectors, as well as academia, are very interested in this research area. Several pharmaceutical companies, notably Ranbaxy and Abbott, have taken the initiative to provide a variety of nutraceutical products for customers of various ages. Researchers have discovered that these products play a significant preventive role; hence, more in-depth academic and pharmaceutical sector research is required regarding their safety and efficacy. Additionally, the application of cutting-edge and high-throughput technologies can aid in our understanding of the underlying mechanisms of action and expand the boundaries of this fascinating field of study for the benefit of humanity in terms of both economic and health results.

REFERENCES

Aboul-Enein, H. Y., P. A. Berczynski, and I. Kruk. 2013. "Phenolic compounds: The role of redox regulation in neurodegenerative disease and cancer." *Mini-Reviews in Medicinal Chemistry* 3: 385–398.

Ajayi, A. M., S. Umukoro, B. Ben-Azu, B. Adzu, and O. G. Ademowo. 2017. "Toxicity and protective effect of phenolic-enriched ethylacetate fraction of Ocimum gratissimum (Linn.) leaf against acute inflammation and oxidative stress in rats." *Drug Development Research* 78(3–4): 135–145.

Anadón, A., M. R. Martínez-Larrañaga, I. Ares, and M. A. Martínez. 2016. "Interactions between nutraceuticals/nutrients and therapeutic drugs." In: *Nutraceuticals*, 1st ed., Gupta, R.C., Ed.; Academic Press, Cambridge, MA, 855–874.

Andlauer, W., and P. Furst. 2002. "Nutraceuticals: A piece of history, present status and outlook." *Food Research International* 35(2–3): 171–176.

Arslan Burnaz, N., M. Küçük, and Z. Akar. 2017. "An on-line HPLC system for detection of antioxidant compounds in some plant extracts by comparing three different methods." *Journal of Chromatography. Part B* 1052: 66–72.

Augustin, M. A., and L. Sanguansri. 2008. "Encapsulation of bioactives." In: *Food Materials Science: Principles and Practice*, Aguilera, J.M., P.J. Lilford, Eds.; Springer, New York, pp. 577–601.

Augustin, M. A., and Y. Hemar. 2009. "Nano-structured assemblies for encapsulation of food ingredients." *Chemical Society Reviews* 38: 902–912.

Augustin, M. A., and L. Sanguansri. 2012. "Challenges in developing delivery systems for food additives, nutraceuticals and dietary supplements." In: *Encapsulation Technologies and Delivery Systems for Food Ingredients and Nutraceuticals*, Garti, N., D.J. McClements, Eds.; Woodhead, Cambridge, pp. 19–48.

Augustin, M. A., L. Sanguansri, and T. Lockett. 2013. "Nano- and micro-encapsulated systems for enhancing the delivery of resveratrol." *Annals of the New York Academy of Sciences* 1290: 107–112.

Ayaz, M., F. Ullah, A. Sadiq, F. Ullah, M. Ovais, J. Ahmed, and H. P. Devkota. 2019. "Synergistic interactions of phytochemicals with antimicrobial agents: Potential strategy to counteract drug resistance." *Chemico-Biological Interactions* 308: 294–303.

Barnett, R. E., D. C. Bailey, H. E. Hatfield, and V. A. Fitsanakis. 2016. "Caenorhabditis elegans: A model organism for nutraceutical safety and toxicity evaluation." In: *Nutraceuticals*, Gupta, R.C., Ed.; Academic Press, Cambridge, MA, 341–354.

Barrow, C. J., C. Nolan, and B. J. Holub. 2009. "Bioequivalence of encapsulated and microencapsulated fish-oil encapsulation." *Journal of Functional Foods* 1(1): 38–43.

Bassaganya-Riera, J., R. Hontecillas, and M. J. Wannemuehler. 2002. "Nutritional impact of conjugated linoleic acid: A model functional food ingredient." *In vitro cellular and Development Biology - Plant* 38: 241–246.

Bian, W. P., and D. S. Pei. 2016. "Zebrafish model for safety and toxicity testing of nutraceuticals." In: *Nutraceuticals*, Gupta, R.C., Ed.; Academic Press, Cambridge, MA, 333–339.

Borges, G., M. E. J. Lean, S. A. Roberts, and A. Crozier. 2013. "Bioavailability of dietary (poly)phenols: A study with ileostomists to discriminate between absorption in small and large intestine." *Food and Function* 4(5): 754–762.

Bourassa, P., J. Bariyanga, and H. A. Tajmir-Riahi. 2013. "Binding sites of resveratrol, genistein, and curcumin with milk α- and β-caseins." *Journal of Physical Chemistry. Part B* 5(5): 1287–1295.

Bugno, A., A. A. B. Almodovar, T. C. Pereira, T. de J. A. Pinto, and M. Sabino. 2006. "Occurrence of toxigenic fungi in herbal drugs." *Brazilian Journal of Microbiology* 37(1): 47–51.

Burgain, J., C. Giani, M. Linder, and J. Scher. 2011. "Encapsulation of probiotic living cells: from laboratory to industrial applications." *Journal of Food Engineering* 104: 467–483.

Chen, B., D. J. Mc. Clements, and E. A. Decker. 2013. "Design of foods with bioactive lipids for improved health." *Annual Reviewof Food Science and Technology* 4: 35–56.

Chen, S., L. Wan, L. Couch, H. Lin, Y. Li, V. N. Dobrovolsky, N. Mei, and L. Guo. 2013. "Mechanism study of goldenseal-associated DNA damage." *Toxicology Letters* 221(1): 64–72.

Chiang, J. H., J. S. Yang, C. Y. Ma, M. D. Yang, H. Y. Huang, T. C. Hsia, H. M. Kuo, P. P. Wu, T. H. Lee, and J. G. Chung. 2011. "Danthron, an anthraquinone derivative, induces DNA damage and caspase cascades-mediated apoptosis in SNU-1 human gastric cancer cells through mitochondrial permeability transition pores and Bax-triggered pathways." *Chemical Research in Toxicology* 24(1): 20–29.

Cooper, D. A. 2004. "Carotenoids in health and disease: Recent scientific evaluations, research recommendations and the consumer." *Journal of Nutrition* 134(1): 221–224.

Crittenden, R. C., R. Weerakkody, L. Sanguansri, and M. A. Augustin. 2006. "Synbiotic microcapsules that enhance microbial viability during nonrefrigerated storage and gastrointestinal tract transit." *Applied and Environmental Microbiology* 72(3): 2280–2282.

Del Cuéllar-Villarreal, M. R., E. Ortega-Hernández, A. Becerra-Moreno, J. Welti-Chanes, L. Cisneros-Zevallos, and D. A. Jacobo-Velázquez. 2016. "Effects of ultrasound treatment and storage time on the extractability and biosynthesis of nutraceuticals in carrot (Daucus carota)." *Postharvest Biology and Technology* 149: 18–26.

Del Rio, D., A. Rodriquez-Mateos, J. P. E. Spencer, M. Tognolini, G. Borges, and A. Crozier. 2013. "Dietary (poly)phenolics in human health: Structures, bioavailability, and evidence of protective effects against chronic diseases." *Antioxidants and Redox Signaling* 18(14): 1818–1892.

Desmond, C., C. Stanton, G. F. Fitzgerald, K. Collins, and R. P. Ross. 2001. "Environmental adaption of probiotic lactobacilli towards improvement of performance during spray drying." *Intnational Dairy Journal* 14: 801–808.

Dlugaszewska, J., M. Ratajczak, D. Kamin´ska, and M. Gajecka. 2019. "Are dietary supplements containing plant-derived ingredients safe microbiologically?" *Saudi Pharmaceutical Journal* 27(2): 240–245.

Dorman, G., B. Flachner, I. Hajdu, and C. D. Andras. 2016. "Target identification and polypharmacology of nutraceuticals." In: *Nutraceuticals*, Gupta, R.C., Ed.; Academic Press, Cambridge, MA, 315–343.

Drusch, S., and S. Mannino. 2009. "Patent-based review on industrial approaches for the microencapsulation of oils rich in polyunsaturated fatty acids." *Trends in Food Science and Technology* 20: 237–244.

Dube, A., K. Ng, J. A. Nicolazzo, and I. Larson. 2010. "Effective use of reducing agents and nanoparticle encapsulation in stabilizing catechins in alkaline solution." *Food Chemistry* 122(3): 622–667.

Dürr, D., B. Stieger, G. A. Kullak-Ublick, K. M. Rentsch, H. C. Steinert, P. J. Meier, and K. Fattinger. 2000. "St John's Wort induces intestinal P-glycoprotein/MDR1 and intestinal and hepatic CYP3A4." *Clinical Pharmacology and Therapeutics* 68(6): 598–604.

Ezhilarasi, P. N., P. Karthik, N. Chhanwal, and C. Anandharamakrishnan. 2013. "Nanoencapsulation techniques for food bioactive components: A review." *Food and Bioprocess Technology* 6(3): 628–647.

Fang, Z. X., and B. Bhandari. 2010. "Encapsulation of polyphenols – A review." *Trends in Food Science and Technology* 21(10): 510–523.

Favaro-Trindade, C. S., A. S. Santana, E. S. Monterrey-Quintero, M. A. Trindade, and M. N. Netto. 2010. "The use of spray drying technology to reduce bitter taste of casein hydrolysate." *Food Hydrocolloids* 24(4): 336–340.

Fu, S., Z. Zhiping, S. Ajlouni, K. Ng, L. Sanguansri, and M. A. Augustin. 2014. "Interactions of buttermilk with curcuminoids." *Food Cheistry* 149: 47–53.

Gaudineau, C., R. Beckerman, S. Welbourn, and K. Auclair. 2004. "Inhibition of human P450 enzymes by multiple constituents of the Ginkgo biloba extract." *Biochemical and Biophysical Research Communications* 318(4): 1072–1078.

Gil, F., A. F. Hernández, and M. C. Martín-Domingo. 2016. "Toxic contamination of nutraceuticals and food ingredients." In: *Nutraceuticals*, Gupta, R.C., Ed.; Academic Press, Cambridge, MA, 825–837.

Girdhar, S., D. Pandita, A. Girdhar, and V. Lather. 2017. "Safety, quality and regulatory aspects of nutraceuticals." *Applied Clinical Research, Clinical Trials and Regulatory Affairs* 4(1): 36–42.

Goey, A. K. L., K. D. Mooiman, J. H. Beijnen, J. H. M. Schellens, and I. Meijerman. 2013. "Relevance of in vitro and clinical data for predicting CYP3A4-mediated herb–drug interactions in cancer patients." *Cancer Treatment Reviews* 39(7): 773–783.

Gonçalves, R. F. S., J. T. Martins, C. M. M. Duarte, A. A. Vicente, and A. C. Pinheiro. 2018. "Advances in nutraceutical delivery systems: From formulation design for bioavailability enhancement to efficacy and safety evaluation." *Trends in Food Science and Technology* 78: 270–291.

Gonzalez-Suarez, I., F. Martin, J. Hoeng, and M. C. Peitsch. 2016. "Mechanistic network models in safety and toxicity evaluation of nutraceuticals." In: *Nutraceuticals*, Gupta, R.C., Ed.; Academic Press, Cambridge, MA, 237–304.

Gouin, S. 2004. "Microencapsulation: Industrial appraisal of existing technologies and trends." *Trends in Food Science and Technology* 15(7–8): 330–347.

Gulati, K., R. Anand, and A. Ray. 2016. "Nutraceuticals as adaptogens: Their role in health and disease." In: *Nutraceuticals*, Gupta, R.C., Ed.; Academic Press, Cambridge, MA, 193–205.

Guo, X., and N. Mei. 2016. "Aloe vera: A review of toxicity and adverse clinical effects." *Journal of Environmental Science and Health: Part C, Environmental Carcinogenesis and Ecotoxicology Reviews* 34(2): 77–96.

Gupta, R. C. 2016. *Nutraceuticals: Efficacy, Safety and Toxicity*; Academic Press, Cambridge, MA, p. 1042.

Gupta, R. C., A. Srivastava, and R. Lall. 2018. "Toxicity potential of nutraceuticals." In: *Computational Toxicology: Methods and Protocols (Methods in Molecular Biology)*, Nicolotti, O., Ed.; Springer, New York, pp. 367–394.

Gul, K., A. K. Singh, and R. Jabeen. 2016. "Nutraceuticals and functional foods: The foods for the future world." *Critical Reviews in Food Science and Nutrition* 56(16): 2617–2627.

Hasler, C. M. 2005. *Regulation of Functional Foods and Nutraceuticals: A Global Perspective*; John Wiley & Sons, Hoboken, NJ, p. 425.

Heidebach, T., P. Forst, and U. Kulozik. 2012. "Microencapsulation of probiotic cells for food applications." *Critical Reviews in Food Science and Nutrition* 52(4): 291–314.

Herr, M., H. Grondin, S. Sanchez, D. Armaingaud, C. Blochet, A. Vial, P. Denormandie, and J. Ankri. 2017. "Polypharmacyandpotentially inappropriate medications: A cross-sectional analysis among 451 nursing homes in France." *European Journal of Clinical Pharmacology* 73(5): 601–608.

Iannitti, T., and B. Palmieri. 2010. "Therapeutical use of probiotic formulations in clinical practice." *Clinical Nutrition* 29(6): 701–725.

Kadakkuzha, B. M., X. Liu, S. Swarnkar, and Y. Chen. 2016. "Genomic and proteomic mechanisms and models in toxicity and safety evaluation of nutraceuticals." In: *Nutraceuticals*, Gupta, R.C., Ed.; Academic Press, Cambridge, MA, 227–337.

Kim, M., and Y. Kim. 2010. "Hypocholesterolemic effects of curcumin via up-regulation of cholesterol 7a-hydroxylase in rats fed a high fat diet." *Nutrition Research and Practice* 4(3): 191–195.

Krishna, G., and G. Gopalakrishnan. 2016. "Alternative in vitro models for safety and toxicity evaluation of nutraceuticals." In: *Nutraceuticals*, Gupta, R.C., Ed.; Academic Press, Cambridge, MA, 355–385.

Lin, H., X. Guo, S. Zhang, S. L. Dial, L. Guo, M. G. Manjanatha, M. M. Moore, and N. Mei. 2014. "Mechanistic evaluation of Ginkgo biloba Leaf extract-induced genotoxicity in L5178Y cells." *Toxicological Sciences* 139(2): 338–349.

Maeda, J., A. Kijima, K. Inoue, Y. Ishii, R. Ichimura, S. Takasu, K. Kuroda, K. Matsushita, Y. Kodama, N. Saito, T. Umemura, and M. Yoshida. 2014. "In vivo genotoxicity of Ginkgo biloba Extract in gpt Delta Mice and constitutive androstane receptor knockout mice." *Toxicological Sciences* 140(2): 298–306.

Markets, and Markets. 2014. *Nutraceutical Ingredients Market by Type & Application-Global Trends & Forecasts to 2018 Rep. FB 2214, MarketsandMarkets, Jan.* http://www.marketsandmarkets.com/Market-Reports/nutraceutical-ingredient-market-1319.html.

Marques, C., M. M. Amorim, J. O. Pereira, M. E. Pintado, D. Moura, C. Calhau, and H. Pinheiro. 2012. "Bioactive peptides: Are there more antihypertensive mechanisms beyond ACE inhibition?." *Current Pharmaceutical Design* 18(30): 4706–4713.

McClements, D. J., E. A. Decker, and J. Weiss. 2007. "Emulsion-based delivery systems for lipophilic bioactive components." *Journal of Food Science* 72(8): 109–124.

McClements, D. J., E. A. Decker, Y. Park, and J. Weiss. 2009. "Structural design principles for delivery of bioactive components in nutraceuticals and functional foods." *Critical Reviews in Food Science and Nutrition* 49(6): 577–606.

McClements, D. J. 2012. "Requirements for food ingredient and nutraceutical delivery systems." In: *Encapsulation Technologies and Delivery Systems for Food Ingredients and Nutraceuticals*, Garti, N., D.J. McClements, Eds.; Woodhead, Cambridge, pp. 3–18.

Mendanha, D. V., S. E. M. Ortiz, C. S. Favaro-Trindade, A. Mauri, E. S. Monterrey-Quintero, and M. Thomazini. 2009. "Microencapsulation of caseinhydrolysate by complex coacervation with SPI/pectin." *Food Research International* 42(8): 1099–1404.

Meng, X. C., C. Stanton, G. F. Fitzgerald, C. Daly, and R. P. Ross. 2008. "Anhydrobiotics: The challenges of drying probiotic cultures." *Food Chemistry* 106(4): 1406–1416.

Merz, K. H., and D. Schrenk. 2016. "Interim relative potency factors for the toxicological risk assessment of pyrrolizidine alkaloids in food and herbal medicines." *Toxicology Letters* 263: 44–57.

Mooiman, K. D., R. F. Maas-Bakker, J. J. M. A. Hendrikx, P. C. D. Bank, H. Rosing, J. H. Beijnen, J. H. M. Schellens, and I. Meijerman. 2014. "The effect of complementary and alternative medicines on CYP3A4-mediated metabolism of three different substrates: 7-benzyloxy-4trifluoromethyl-coumarin, midazolam and docetaxel." *Journal of Pharmacy and Phramacology* 66: 865–874.

Mouly, S., C. Lloret-Linares, P. O. Sellier, D. Sene, and J. F. Bergmann. 2017. "Is the clinical relevance of drug-food and drug-herb interactions limited to grapefruit juice and Saint-John's Wort?." *Pharmacological Research* 148: 82–92.

Oidtmann, J., M. Schantz, K. Mader, M. Baum, S. Berg, M. Betz, U. Kulozik, S. Leick, H. Rehage, K. Schwarz, and E. Richling. 2012. "Preparation and comparative release characteristics of three anthocyanin encapsulation systems." *Journal of Agricultural and Food Chemistry* 60(3): 844–851.

Oh, H. A., H. Lee, D. Kim, and B. H. Jung. 2017. "Development of GC-MS based cytochrome P450 assay for the investigation of multi-herb interaction." *Analytical Biochemistry* 519: 71–83.

Onwulata, C.I. 2012. "Encapsulation of new active ingredients." *Annual Review of Food Science and Technology 3*: 183–202.

Paes-Leme, A. A., E. S. Motta, J. C. P. De Mattos, F. J. S. Dantas, R. J. A. C. Bezerra, and A. Caldeira-de-Araujo. 2005. "Assessment of Aloevera (L.) genotoxic potential on Escherichia coli and plasmid DNA." *Journal of Ethnopharmacology* 102(2): 197–201.

Pandiri, A. R., R. C. Sills, M. J. Hoenerhoff, S. D. Peddada, T. V. T. Ton, H. H. L. Hong, G. P. Flake, D. E. Malarkey, G. R. Olson, and I. P. Pogribny. 2011. "Aloe vera non-decolorized whole leaf extract-induced large intestinal tumors in F344 rats share similar molecular pathways with human sporadic colorectal tumors." *Toxicology Pathology* 39: 1065–1074.

Panter, K. E., K. D. Welch, and D. R. Gardner. 2019. "Poisonous Plants: Biomarkers for Diagnosis." In: *Nutraceuticals*, Gupta, R.C., Ed.; Academic Press, Cambridge, MA, 627–652.

Peterson, J. D. 2016. "Noninvasive in vivo optical imaging models for safety and toxicity testing." In: *Nutraceuticals*, Gupta, R.C., Ed.; Academic Press, Cambridge, MA, 305–317.

Pirillo, A., and A. L. Catapano. 2015. "Berberine, a plant alkaloid with lipid- and glucose-lowering properties: From in vitro evidence to clinical studies." *Atherosclerosis* 243(2): 449–461.

Prabu, S. L., T. N. K. SuriyaPrakash, C. D. Kumar, S. SureshKuma, and T. Ragavendran. 2012. "Nutraceuticals: A review." *Elixir International Journal* 46: 8372–8377.

Prado, G., A. F. Altoe, T. C. B. Gomes, A. S. Leal, V. A. D. Morais, M. S. Oliveira, M. B. Ferreira, M. B. Gomes, F. N. Paschoal, R. V. S. Souza, D. A. Silva, and J. E. Cruz Madeira. 2012. "Occurrence of aflatoxin B1 in natural products." *Brazilian Journal of Microbiology* 43(4): 1428–1435.

Preliasco, M., D. Gardner, J. Moraes, A. C. González, G. Uriarte, and R. Rivero. 2017. "Senecio grisebachii Baker: Pyrrolizidine alkaloids and experimental poisoning in calves." *Toxicon* 133: 68–73.

Research and Markets. 2013. "Global functional food and nutraceuticals market (2013–2018). Rep., IndustryARC, Oct. http://www.researchandmarkets.com/reports/2788610.

Resende, F. A., W. Vilegas, L. C. Dos Santos, and E. A. Varanda. 2012. "Mutagenicity of flavonoids assayed by bacterial reverse mutation (Ames) test." *Molecules* 17(5): 5255–5268.

Riccioni, G., M. A. Gammone, W. Currenti, and N. D'Orazio. 2018. "Effectiveness and safety of dietetic supplementation of a new nutraceutical on lipid profile and serum inflammation biomarkers in hypercholesterolemic patients." *Molecules* 23(5): 1468.

Rokka, S., and P. Rantam¨aki. 2010. "Protecting probiotic bacteria by microencapsulation: Challenges for industrial applications." *European Food Research and Technology* 231(1): 1–12.

Ruxton, C. H. S., S. C. Reed, M. J. A. Simpson, and K. J. Millington. 2007. "The health benefits of omega-3 polyunsaturated fatty acids: A review of evidence." *Journal of Human Nutrition and Dietetics* 20(3): 275–285.

Sanders, M. E., and M. Marco. 2010. "Food formats for effective delivery of probiotics." *Annual Review of Food Science and Technology* 1: 65–85.

Sanguansri, L., Z. Shen, R. Weerakkody, M. Barnes, T. Lockett, and M. A. Augustin. 2013. "Omega-3 fatty acids in ileal effluent after consuming different foods containing microen capsulated fish oil powder–anileostomy study." *Food and Function* 4(1): 74–82.

Sarma, H., S. Deka, H. Deka, and R. R. Saikia. 2011. "Accumulation of heavy metals in selected medicinal plants." In: *Reviews of Environmental Contamination and Toxicology*, Whitacre, D.M., Ed.; Springer, New York, pp. 63–86.

Scalbert, A., and G. Williamson. 2000. "Dietary intake and bioavailability of polyphenols." *Journal of Nutrition* 130(8S Suppl): 2073–2085.

Sehgal, I., W. D. Winters, M. Scott, and K. Kousoulas. 2013. "An in vitro and in vivo toxicologic evaluation of a stabilized Aloe vera gel supplement drink in mice." *Food and Chemical Toxicology* 55: 363–370.

Shahidi, F. 2012. "Nutraceuticals, functional foods and dietary supplements in health and disease." *Journal of Food and Drug Analysis* 20(1): 226–230.

Shao, A., A. Broadmeadow, G. Goddard, E. Bejar, and V. Frankos. 2013. "Safety of purified decolorized (low anthraquinone) whole leaf aloe vera (L) Burm. f. juice in a 3-month drinking water toxicity study in F344 rats." *Food and Chemical Toxicology* 57: 21–31.

Shao, F., H. Zhang, L. Xie, J. Chen, S. Zhou, J. Zhang, J. Lv, W. Hao, Y. Ma, Y. Liu, N. Ou, and W. Xiao. 2017. "Pharmacokinetics of ginkgolides A, B and K after single and multiple intravenous infusions and their interactions with midazolam in healthy Chinese male subjects." *European Journal of Clinical Pharmacology* 73(5): 537–546.

Sobrinho, A. P., A. S. Minho, L. L. C. Ferreira, G. R. Martins, F. Boylan, and P. D. Fernandes. 2017. "Characterization of anti-inflammatory effect and possible mechanism of action of Tibouchina granulosa." *Journal of Pharmacy and Phramacology* 69: 706–713.

Taneja, A., and H. Singh. 2012. "Challenges for the delivery of long-chain n-3 fatty acids in functional foods." *Annual Review of Food Science and Technology* 5: 105–123.

Tong, M., W. Gao, W. Jiao, J. Zhou, Y. Li, L. He, and R. Hou. 2017. "Uptake, translocation, metabolism, and distribution of glyphosate in nontarget tea plant (Camellia sinensis L.)." *Journal of Agricultural and Food Chemistry* 65(35): 7638–7646.

Ubbink, J., and J. Kruger. 2006. "Physical approaches for the delivery of active ingredients in foods." *Trends in Food Science and Technology* 17(5): 244–254.

Unger, M., and A. Frank. 2004. "Simultaneous determination of the inhibitory potency of herbal extracts on the activity of six major cytochrome P450 enzymes using liquid chromatography/mass spectrometry and automated online extraction." *Rapid Communication in Mass Spetrometry* 18(19): 2273–2281.

Visioli, F., C. A. De La Lastra, C. Andres-Lacueva, M. Aviram, and C. Calhau. 2014. "Polyphenols and human health: A prospectus." *Critical Reviews in Food Science and Nutrition* 51: 524–546.

Wada, Y., and B. L¨onnerdal. 2014. "Bioactive peptides derived from human milk proteins—Mechanisms of action." *Journal of Nutrition Biochemistry* 25: 503–514.

Wang, K. 2016. "Adverse reaction prediction and pharmacovigilance of nutraceuticals: Examples of computational and statistical analysis on big data." In: *Nutraceuticals*, Gupta, R.C., Ed.; Academic Press, Cambridge, MA, 238–45.

Wani, S. A., J. Kour, R.-U. Haq, and M. S. Elshikh (Eds.). 2023. *Fruits and Their Roles in Nutraceuticals and Functional Foods*, 1st ed.; CRC Press. ISBN 9781032194462. https://doi.org/10.1201/9781003259213.

Williams, L. D., G. A. Burdock, E. Shin, S. Kim, T. H. Jo, K. N. Jones, and R. A. Matulka. 2010. "Safety studies conducted on a proprietary high-purity Aloe vera inner leaf fillet preparation, Qmatrix." *Regulatory Toxicology and Pharmacology* 57(1): 90–98.

Wu, J., and L. Lin. 2002. "Ultrasound-induced stress responses of panax ginsengCells: Enzymatic browning and phenolics production." *Biotechnology Progress* 18(4): 862–866.

Wu, L., L. D. Melton, L. Sanguansri, and M. A. Augustin. 2014. "The batch adsorption of epigallocatechin gallateonto apple pomace." *Food Chemistry* 169: 260–265.

Yang, X., W. Li, Y. Sun, X. Guo, W. Huang, Y. Peng, and J. Zheng. 2017. "Comparative study of hepatotoxicity of pyrrolizidine alkaloids retrorsine and monocrotaline." *Chemical Research in Toxicology* 30(2): 532–539.

Yao, M., H. Xiao, and D. J. Mc. Clements. 2014. "Delivery of lipophilic bioactives: Assembly, disassembly, and reassembly of lipid nanoparticles." *Annual Review of Food Science and Technology* 5: 53–81.

Ye, J. H., E. Thomas, L. Sanguansri, Y. R. Liang, and M. A. Augustin. 2013. "Interaction between whole buttermilk and resveratrol." *Journal of Agricultural and Food Chemistry* 29: 7096–7101.

Ying, D. Y., S. Schwander, R. Weerakkody, L. Sanguansri, C. Gantenbein-Demarchic, and M. A. Augustin. 2013. "Microencapsulated Lactobacillus rhamnosus GG in whey protein and resistant starch matrices: Probiotic survival in fruit juice." *Journal of Functional Foods* 5(1): 98–105.

Yu, J., N. J. Engeseth, and H. Feng. 2016. "High intensity ultrasound as an abiotic elicitor—Effects on antioxidant capacity and overall quality of romaine lettuce." *Food and Bioprocess Technology* 9(2): 262–273.

Zayed, G., and Y. H. Roos. 2004. "Influence of trehalose and moisture content on survival of Lactobacillus salivarius subjected to freeze-drying and storage." *Process Biochemistry* 39(9): 1081–1086.

Zhang, L., and A. Sparreboom. 2017. "Predicting transporter-mediated drug interactions: Commentary on: 'Pharmacokinetic evaluation of a drug transporter cocktail consisting of digoxin, furosemide, metformin and rosuvastatin' and 'Validation of a microdose probe drug cocktail for clinical drug interaction assessments for drug transporters and CYP3A'." *Clinical Pharmacology and Therapeutics* 101(4): 447–449.

Index

P

Q

R

S

T

For Product Safety Concerns and Information please contact our
EU representative GPSR@taylorandfrancis.com Taylor & Francis
Verlag GmbH, Kaufingerstraße 24, 80331 München, Germany